Energy-Efficient Underwater Wireless Communications and Networking

Nitin Goyal
Chitkara University Institute of Engineering and Technology, Chitkara University, Punjab, India

Luxmi Sapra
Chitkara University Institute of Engineering and Technology, Chitkara University, Punjab, India

Jasminder Kaur Sandhu
Chitkara University Institute of Engineering and Technology, Chitkara University, Punjab, India

A volume in the Advances in Environmental Engineering and Green Technologies (AEEGT) Book Series

IGI Global
PUBLISHER of TIMELY KNOWLEDGE

Published in the United States of America by
IGI Global
Engineering Science Reference (an imprint of IGI Global)
701 E. Chocolate Avenue
Hershey PA, USA 17033
Tel: 717-533-8845
Fax: 717-533-8661
E-mail: cust@igi-global.com
Web site: http://www.igi-global.com

Library of Congress Cataloging-in-Publication Data

Names: Goyal, Nitin, 1986- editor. | Sapra, Luxmi, 1978- editor. | Sandhu, Jasminder Kaur, 1984- editor.
Title: Energy-efficient underwater wireless communications and networking / Nitin Goyal, Luxmi Sapra, and Jasminder Kaur Sandhu, editors.
Description: Hershey, PA : Engineering Science Reference, [2020] | Includes bibliographical references and index. | Summary: "This book covers existing and future possibilities of the area as well as the current challenges presented in the implementation of underwater sensor networks"-- Provided by publisher.
Identifiers: LCCN 2019057525 (print) | LCCN 2019057526 (ebook) | ISBN 9781799836407 (hardcover) | ISBN 9781799836414 (paperback) | ISBN 9781799836421 (ebook)
Subjects: LCSH: Wireless sensor networks--Energy conservation. | Signals and signaling, Submarine.
Classification: LCC TK7872.D48 E535 2020 (print) | LCC TK7872.D48 (ebook) | DDC 006.2/5--dc23
LC record available at https://lccn.loc.gov/2019057525
LC ebook record available at https://lccn.loc.gov/2019057526

This book is published in the IGI Global book series Advances in Environmental Engineering and Green Technologies (AEEGT) (ISSN: 2326-9162; eISSN: 2326-9170)

British Cataloguing in Publication Data
A Cataloguing in Publication record for this book is available from the British Library.

For electronic access to this publication, please contact: eresources@igi-global.com.

Advances in Environmental Engineering and Green Technologies (AEEGT) Book Series

Sang-Bing Tsai
Zhongshan Institute, University of Electronic Science and Technology of China, China & Wuyi University, China
Ming-Lang Tseng
Lunghwa University of Science and Technology, Taiwan
Yuchi Wang
University of Electronic Science and Technology of China Zhongshan Institute, China

ISSN:2326-9162
EISSN:2326-9170

Mission

Growing awareness and an increased focus on environmental issues such as climate change, energy use, and loss of non-renewable resources have brought about a greater need for research that provides potential solutions to these problems. Research in environmental science and engineering continues to play a vital role in uncovering new opportunities for a "green" future.

The **Advances in Environmental Engineering and Green Technologies (AEEGT)** book series is a mouthpiece for research in all aspects of environmental science, earth science, and green initiatives. This series supports the ongoing research in this field through publishing books that discuss topics within environmental engineering or that deal with the interdisciplinary field of green technologies.

Coverage

- Pollution Management
- Sustainable Communities
- Cleantech
- Industrial Waste Management and Minimization
- Waste Management
- Contaminated Site Remediation
- Renewable Energy
- Policies Involving Green Technologies and Environmental Engineering
- Alternative Power Sources
- Water Supply and Treatment

IGI Global is currently accepting manuscripts for publication within this series. To submit a proposal for a volume in this series, please contact our Acquisition Editors at Acquisitions@igi-global.com or visit: http://www.igi-global.com/publish/.

The Advances in Environmental Engineering and Green Technologies (AEEGT) Book Series (ISSN 2326-9162) is published by IGI Global, 701 E. Chocolate Avenue, Hershey, PA 17033-1240, USA, www.igi-global.com. This series is composed of titles available for purchase individually; each title is edited to be contextually exclusive from any other title within the series. For pricing and ordering information please visit http://www.igi-global.com/book-series/advances-environmental-engineering-green-technologies/73679. Postmaster: Send all address changes to above address.

Titles in this Series

For a list of additional titles in this series, please visit:
http://www.igi-global.com/book-series/advances-environmental-engineering-green-technologies/73679

Development Challenges and Strategies in Global Arid Regions
Umar Garba Benna (Benna Associates, Nigeria)
Engineering Science Reference • © 2021 • 300pp • H/C (ISBN: 9781799837930) • US $195.00

Molecular Plant Breeding and Genome Editing Tools for Crop Improvement
Pradip Chandra Deka (Sir Padampat Singhania University, India)
Engineering Science Reference • © 2021 • 489pp • H/C (ISBN: 9781799843122) • US $195.00

Recent Advancements in Bioremediation of Metal Contaminants
Satarupa Dey (Shyampur Siddheswari Mahavidyalaya, India) and Biswaranjan Acharya (School of Computer Engineering, KIIT University (Deemed), India)
Engineering Science Reference • © 2021 • 363pp • H/C (ISBN: 9781799848882) • US $195.00

Global Environmental Impacts on Food Security, Nutrition, and Human Health
Hicham Chatoui (Private University of Marrakech, (UPM), Morocco & Cady Ayyad University, Morocco) Mohamed Merzouki (Faculty of Science and Technology, Sultan Moulay Slimane University, Beni Mellal, Morocco) Hanane Moummou (Private University of Marrakech (UPM), Marrakech, Morroco) Mounir Tilaoui (Private University of Marrakech (UPM), Sultan Moulay Slimane University, Beni Mellal, Morocco) and Nabila Saadaoui (Private University of Marrakech (UPM), Marrakech, Morroco)
Engineering Science Reference • © 2020 • 300pp • H/C (ISBN: 9781799837480) • US $195.00

Spatial Information Science for Natural Resource Management
Suraj Kumar Singh (Suresh Gyan Vihar University, Jaipur, India) Shruti Kanga (Suresh Gyan Vihar University, Jaipur, India) and Varun Narayan Mishra (Suresh Gyan Vihar University, Jaipur, India)
Engineering Science Reference • © 2020 • 355pp • H/C (ISBN: 9781799850274) • US $195.00

Innovations, Models, and Policies of Green Engineering Technologies
Daniel Dacuma Dasig Jr (De La Salle University Dasmarinas, Philippines) and Mirzi Llego Betasolo (Papua New Guinea University of Technology, Papua New Guinea)
Engineering Science Reference • © 2020 • 300pp • H/C (ISBN: 9781799840336) • US $195.00

Energetics Perspective on the Environmental and Human Impact of Buildings
Teodora Melania Şoimoşan (Technical University of Cluj-Napoca, Romania) and Ligia Mihaela Moga (Technical University of Cluj-Napoca, Romania)
Engineering Science Reference • © 2020 • 300pp • H/C (ISBN: 9781799835721) • US $195.00

IGI Global
PUBLISHER of TIMELY KNOWLEDGE

701 East Chocolate Avenue, Hershey, PA 17033, USA
Tel: 717-533-8845 x100 • Fax: 717-533-8661
E-Mail: cust@igi-global.com • www.igi-global.com

Editorial Advisory Board

Table of Contents

Detailed Table of Contents

An underwater wireless sensor network (UWSN) is a group of spatially distributed sensors for monitoring the physical changes of underwater environment and organize the measured data at a base station. Out of many big challenges in this field, network protocol design is of great concern due to the idiosyncratic nature of underwater environment. As underwater sensor networks are very much distinct in nature from terrestrial network so as the challenges, some of the major issues of concern in underwater environment are long propagation delay, available bandwidth, error rate, restricted battery power, movement of sensor nodes, etc. Because of all these issues and challenges, the routing protocols for terrestrial network can't be implemented in UWSNs. Hence, there is demand for new and different routing protocols for UWSNs. In this chapter, an overall survey of various routing protocols along with pros and cons are explained with challenges for designing new routing protocols for vast and enormous UWSNs.

Due to widespread applications and an orientation towards dimensionality enhancement, the underwater sensor networks (UWSNs) are going to be in demand for today's researchers. The core idea to design a strong and reliable build under the water revolves around the master plan of sensor deployment. The quality parameters like coverage, connectivity, and network lifespan, etc. are considered as main pillars for these deployment plans. There is always a trade-off between these quality parameters. So, it is quite necessary to maintain a balance among these parameters in order to establish a successful communication system. In this chapter, the authors have categorized the node deployment plans on the basis of the above-stated quality parameters. During this categorization, the view of dimensionality enhancement from two dimensions to three dimensions is also kept in the mind.

The chapter focuses on the recent development in the field of the sensor node deployment in the UWSN (under water wireless sensor network). In the chapter, the technical challenges during the node deployment of the sensor nodes in the UWSN (under water wireless sensor network) are represented with prefacing the background. The chapter focuses on the different methods of node deployment and presents a generalized model for ensure the reliability. A view of analyzing the deployment of sensor nodes is also shown in the example by following the recent researches in the domain. Finally, the future scope and conclusion is represented with the idea of new paradigms in the deployment of sensor nodes in the UWSN.

Underwater acoustic communication uses sound waves to trans-receive information, diving deep inside water, environment scanning, undersea explorations, disaster prevention, etc. In this chapter, an attempt has been made to cover stationary and mobile localization algorithm. They are further subdivided into distributed and centralized. Each one is further subcategorized into estimation-based and prediction-based schemes. The category therefore extends on the basis of ranging method, communication, and synchronization, some of which are area localization, sensor-based localization, forming a sensor array, motion-aware self-localization, silent localization. Each one will be discussed in detail in this chapter. At last, hybrid technique is also discussed, which combines stationary and mobile techniques. The discussion includes various nodes including anchor node, unknown node, sink node, and reference node. Various methods to follow the techniques are also discussed, which include anchor-based method, ranging method, and message communication.

In this technological arena, a firewall is a major tool for network security system to predetermine its basic rules so that it can control and monitor incoming and outgoing network traffic. The firewall generally restricts the faulty and unusual data entering the system. It can be categorized as host-based firewalls and network-based firewalls. There are various applications of firewall and how these firewalls can help the underwater wireless sensor network (UWSN) in protecting the sensor nodes where a huge amount of sensitive data is communicated among different countries. Firewall mainly aims to secure the network from different attacking devices and the attackers in a very efficient way. Firewall creates a barrier in which only the authorized data can be passed through after continuously checking by the firewall policies. To protect a local system and network of system from threats which are generally network based firewall is the most effective way. The design of the firewall is required to be accurate as it acts as a filter at higher protocol layer and some levels of IP-packets.

Underwater wireless sensor networks (UWSNs) contains many components such as vehicles and sensors that are deployed for cooperative monitoring and data collection tasks in a particular acoustic environment various nodes and ground-based stations use these networks interactively. Presently, UWSNs face the problems and obstacles regarding limited bandwidth, high propagation delay, 3D topology, media access control, routing, resource utilization, and power constraints. The research community has developed different methodologies over the past few decades to address these issues and challenges; but, due to complex characteristics of the underwater environment, some of them are still open to research. The main drawback of the traditional approach is the lack of direct interaction between different ends, recorded information can never be accessed during any mission, and data recorded will be lost in the event of any failure.

It is known that for underwater communication systems, lower frequency bands are used, which also affects the data rate transmission at lower rate. This is due to different physical phenomena such as reflections, refraction, and energy dispersion. This can be mitigated by installing wireless sensors that are placed close to each other, and hence, for accurate measurements, higher communication bandwidth is required. By using real scenario for measurement at higher data rate, 2.4GHz ISM band is used efficiently for underwater communication system with efficient use of energy.

Earth is solitary among the rocky planets revolving in the solar system. It is the only planet that is 3/4 covered with oceans of liquid water. In today's era, wireless sensor network is used in almost places and also is an interesting topic for researchers too. But with the advancement in technology, a network similar to wireless sensor network can be deployed under the water with acoustic signals for data transmission. Because of harsh environment and some more challenges, underwater communication becomes tricky. Underwater wireless sensor networks (UWSNs) is an advanced technology for underwater explorations and have demonstrated their quality in underwater applications. UWSN is a blend of wireless technology with sensor technology that has some extraordinary features like smart sensing, intelligent computing, and communication capabilities. Further, to maintain the secure communication underwater, it becomes more and more tricky. This chapter presents an expansive analysis of challenges, attacks, and security techniques of UWSNs.

This chapter is structured around the concepts of risk analysis due to underwater deployment of the cables for data transfer. Most of the organizations have deployed their networks below the water for shortening the distances between peers and also to reduce the physical destruction cost of cables. Furthermore, some organizations like Google kept their servers below the water because it reduced the cost of getting it cool, which in turn increases the efficiency. However, security consultants around the world in recent past expressed their considerations that a cyber-attack on these servers or cables might result in miserable economic collision. This might be overstated but there are infrequent situations in which cable breakage could be riotous. Although organizations cannot rule the threat of attacks on these apparatuses, there are fewer check measures that could reduce the possible attack chances in underwater communication.

With the enormous growth of sensor networks, information seeking from such networks has become an invaluable source of knowledge for various organizations to enhance the comprehension of people interests. Not only wireless sensor networks (WSNs) but its various classes also remain the hot topics of research. In this chapter, the primary focus is to understand the concept of sensor network in underwater scenario. Various mechanisms are used to recognize the activities underwater using sensor which examines the real-time events. With these features, a few challenges are also associated with sensor networks, which are addressed here. Machine learning (ML) techniques are the perfect key of success to resolve such issues due to their feasibility and adaption in complex problem environment. Therefore, various ML techniques have been explained to enhance the operational performance of WSNs, especially in underwater WSNs (UWSNs). The main objective of this chapter is to understand the concepts of UWSNs and role of ML to address the performance issues of UWSNs.

Ahmed Ali Saihood, University of Thi-Qar, Iraq
Laith Alzubaidi, University of Information Technology and Communications, Iraq

The wireless sensor networks have been developed and extended to more expanded environments, and the underwater environment needs to develop more applications in different fields, such as sea animals monitoring, predict the natural disasters, and data exchanging between underwater and ground environments. The underwater environment has almost the same infrastructure and functions with ground environment with some limitations, such as processing, communications, and battery limits. In terms of battery limits, many techniques have been proposed; in this chapter, the authors will focus in deep reinforcement learning techniques.

Gaurav Sharma, Seth Jai Parkash Mukand Lal Insitute of Engineering and Technology, India
Shilpi Harnal, Seth Jai Parkash Mukand Lal Insitute of Engineering and Technology, India
Neha Miglani, National Institute of Technology, Kurukshetra, India
Savita Khurana, Seth Jai Parkash Mukand Lal Insitute of Engineering and Technology, India

Underwater wireless sensor networks (UWSNs) have gained importance as well as diverted attention of many researchers, domain experts to a great extent in recent past. The devices used for UWSN deployment are resource-constrained like storage issue, low processing speed, as well are vulnerable to a wide class of security threats and malicious attacks, which affect reliable communication. For reliable data delivery, a system should include packet delivery ratio, battery life, delays incurred, and energy consumption, etc. Numerous reliability models for underwater networks have been designed to incorporate the parameters and performance metrics in optimized manner. The chapter deals with focusing on such models and their efficiency in terms of battery life, packet loss, error handling mechanism, and network delays. Further, it is also explained how and why the error-controlled schemes should be designed and implemented in order to incorporate reliable data delivery in limited resources-constraints of UWSN along with the consideration of efficiency and performance concerns.

Renu Popli, Chitkara University Institute of Engineering and Technology, Chitkara University, Punjab, India
Isha Kansal, Chitkara University Institute of Engineering and Technology, Chitkara University, Punjab, India
Kanwal Garg, Department of Computer Science and Applications, Kurukshetra University, India

Due to immense application scenarios emerged by underwater communication, the researchers have shown

The earth is covered 71% by water and the human utilizes the remaining 29% for their shelter and living. The idea of living under the water is possible only in fiction movies for our ancestors. But in this century, the idea of living in an underwater city has become a reality with the development of the existing technology. The exploration of the undersea is booming in the science community, which shows the path for underwater cities, underwater museum, and underwater hotels. This chapter contributes the information related to underwater smart cities in three folds: (1) discusses the major challenges in developing the underwater infrastructure, (2) introduces the internet of underwater things components involved in interconnecting the devices for underwater acoustic communication, (3) list the examples of the existing masterpiece architecture constructed underwater.

Foreword

Investigators have contributed greatly in the field of terrestrial sensor networks and nowadays the investigators are pioneering their efforts towards Underwater Wireless Sensor Networks (UWSNs) being a forthcoming area of research. The major similarity between these two sensor-based networks being the composition of network consisting of sensor nodes to collect information and forwarding it to the sink. The major difference lies in terms of communication method of acoustic signal, cost of expensive sensors, greater memory space to save data, power for communication, and dense deployment of sensors. Further, realization of UWSN has fair amount of challenges like large propagation delays, limited link capacity, narrow bandwidth, shadow zones, harsh geographical atmosphere, attenuation, comparatively smaller network scale, high bit error rates, limited energy and temporary losses of connectivity. Since many years various UWSN protocols have been designed and/or existing protocols have been improvised for effective and efficient communication. This book indeed provides very useful insights of the communication issues in underwater sensor networks.

Chapter 1 describes about issues and challenges while designing routing protocols in UWSN. In this chapter, an overall survey of various routing protocols along with pros and cons are explained with challenges for designing new routing protocols for vast and enormous UWSNs. Chapter 2 details the node deployment plans on the basis of various quality parameters. The core idea in this chapter is to design a strong and reliable model under the water that revolves around the master plan of sensor deployment. Chapter 3 focuses on the different methods of node deployment and present a generalized model for ensure the reliability while node deployment. A view of analyzing the deployment of sensor nodes is also shown in the example by following the recent researches in the domain. Chapter 4 tries to cover stationary and mobile localization algorithms along with hybrid methods which are further subdivided into Distributed and centralized. The discussion includes various nodes that is anchor node, unknown node, sink node, and reference node. Chapter 5 elaborate challenging aspect of underwater data fusion for intelligent traffic control and underwater vehicle navigation systems. This chapter provides an insight into basic concepts about multi-sensor data fusion and detailed review of most popular data fusion architectural models available for UWSN.

Chapter 6 analyses in the process of transmission and reception of acoustic waves that it is the only best method to communicate under the water. In this chapter, the congestion control mechanism for sensor nodes in UWSN is described. Chapter 7 presents a survey of available fault management techniques and their pros and cons for further advancement in underwater sensor networks to highlight new research trends. Chapter 8 demonstrated firewall as a major tool for network security system to control and monitor incoming and outgoing network traffic. In this chapter, it is shown that how various applications of firewall can help the UWSN in protecting the sensor nodes where a huge amount of sensitive data is communicated. Chapter 9 offer a review over various error control and reliable data delivery schemes in UWSN. The research community has developed different methodologies over the past few decades to address these issues and challenges. Chapter 10 launches the need for higher data rate transmission for UWSN communications using Industrial, Scientific and Medical (ISM) bands. The underwater problems can be mitigated by installing wireless sensors which are placed closed to each other and hence for accurate measurements, higher communication bandwidth is required.

Chapter 11 presents the need for secure communication techniques for UWSN. This chapter presents an expansive analysis of challenges, attacks and security techniques of UWSNs. Chapter 12 highlights the issues those surround the problems due to cyber-attacks and the need of cyber security in underwater communication. Furthermore, some organizations like Google kept their servers below the water because it reduces the cost of getting it cool which in turn increases the efficiency too. Chapter 13 shows the mechanism to use machine learning (ML) techniques for operating collected data from UWSN. Various ML techniques have been explained to enhance the operational performance of WSNs; especially in UWSNs. Chapter 14 intricate the deep reinforcement learning methods for energy efficient underwater wireless networking. The underwater environment has almost the same infrastructure and functions with ground environment with some limitations, such as, processing, communications, and battery limits.

Chapter 15 expands the reliability in underwater wireless sensor networks where numerous reliability models for underwater networks have been designed to incorporate the parameters and performance metrics in optimized manner. The chapter deals with focusing on such models and their efficiency in terms of battery life, packet loss, error handling mechanism and network delays. Chapter 16 explains the importance of real time data in marine environment for plenty of applications and discuss the major issue in efficient data communication. Chapter 17 reviews applications surrounding demur and routing protocols with applications in UWSN for smart city. This chapter investigate how to be profited by the Internet of Underwater Things (IoUT) to gain from, adventure and safeguard the normal submerged assets and we bring up the contrasts among UWSNs. Chapter 18 conclude the concepts of risk analysis due to underwater deployment of the cables for data transfer. Underwater cable deployment facilitates the coverage and speed of internet all over the world for various applications like international trade, various communication transfers, and other daily end user requirements. Chapter 19 finally concludes and presents the application area of UWSN in forming the smart city below the water. The exploration of the undersea is booming in the science community, which shows the path for underwater cities, underwater museum, and underwater hotels.

I appreciate the efforts put in by the authors for bringing out recent trends in the upcoming area of underwater network illustrating variety of applications like pollution monitoring especially chemical waste, monitoring of the population of underwater flora and fauna, the examining of the health of rare marine creatures, mine reconnaissance, disaster prevention, assisted navigation, nutrient production, oil leakage detection, distributed tactical surveillance, oceanographic data collection, target detection, tracking and underwater military applications etc.

I am delighted to see the domain knowledge of Editors being fructified in the form of this book. Their sound knowledge of the subject matter has indeed made it simple to present the text which can be used as a text book for students and a reference guide for researchers. I wish them all the very best in their future endeavors.

Mayank Dave
National Institute of Technology, Kurukshetra, India

Preface

The goal of this book is to share experiences and knowledge about the applications, data and resources deployed under the water. This book describes security attacks, trends and scenarios, along with protocol, topology for reliable data communication; and discusses the applicability of machine learning in UWSN. Nearly 70% portion of the globe is covered with water and most of this is unexplored like rivers, sea, oceans etc. Recently, the exploration by the researchers is ominously growing in this region because of regular incidences or calamities. It's not only rich in resources but also it's playing an important role for underwater exploration like mine exploration, assisted navigation, environment monitoring, military purpose, oceanographic data sampling network for disaster management or prevention, tactical surveillance for attack protection. The most important concern for the researcher is to handle such a big data of naturally existing underwater area. UWSN intended as an aquatic medium for variety of geological resources like gas, oil, detecting and tracking fish banks etc. Inventors analyzed the extensive underwater environment by developing various UWSN protocols as in 1826, Lake, Charles and Daniel evaluated sound velocity in water and in 1906, Lewis invented sonar type listening device. Further in 1915, Paul invented submarine localization and in early 90's, underwater vehicles and surface floats coordinated the data in the form of signals. In diverse aquatic mediums the UWSN protocols are being used for monitoring purpose. However these protocols may not give the effective information in case of oceans, sea etc. as these are larger in area and there are numerous factors like area, water level, pressure etc. that are required to be analyzed at real time. Disasters like earthquake, volcanic eruption, tsunami, or other seismic activities affect the larger area where there are number of hurdles in collection and storage of data in a readily available manner.

The target audience of this book will include Researchers, IT Industry, Government Research Agencies, Naval Forces, Machine Learning experts, concerned with optimized data flow in communication network, securing assets and mitigating security attacks.

ORGANIZATION OF THE BOOK

The book is organized into 19 chapters. A brief description of each of the chapters of this book is as follows:

In Chapter 1, a detailed overview about issues and challenges while designing routing protocols in UWSN is explained. As UWSN is a group of spatially distributed sensors for monitoring the physical changes of underwater environment and organize the measured data at a base station. Out of many big challenges in this field, network protocol design is of great concern due to the idiosyncratic nature of underwater environment. In this chapter, an overall survey of various routing protocols along with pros and cons are explained with challenges for designing new routing protocols for vast and enormous UWSNs.

Chapter 2 categorize the node deployment plans on the basis of various quality parameters. The core idea in this chapter is to design a strong and reliable model under the water that revolves around the master plan of sensor deployment. The quality parameters like coverage, connectivity, and network lifespan, etc. are considered as main pillars for these deployment plans. There is always a trade-off between these quality parameters. So, it is quite necessary to maintain a balance among these parameters in order to establish a successful communication system. In the above-stated categorization, the view of dimensionality enhancement from two dimensions to three dimensions is also taken care.

Chapter 3 identifies the recent development for sensor node deployment in UWSN with technical challenges prefacing the background. The chapter focus on the different methods of node deployment and present a generalized model for ensure the reliability while node deployment. A view of analyzing the deployment of sensor nodes is also shown in the example by following the recent researches in the domain. Finally the idea of new paradigms in the deployment is also shown.

Chapter 4 made an attempt to cover stationary and mobile localization algorithms which are further subdivided into Distributed and centralized. The categories extends on the basis of ranging method, communication and synchronization, some of which are area localization, sensor based localization, forming a sensor array, motion aware self-localization, silent localization. The discussion includes various nodes that is anchor node, unknown node, sink node, and reference node. Various methods to follow the techniques are also discussed, which include, anchor based method, ranging method, and message communication.

Chapter 5 discuss the most critical issues related to UWSN as limited bandwidth, data aggregation, data fusion, data collection, routing, media access control, higher propagation delay and higher power consumption. Underwater data fusion is considered as an equally important and challenging aspect for underwater intelligent traffic control and underwater vehicle navigation systems. This chapter provides an insight into basic concepts about multi-sensor data fusion and detailed review of most popular data fusion architectural models available for UWSN. Finally, the chapter also provides open research directions to researchers for multi-sensor data fusion.

Chapter 6 details the process of transmission and reception of messages by propagating sound in an underwater environment which is known as acoustic communication. Transmission of acoustic waves is the only method to communicate underwater, as radio waves get attenuated severely and there is severe scattering in optical transmission. In this chapter, the congestion control mechanism for sensor nodes in UWSN is described.

Chapter 7 presents a survey of fault management techniques in UWSN regarding types of faults and their classification, environmental factors influencing the identification of faults, fault detection schemes, issues, and future directions is performed. This chapter outlines available fault management techniques and their pros and cons for further advancement in underwater sensor networks to highlight new research trends.

Chapter 8 analyses the firewall as a major tool for network security system to predetermine its basic rules so that it can control and monitor incoming and outgoing network traffic. It can be categorized as host based firewalls and network based firewalls. In this chapter, it is shown that how various applications of firewall can help the UWSN in protecting the sensor nodes where a huge amount of sensitive data is communicated among different countries. Firewall mainly aims to secure the network from different attacking devices and the attackers in a very efficient way.

Chapter 9 presents a review over various error control and reliable data delivery schemes in UWSN. Various nodes and ground-based stations use these networks interactively. Presently, UWSNs face the problems and obstacles regarding limited bandwidth, high propagation delay, 3D topology, resource utilization, and power constraints. The research community has developed different methodologies over the past few decades to address these issues and challenges; but, due to complex characteristics of the underwater environment, some of them are still open to research.

Chapter 10 establishes the need for higher data rate transmission for UWSN communications using Industrial, Scientific and Medical (ISM) bands. It is known fact that for underwater communication system, lower frequency bands are used which also effects the data rate transmission at lower rate. This is due to different physical phenomena such as reflections, refraction and energy dispersion. This above problems can be mitigated by installing wireless sensors which are placed closed to each other and hence for accurate measurements, higher communication bandwidth is required.

Chapter 11 establishes the need for secure communication techniques for UWSN. Because of harsh environment and some more challenges, underwater communication become tricky. UWSN is a blend of wireless technology with sensor technology that has some extraordinary features like smart sensing, intelligent computing, and communication capabilities. This chapter presents an expansive analysis of challenges, attacks and security techniques of UWSNs.

Chapter 12 reviews issues surrounding problems due to cyber-attacks and the need of cyber security in underwater communication. Now a day's most of the organizations have deployed their network below the water for shortening the distances between peers and also to reduce the physical destruction cost of cables. Furthermore, some organizations like Google kept their servers below the water because it reduces the cost of getting it cool which in turn increases the efficiency too. Although organizations cannot rule the threat of attacks on these apparatus, but there is fewer check measures that could reduce the possible attack chances while underwater communication.

Chapter 13 works on machine learning techniques to operate on data collected from UWSN. In this chapter, the primary focus is to understand the concept of sensor network in underwater scenario. Machine learning (ML) techniques are the perfect key of success to resolve such issues due to their feasibility and adaption in complex problem environment. Therefore, various ML techniques have been explained to enhance the operational performance of WSNs; especially in UWSNs. The main objective of this chapter is to understand the role of ML to address the performance issues of UWSNs.

Chapter 14 discuss the deep reinforcement learning methods for energy efficient underwater wireless networking. The underwater environment needs to develop more applications in different fields, such as sea animals monitoring, predict the natural disasters, and data exchanging between underwater and ground environments. The underwater environment has almost the same infrastructure and functions with ground environment with some limitations, such as, processing, communications, and battery limits. In this chapter, authors have focused in deep reinforcement learning techniques.

Chapter 15 elaborates the reliability in underwater wireless sensor networks. Numerous reliability models for underwater networks have been designed to incorporate the parameters and performance metrics in optimized manner. The chapter deals with focusing on such models and their efficiency in terms of battery life, packet loss, error handling mechanism and network delays. Further, it is also explained that how and why the error controlled schemes should be designed and implemented in order to incorporate reliable data delivery in limited resources-constraints of UWSN along with the consideration of efficiency and performance concerns.

Chapter 16 elaborates that in UWSN, sensor nodes are deployed at different locations to collect the information and forward it to the sink node at surface. As a node cannot directly communicate to the sink node at surface, the multi hop communication takes place which possess great challenges under water such as limited resource, bandwidth unavailability, end to end delays, temperature, water pressure etc. This paper presents importance of real time data in marine environment for plenty of applications and discuss the major issue in efficient data communication.

Chapter 17 reviews applications surrounding demur and routing protocols with applications in UWSN for smart city. To help the idea of Internet of Underwater Things (IoUT) is characterized as the system of brilliant consistent submerged items and an overall system of savvy interconnected submerged items that empower to screen tremendous unexplored water zones. This chapter investigate how to be profitted by the IoUT to gain from, adventure and safeguard the normal submerged assets and we bring up the contrasts among UWSNs. Also, the difficulties for IoUT with diverse protocol deployments is also discussed.

Chapter 18 conclude the concepts of risk analysis due to underwater deployment of the cables for data transfer. Underwater cable deployment facilitates the coverage and speed of internet all over the world for various applications like international trade, various communication transfers, and other daily end user requirements. This chapter solves two primary goals. The first is to highlight the issues in underwater cable deployment that may be seen or unseen leads to various potential risks that could interrupt cable-supported services. The second goal is to explore the various possibilities for securing the organization to ensure business continuity.

Chapter 19 finally concludes and presents the application area of UWSN in forming the smart city below the water. The exploration of the undersea is booming in the science community, which shows the path for underwater cities, underwater museum, and underwater hotels. This chapter contributes the information related to underwater smart cities in three folds as by discussing the major challenges in developing the underwater infrastructure and to introduce the IoUT components involved in interconnecting the devices, and to list the examples of the existing masterpiece architecture constructed underwater.

Thus, the aim of this book is to familiarize researchers with the UWSN area starting from routing of data to its applications through fault management and error control for energy efficient reliable communication and networking.

Acknowledgment

First of all, we express our gratitude to the Almighty, who blessed us with the zeal and enthusiasm to complete this book successfully. We are extremely thankful to get the continuous guidance from Prof. Mayank Dave, Computer Engineering Department, National Institute of Technology, Kurukshtera, and Prof. Anil Kumar Verma, Computer Science and Engineering Department, Thapar University, Patiala for their tireless efforts to get in-depth knowledge of underwater wireless sensor networks field. Also we are thankful to Prof. Lalit Kumar Awasthi, Director, Dr. B. R. Ambedkar National Institute of Technology, Prof. Ajay Kumar Sharma, Vice-Chancellor, Inder Kumar Gujral Punjab Technical University, Prof. Awadhesh Kumar Singh, Computer Engineering Department, National Institute of Technology, Kurukshtera, for their support throughout the life cycle of this book.

We are also thankful to Dr. Ashok Kumar Chitkara, Chancellor, and Dr. Madhu Chitkara, Pro-Chancellor, Chitkara University, Punjab, India for their fruitful guidance and every possible support at Chitkara University. We are also thankful to Prof. Archana Mantri, Vice-Chancellor, Chitkara University, Punjab for her continuous motivation. Her wide knowledge and logical way of thinking have been a great value for us. We especially thank our co-authors for contributing their valuable work, without which it was impossible to complete this book.

This book would not have been possible without the heartiest support of our family. Our deepest regards to our Parents for their blessings and affection even during the tough period. Last but not the least, we thank to GOD, for giving us the strength and wisdom to carry out this work successfully.

Nitin Goyal
Chitkara University Institute of Engineering and Technology, Chitkara University, Punjab, India

Luxmi Sapra
Chitkara University Institute of Engineering and Technology, Chitkara University, Punjab, India

Jasminder Kaur Sandhu
Chitkara University Institute of Engineering and Technology, Chitkara University, Punjab, India

Chapter 1
Routing Protocol Design Issues and Challenges in Underwater Wireless Sensor Network

Monika Choudhary
Chitkara University Institute of Engineering and Technology, Chitkara University, Punjab, India

Nitin Goyal
https://orcid.org/0000-0001-7878-363X
Chitkara University Institute of Engineering and Technology, Chitkara University, Punjab, India

ABSTRACT

An underwater wireless sensor network (UWSN) is a group of spatially distributed sensors for monitoring the physical changes of underwater environment and organize the measured data at a base station. Out of many big challenges in this field, network protocol design is of great concern due to the idiosyncratic nature of underwater environment. As underwater sensor networks are very much distinct in nature from terrestrial network so as the challenges, some of the major issues of concern in underwater environment are long propagation delay, available bandwidth, error rate, restricted battery power, movement of sensor nodes, etc. Because of all these issues and challenges, the routing protocols for terrestrial network can't be implemented in UWSNs. Hence, there is demand for new and different routing protocols for UWSNs. In this chapter, an overall survey of various routing protocols along with pros and cons are explained with challenges for designing new routing protocols for vast and enormous UWSNs.

INTRODUCTION

Almost 71% of earth's surface area is covered with water. Due to this large coverage, huge opportunities and way of wireless communication under the water attracted many of the research participants and also experts from the industry. Underwater Wireless Sensor Networks (UWSNs) are the networks, consisting of number of sensor nodes which sense or control various activities that are occurring under the water and the vehicles which are deployed under the water for communication (Kumar et al., 2014). One of

DOI: 10.4018/978-1-7998-3640-7.ch001

the major differences between UWSN and terrestrial networks is the use of signals i.e. acoustic signals for UWSNs communication, whereas WSNs uses radio signals plays an important role in communication. This movement from radio signals to the acoustic channel comes because of the low pursuance of the radio signals in the aqueous environment. Underwater Sensor network have received extensive attention of many research personnel and form the industry experts due to the increasing number of utilization for ample range of usage involved in military, scientific, environmental and commercial sectors. These areas have a huge scope of underwater sensor network applications in it (Goyal et al., 2014). To accomplish all these work and monitoring an efficient communication among the sensor nodes is required which is a challenging task due to the hard and rigid conditions under the water as compare to terrestrial network which are completely different to it. For this purpose we need highly efficient routing protocol that provides shipshape and satisfactory communication between all the nodes and maximizing the performance (Khurana et al., 2012). Many of researchers have worked on it and designed superior quality routing algorithms but still it needs improvement (Gaba et al., 2013; Goyal et al., 2013). The major challenges in aqueous environment are float mobility, long propagation delays that is due to the acoustic signal used for communication that is almost 1500 meter per second, bandwidth limitations (approximately less than 100 KHz) and water pressure (Goyal et al., 2016). Due to this, UWSN degrades the output in terms of the throughput and life of nodes. The design of such routing protocol faces major challenges because of the idiosyncratic properties of the acoustic communication channel, an efficient routing protocol for data forwarding is crucial to UWSNs.

The dynamic network topology is also one of the challenges in underwater environment. Every algorithm works according to their routing techniques and the problem area it handles and solves in that particular domain. In UWSN those protocols are required that are more smart, reliable and perform well with respect to the covet parameters (Goyal et al., 2017). In this chapter, all the UWSN requirement with its architecture and its advantages and disadvantages are discussed which can provide basic concepts to researchers with clear and unambiguous perspicacity for further research. The main contribution of this article is towards the challenges and design issues regarding the routing protocol which is of high demanding area of research. Routing is a technique that enables the movement of packets or messages from source to sink node (Goyal et al., 2018). Whenever there is any kind of communication involved in the networks there is need of routing as it is also defined as the method of selecting or finding best route for routing or forwarding data packets. Routing always involves the concept of network topology which is the topological structure or arrangement of network physically and logically. Here in UWSNs, the sensor nodes having different capabilities are spread out in aqueous environment.

All the sensor nodes are connected using any network topologies which construct its network structure. But Routing packets, in UWSN face numerous challenges; one of them is that dealing with void communication areas (Goyal et al., 2019a; 2019b). The communication takes place by broadcasting packets from bottom of the level to the seashore, in this all the deployed sensor nodes communicates with other nodes to determine and uses routes that performed well according to the condition for selection which are followed for routing. Routing protocols used in UWSNs chooses such routes to forward data packets for reliable and efficient communication. Newly, it draws an attention of many of the researcher participants, practioners and engineers and they have utilized these protocols to monitor and investigate the aqueous medium for highest applications. These applications include monitoring and controlling of the under the water conditions for different purposes like army and habitat monitoring, disaster forecasting and its management and general exploration of underwater environment. As an example of UWSN there is large number of application areas one of them is Sea-web. In sea web there is a large network that is deployable under the sea for many applications in the area of military. The main objective of sea web was to examine the technique that is appropriate for communication with and apprehension of submarines that are present under the water. All these distributions were in coastal sea zones for long periods. UWSN is advantageous because of

- **Low Complexity:** Underwater sensor network is an example of wireless network that avoids lot of wiring which is used to connect various components and sensor nodes of UWSN. It further saves cost and also decreases network complexity.
- **High Scalability:** There is high degree of scalability in UWSNs as new devices can be easily added to the network. It can accommodate N number of devices at any time which improves the overall performance of the network.
- **Centralized Monitoring:** There are many applications like ocean monitoring where underwater sensor network can be easily accessed through a centralized monitor.
- **Flexibility:** UWSN is very flexible as it passes through physical partition that is aimed at the message passing among the sensor nodes and different components.

But advantages comes with some disadvantages as well, so some of the disadvantages of UWSN are

- **Monitoring:** UWSNs are not capable to do real time monitoring, as the data that is recorded through the devices is accessed after the devices regained from water which can take considerable amount of time.
- **Onshore-Offshore Communication:** The interaction between all the onshore control system and rest of the monitoring and controlling devices that are offshore can't be carried out easily.
- **Limited Storage:** The devices that are used for communication like sensors have limited storage space to store data due to their design implications.
- **Fault Tolerance-**when sensor system has got failure and misconfigurations occurs, it is very crucial to detect the damage sensor where actual problem lies.

UWSN do have applications in various unexplored emerging fields in the domain of underwater environment. On broad level we can have the UWSN applications such as monitoring, military, navigation, and natural disaster forecasting that can significantly affect the habitat living under the water. Different application are explained as follows-

- **Ocean Monitoring:** This particularly refers to the underwater network which is established to monitor the underwater habitat, their attributes and properties.
- **Quality of Water:** The habitat living under the water is highly conscious towards the quality of water as it is biggest factor affecting their life.
- **Water Resources**: This application area included the exploration of various natural resources like we can explore, management crust available under the water and like this we can also explore some other resources.
- **Oil Monitoring**: The spillage of oil causes harmful effects to the marine life. So by using UWSN, we can monitor the thickness of oil spill in water which will be further helpful in the cleaning process.

- **Earthquake and Tsunami**: There are the natural disasters which can came at any time or at any place, so same is the case with the environment in the water also.
- **Flood Occurrence**: This is also kind of natural disaster that can happen at any place and also happens frequently so it requests timely alerts. Various resources have proposed solve for these kinds of problems using various hardware in collection with sensor nodes.
- **Military**: In military surveillance UWSN presents a great help and it is implemented with the help of sensor nodes function activities.

UWSN System Architecture i.e. basic requirement for the design of new application in this field can be divided into four parts. The techniques of optical, Radio Frequency links and acoustic communication channel can be used for all kind of communications.

- **1D UWSN Architecture:** This architecture basically refers to interconnection of nodes creating a network where all the nodes are spread out in self-governing fashion. Each of the sensor nodes is itself a stand-alone network that transmits the information to the remote node, sensing and processing. In this architecture, the communication between the nodes occurs using various techniques like acoustic channel, Radio Frequency waves or optical waves as communication method. The transmission among the different sensor node and other node is carried out hop by hop, it means on single hop distance.
- **2D UWSN Architecture:** In this architecture, a network is created in group of nodes also called as cluster that is spread out under the water. Every cluster is maintained by a cluster head (CH) (as anchor node).Each member node that is part of one cluster gathers the information and forwards it to respective CH. Similarly the CH gathers the data from all of its respective members and forwards it to the node present on the surface named as surface buoyant nodes. In it, various communication medium techniques like acoustic channel, optical, and Radio Frequency (RF) communication can be used.
- **3D UWSN Architecture:** All the nodes are spread out under the water that make clusters and are anchored at various heights or levels. Due to the deployment of sensors at altered heights or depths, the entire message passing occurring between the sensors goes from2D to the next higher level that is one dimension further. There exist 3 communication scenarios or three dimensions: inter-cluster, intra-cluster, and CH - surface buoyant node communication methods.
- **4D UWSN Architecture:** 4D architecture worked with the collaborations of fixed and mobile UWSNs. This architecture uses the features of both the architecture by mixing them according to the application and use. As the name suggests, the mobile UWSN nodes are remote in nature that works as remote operative underwater vehicles (abbreviated as ROVs) that collects the information from CHs and passes the information to different remote stations.

RELATED WORK

After the fruitful analysis and research on the Wireless Sensor Network (WSN) diverts the inclination headed for research on UWSN. Mukhtiar Ahmed et al. (2017) have described many research challenges, out of them design of routing protocol is also of significant importance. Data forwarding is one of problem for forwarding data packets along with the route development, route maintenance of the routing protocols. Abdul Wahid et al. (2010) presented an energy effective routing protocol EEDBR that handles the problem of battery replacement and battery recharge problem of nodes. The protocol, for forwarding the message packets mainly used the depth parameter of the sensor nodes.

Hao Qin et al. (2014) focuses on the unbalanced energy consumption that tends to deteriorate the lifetime of network. To address this problem, they propose a combination of Depth-Controlled and Energy-Balanced Routing Protocol. The proposed routing protocol reinstate the nodes with low-energy, having less capability efficiently with the nodes having high-energy that are very good in performance after adjusting their depths, for achieving uniformly consumption of energy throughout the whole network which improves the lifetime and functioning of overall network. Peng Xie et al. (2006) presented the concept of a geographic routing protocol known as VBF. In VBF, source to destination path is truly given by a vector, there is no requirement of state information of the sensors, and for routing merely sensors are involved in routing. For improving the reliability, the packets are broadcasted from source to target on redundant and interleaved paths. Tariq Ali et al. (2014) projected a unique routing protocol for controlling one of the main drawbacks of underwater sensor that is network issues of continuous node movements, inefficient energy consumption, and delays, known asL2-ABF. In this protocol every node calculates its flooding angle that is required for furthering the data packets headed for sink node. Awan et al. projected an energy-efficient adaptive clustering algorithm for UWSN. This algorithm uses a hierarchical structure of network at multi-level that is completely based on the hop distance between different CHs, main entity responsible for maintenance of whole cluster and the destination node.

Now days the underwater communication utilizes mainly acoustic channels, electromagnetic, and optics as data transmission techniques for sending data at various positions. The underwater communication is mostly affected by the conducting feature of seawater and optical waves as they can be applied on very short distances and optical waves are easily absorbed by the seawater. Sound's speed does not remain constant all the time but, it variates from position to position. The main concept that is used under the water is acoustic channel communication that performed better for communication in underwater because of less attenuation property. Acoustic communication also performed better in deep sea water as it has less attenuation in it. The performance of acoustic communication is affected by the temperature gradients, multipath effect and ambient noises regarding surface.

Due to unique challenges of UWSN such that movement of sensor nodes with water currents in ocean, costly senor nodes, limited battery power and inapplicability of global position system, they cannot be employed underwater straight forward. Guo et al. (2013) deliver an algorithm AFLA. Its main purpose is for making an effective localization based technique for UWSN. This algorithm presents a property of self-localization or self-configuration for anchor-free nodes. AFLA is powerful localization schemes that can be utilized in all kind of network conditions. Efficient energy consumption is the key characteristics that are mainly required for underwater environment. Such protocols are easy to manage dynamic asymmetric links and large propagation end to end delay.

Du et al. (2014) presented Level-Based LB-AGR protocol that categorizes traffic basically into the four parts of communication-

- Up-stream to destination
- Down-stream to different sensor nodes
- Down-stream to some particular nodes
- Down-stream to all type of nodes that are involved in communication

Presently, for all routing protocols that are working for UWSN, their major requirements are minimum propagation delays and high power efficiency. Huang et al. (2014) presented a routing technique based on energy consumption, with the concept of fuzzy logic and decision tree. This routing protocol focuses on effective consumption of battery power, so that the expenditures for energy get reduced. This protocol successfully reduces the main problem of traffic overloading in acoustic communication and also effectively decreases the energy requirements of different nodes.

Basagni et al. (2012) presented the technique of data fragmentation that focuses on minimizing the problem of collision by partitioning or dividing the long data packets into very short parts or fragments. UWSN mostly uses the half-duplex communication technique i.e. easily implemented by selecting optimum data packet size. In the suggested system communication for underwater; all sensor nodes work in a team for implementing all necessary actions or decision. The main advantage of performing fragmentation is that it decreases retransmissions that occur due to packet loss during collisions, energy consumption, and high overhead. Basagni et al. (2012) ignored the factor of variability in bandwidth and interference caused during communication, hence as a effect of performing fragmentation the overall traffic load increases and the communication channel gets congested.

UWSN ROUTING PROTOCOLS

Under this section layout of various routing protocols, proposed for UWSNs is presented by classifying them into various groups. Here various routing schemes that have been suggested in the area of literature are elaborated.

The protocols can be mainly classified into 4categories i.e. nodes location, level based (Depth based), energy efficiency based and also based on reliability. The prime differences between all these categories are based on their application area. The requirement of the application of the network and the optimal routes are the factors which are required for designing of various routing protocols along with other features and parameters. Here the routing protocols directly could not be used for UWSN as there is lot of difference between the environment of both the networks that is UWSN and terrestrial network. As terrestrial network works onto other parameters in which all the factors vary according to use.

The location-based categorized routing protocols, works on the principle that mostly all the times nodes are awake of all their positioning coordinates with the guidance of some of the localization enabled services.

Level-based routing protocols or sometimes called depth-based which uses the depth information of particular nodes to route the packets. There exists a grouping of vector-based (VB) and location-based (LB) routing protocols like VBF, Hop-by-Hop VBF (HHVBF), and Adaptive HHVBF (AHHVBF), which restricted the grouping of intermediate nodes inside the virtually created pipeline goes towards the only sink on the surface of water. The VBVA is also vector-based protocol. It is mainly designed with a void-handling technique. Also, this type of protocol is very arduous to be accepted in an aqueous environment.

Another category is based on one of the important parameter that is energy efficiency of the nodes involved in the communication. Under this category one protocol is Energy Efficient Routing Protocol. The depth of node at which it is deployed, is used as a basis for reducing the energy consumption. In this protocol, mainly hop by hop broadcasting of packets implemented. Another protocol is Temporary Cluster Based Routing (TCBR) protocol, does not implemented the concept consumption of energy should be equal for all the sensor nodes deployed throughout the whole network. It works on assigning

Figure 1. Classification of UWSN routing protocol

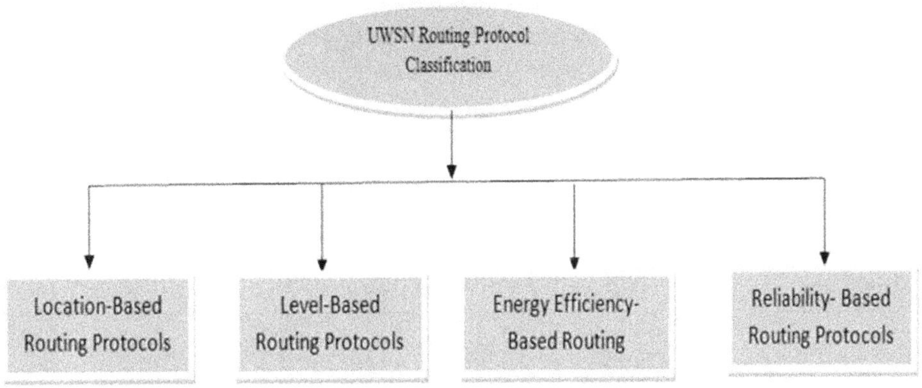

the head for one particular cluster that is known as CH for that particular cluster and responsible for all the functioning of that cluster. This protocol also helps to save one of the important parameter that is energy consumption, as the main concept used in this is to broadcast the sensed data through the short distances. TCBR also not requires any information of location of sensor nodes of the network and in the routing process only a less quantity of sensors are involved throughout the communication.

Next protocol under this category is ERP2R that is based on distance and residual energy. The key concept behindhand is to utilize the important routing metrics or parameter as -

- The physical distance among the nodes involved in the communication
- To balance consumption of energy among all the sensor nodes.

Furthermore, for selecting the forwarding nodes or intermediate node, their link quality to intermediate nodes is also measured to deliver reliability for communication and residual energy of nodes to enhance lifetime of the network. This protocol comes under the reliability based routing protocols as it also works for maintaining the reliability of the communication in the network. DFR is another location-based flooding routing scheme equipped with a void-controlling or maintaining technique. DFR based on the concept of applying flooding approach to communicate the packets to other nodes. FBR is focused beam routing protocol which applies precautionary void-controlling techniques for handling the problem of void-transactions. The main focus of FBR is to minimize consumption of energy by controlling transmission power of the forwarding nodes. Further, receiving and sending the packets at each distance to create and maintain a session is very long process.

DBMR and EEDBR protocol are Level-based or Depth- based protocols without any technique to handle the problem of void. In both the protocols, every node delivers the control packet to a neighboring node with lower depth till the control packet is finally delivered to any of the sink nodes on the seashore. To handle the problem of void, the solution is Hydro Cast that is of two parts:

- Greedy pressure-based routing algorithm
- Lower-depth-first method of recovery.

In this, the recovery paths and void nodes can be identified at early stage, by using the depth-based characteristics of all the deployed sensor nodes that are dispersed in aqueous environment.

Many of the routing protocols have been extensively compared on the basis of different parameters and presented in Table 1. The protocols have also been compared based on the problem they address, the parameters and points that are used to make the routing decision.

Table 1. Routing protocols comparison for UWSN

Protocol	Routing Technique	Area Covered	Pros	Cons	Reference
VBF	Choose a node for forwarding within a pipeline from source to destination	Mobility of nodes	Energy consumption is less, Improved scalability	Dead nodes problem	SM Ghoreyshi et. al (2017)
HH-VBF	Multiple pipelines are used for forwarding from source to destination	Efficient load management as load is distributed	Minimized dead node problem as load is managed through different pipelines	Computational delay gets high due to separate pipe for each node	PengXie et al., 2006
AHH-VBF	Void regions are considered and improves VBF performance	Energy Consumption and void region problem	Energy consumption is comparatively less	Some packet are lost to save energy consumption	Mukhtiar Ahmed et al., 2017
DFR	Restricted forwarded zone is created for flooding between the source, intermediate and destination nodes	Harsh Channel conditions	Extenuation of reluctant conditions	Redundancy in Packet transmission due to flooding causes high energy consumption	Hao Qin et al., 2014
FBR	Retransmit the selected from source to destination	Main problem of energy consumption	Lower the end to end delay and energy efficient	Low throughput when density is low	Faiza Al Salti et al., 2017
DBMR	Make use of lowest depth sensor node from bottom to top during flooding	Mobility of nodes	Does not require the full dimensional information about the node and high throughput	Life of low depth nodes is short as they are always used as forwarder nodes.	Anwar Khan et al., 2018
EEDBR	Handle the problem of short life span of low depth nodes	Balance the energy level and control the short life span of low depth nodes	Energy consumption is balanced and low depth node problem is solved	Less reliability as sender node selects the forwarder nodes always	Abdul Wahid et al.,2010
DTMR	Direct transmission from Source to destination	Energy consumption is high due to the signaling overhead	Lower delay and throughput is high	Reliability is poor in case of direct transmission	Anwar Khan et al.,2018
TCBR	Broadcast the sensed data through the short distances	Consumption of Energy is reduced	Reduced overhead of traffic and does not require information of all the nodes, all the time	All the communication dependent on the performance of CH	Awan et al.,2019
ERP2R	Utilizes physical distance among the nodes	To balance consumption of energy	Provide reliability and enlarge the network lifetime	Sometimes traffic overhead increases	Al Salti et al.,2017

The design of routing protocols for UWSNs is of utmost importance. These protocols analyze paths from the bottom to the surface of the water to assure performance of UWSNs in accordance with the required parameters. These routing protocols cope with the battery power limitation of sensor nodes, high noise and interference, water pressure, mobility of the sensor nodes with water currents, long propagation delay due to the use of acoustic signal for communication, reliable delivery of data packets during unfavorable channel conditions.

Acoustic signals which are used underwater networks for communication have many limitations comparatively to radio waves which are used in wireless sensor nodes, such as limited bandwidth, higher end-to-end delay, loss of network path, longer propagation delay, and highly dynamic topology. Usually, all these limitations result in high amount of energy consumption with a less number of packets delivered. The main aim for designing the powerful and adequate UWSN routing protocols is efficient energy consumption. Some of the routing protocols take into the account of the load balancing technique to enlarge the life time of network. The sensed data for data centric networks in aqueous environment must be precisely delivered or transferred in the bounded latency to destination as higher propagation delays in these networks that use acoustic signals as compare to radio waves.

UWSN ROUTING PROTOCOL DESIGN ISSUES AND CHALLENGES

The process of relaying data from source nodes to other control stations or other sensor nodes efficiently is very challenging task in UWSNs. In this, energy consumption that is saving energy is a major concern. Underwater sensors are quite costly due to housing of additional components like acoustic transceivers used in communication.

This section mainly focuses on the issues and challenges regarding underwater sensor networks that make underwater communication rigid and sometimes crucial as compared to communication in terrestrial sensor network. In this issues and challenges, like selection of optimal packet size for communication, efficient utilization of acoustic channel for communication and reliability, power consumption of nodes, environmental effects on underwater habitat, distributed localization, media access control (MAC) protocols and network routing protocols are still needed more attention. All these problems that is mainly faced during communication require attention from researchers and academia community.

Mobility of nodes during the communication is another challenge, as nodes moves with the flow of water in UWSN. However, due to rapid change in "network topology" and the mobility of nodes and also the environment difference make the existing terrestrial routing protocols inadequate for underwater environment. Hence UWSN requires different routing protocols that fulfill its requirements and challenges. Following are some of the challenges faced in underwater sensor networks like-

- **Limited Bandwidth:** Acoustic channels specially used in UWSN offers very restricted amount of bandwidth, as radio transmission is not suitable to use for underwater communication. Acoustic communication consumes more energy for sending even a small amount of data, due to its poor bandwidth.
- **Energy Consumption**: Due to design implications of the nodes that are used in underwater communication like they are larger in size; hence consumes larger amount of energy for any communication. Furthermore, acoustic channels that are mainly used for UWSN also required more energy for communication than the terrestrial network. Battery replacement is also a challenge means

Figure 2. Routing protocol issues and challenges

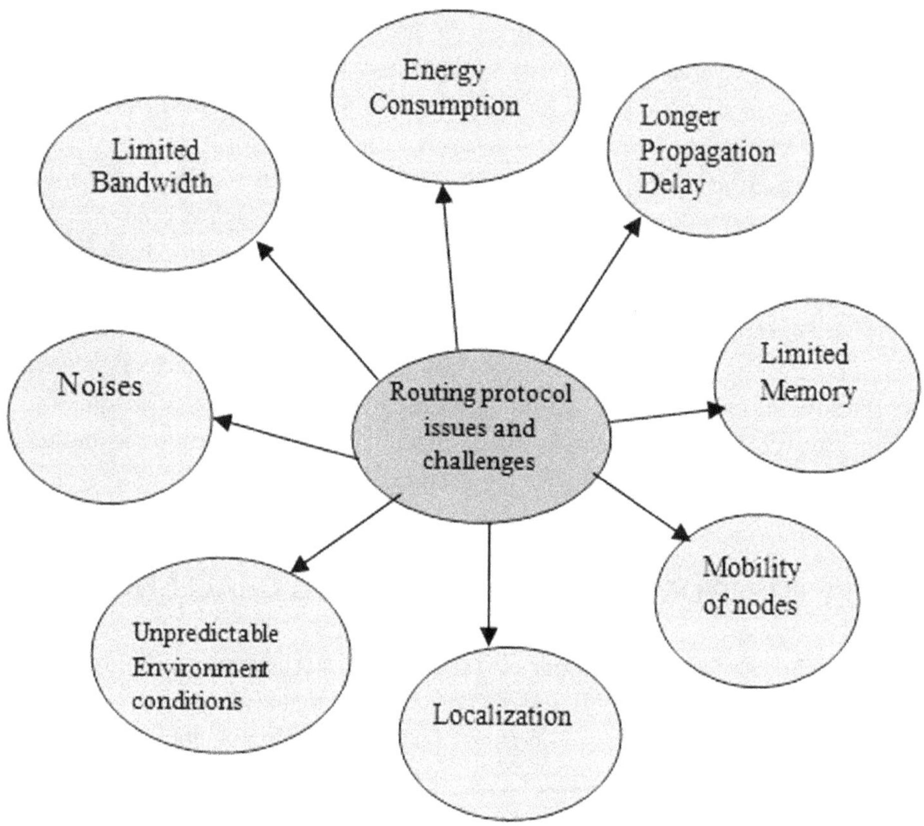

batteries in UWSN cannot be recharged or replaced easily; therefore, use of energy-efficient communication is always a need to provide good network.

- **Longer Propagation Delays:** Due to use of acoustic communication as compare to RF Communication, propagation speed becomes five times slower, that is, 1500m/sec, which ultimately results in longer propagation delays in the network.
- **Limited Memory:** In UWSN, design and features of sensor nodes is itself a bottleneck. The small size, limited amount of storage and processing capacity, high cost are major constraints in terms of the devices used.
- **Mobility of Nodes:** UWSN nodes does not have a fixed or static topology for the communication as flow of water makes it a difficult task for node to remain static in one place; therefore, node moves randomly.
- **Localization:** Localization is one of the challenging factors that is required for data labeling while some of the critical applications require data without any time delay which is of great concerns in the UWSNs.
- **Unpredictable Environment Condition Problem:** The challenge in aqueous environment is different from the terrestrial application environment. The water, a main component, will itself be a significant source for signal interference. The deployment showed, the variability in type of water,

dissolved impurities, water composition and its temperature has a major effect on sound propagation conditions in environmental area and the depth pressure.

- **Noises:** The noises can be man-made noise or diffusive noise. The man-made noise are caused by the machinery like pumps, power plants, reduction gears, etc. and shipping activity. The diffusive noise is associated to hydrodynamics that is movement of water including tides, storms, wind, rain, currents and phenomena of seismic and biological.
- **Physical Damage to Devices:** The sensors that are used under the water are vulnerable to routine underwater challenges, like, salt accumulation algae collection on the lens of camera and deteriorate the working of sensors.

SOLUTIONS TO UWSN ROUTING PROTOCOL CHALLENGES

The routing in under water environment is a challenging task as compare to terrestrial networks; mainly quality of service (QoS) gets degraded. Underwater environment issues like limited bandwidth, longer propagation delays and power constraints of the sensor nodes as well as 3-D topology are some of the challenges for the successful routings. The solutions for all these problems should be local, so as to avoid communication with most of the surface station and of the network, and minimizing the signaling overhead.

- The development of good algorithms and protocols is required that detect and deal with disconnections or failures, battery depletion, unforeseen mobility of nodes or underwater device constraints. For example in case of geographical routing protocols, it is more important to devise efficient underwater location discovery techniques that helps or improves the quality of communication under the water.
- The networking solutions towards the efficient and reliable underwater data collection required to be examined to enable the deployments of UWSNs. Under this, the topology control algorithms used for suitable, autonomous, and on-the-fly organization. UWSN topology might reduce the unwanted effects of underwater network communications and also enhances the performance of services of networking and protocols are designed for UWSNs

CONCLUSION AND FUTURE WORK

UWSN has attracted focuses of many researchers and industry persons in spite of the many challenges faced by it. Although UWSNs have seen an extraordinary growth in the past few years, but as always there is still a room for ample contributions particularly in routing and in the physical deployments of the systems under the water. As it is a well-known fact that routing is a big challenging task in underwater networks. Some of the challenges like low bandwidth, higher propagation delay, battery power or battery replacement etc. are the major bottlenecks in these networks. To conquer these issues in UWSN, the routing protocols are still needed to be strengthened. A better communication technique can be proposed by considering all the environmental effect that occurs during communication. In the development of underwater communication technique, the utmost care must be taken regarding habitat or the life of marine animals and their communication with others.

Some of the major challenges for designing the cognitive acoustic network are dynamic power control, spectrum related issues like sensing strategy, spectrum decision and it's sharing. Therefore, MAC and routing protocols needs to be design by focusing on maximizing the channel utilization during communication. Additionally, the standards for UWSN are also one of the challenges for researchers and industry experts. Much of the investigation is demanded in localization for underwater sensor networks, always plays an important role in design of routing protocol. One of the highly used technology GPS, used for terrestrial network for providing localization services is still not discovered for underwater sensor networks. This technology will be beneficial for UWSN also. From considering environmental effect on the underwater habitat during communication, any good communication technique can also be proposed. As in research and development of UWSN, the center of research should also be the life of marine animals and their inter or intra communication. Cognitive networks area and efficient underwater spectrum can also be future research exploration areas, considering many of the challenges of underwater networks.

REFERENCES

Ahmed, M., Salleh, M., & Channa, M. I. (2017). Routing protocols based on node mobility for Underwater Wireless Sensor Network (UWSN): A survey. *Journal of Network and Computer Applications*, *78*, 242–252. doi:10.1016/j.jnca.2016.10.022

Ahmed, M., Salleh, M., & Channa, M. I. (2018). Routing protocols based on protocol operations for underwater wireless sensor network: A survey. *Egyptian Informatics Journal*, *19*(1), 57–62. doi:10.1016/j.eij.2017.07.002

Akyildiz, I. F., Pompili, D., & Melodia, T. (2005). Underwater acoustic sensor networks: Research challenges. *Ad Hoc Networks*, *3*(3), 257–279. doi:10.1016/j.adhoc.2005.01.004

Al Salti, F., Alzeidi, N., & Arafeh, B. R. (2017). EMGGR: An energy-efficient multipath grid-based geographic routing protocol for underwater wireless sensor networks. *Wireless Networks*, *23*(4), 1301–1314. doi:10.100711276-016-1224-0

Ali, T., Jung, L. T., & Faye, I. (2014). Classification of routing algorithms in volatile environment of underwater wireless sensor networks. *International Journal of Communication Networks and Information Security*, *6*(2), 129.

Ali, T., Jung, L. T., & Faye, I. (2014). Diagonal and vertical routing protocol for underwater wireless sensor network. *Procedia: Social and Behavioral Sciences*, *129*, 372–379. doi:10.1016/j.sbspro.2014.03.690

Awan, K. M., Shah, P. A., Iqbal, K., Gillani, S., Ahmad, W., & Nam, Y. (2019). Underwater wireless sensor networks: A review of recent issues and challenges. *Wireless Communications and Mobile Computing*, *2019*, 2019. doi:10.1155/2019/6470359

Basagni, S., Petrioli, C., Petroccia, R., & Stojanovic, M. (2012). Optimized packet size selection in underwater wireless sensor network communications. *IEEE Journal of Oceanic Engineering*, *37*(3), 321–337. doi:10.1109/JOE.2012.2197271

Bhardwaj, M. (2017). Faulty Link Detection in Cluster based Energy Efficient Wireless Sensor Networks. *International Journal of Scientific Research in Network Security and Communication*, *5*(3), 1–8.

Coutinho, R. W., Boukerche, A., Vieira, L. F., & Loureiro, A. A. (2018). Underwater wireless sensor networks: A new challenge for topology control–based systems. *ACM Computing Surveys*, *51*(1), 19. doi:10.1145/3154834

Domingo, M. C., & Prior, R. (2008). Energy analysis of routing protocols for underwater wireless sensor networks. *Computer Communications*, *31*(6), 1227–1238. doi:10.1016/j.comcom.2007.11.005

Du, X. J., Huang, K. J., Ian, S. L., Feng, Z. X., & Fan, L. I. U. (2014). LB-AGR: Level-based adaptive geo-routing for underwater sensor network. *Journal of China Universities of Posts and Telecommunications*, *21*(1), 54–59. doi:10.1016/S1005-8885(14)60268-5

Felemban, E., Shaikh, F. K., Qureshi, U. M., Sheikh, A. A., & Qaisar, S. B. (2015). Underwater sensor network applications: A comprehensive survey. *International Journal of Distributed Sensor Networks*, *11*(11), 896832. doi:10.1155/2015/896832

Gaba, A., & Goyal, N. (2013). Review over Diverse Location Aided Routing. *Global Journal for Current Engineering Research*, *2*(2), 141–144.

Ghoreyshi, S. M., Shahrabi, A., & Boutaleb, T. (2017). Void-handling techniques for routing protocols in underwater sensor networks: Survey and challenges. *IEEE Communications Surveys and Tutorials*, *19*(2), 800–827. doi:10.1109/COMST.2017.2657881

Goyal, N., Dave, M., & Verma, A. K. (2014). Fuzzy based clustering and aggregation technique for under water wireless sensor networks. In *2014 International Conference on Electronics and Communication Systems (ICECS)* (pp. 1-5). IEEE. 10.1109/ECS.2014.6892804

Goyal, N., Dave, M., & Verma, A. K. (2017). Data aggregation in underwater wireless sensor network: Recent approaches and issues. *Journal of King Saud University-Computer and Information Sciences.*

Goyal, N., Dave, M. & Verma, A. K. (2017a). Improved data aggregation for cluster based underwater wireless sensor networks. *Proceedings of the National Academy of Sciences, India Section A: Physical Sciences, 87*(2), 235-245.

Goyal, N., Dave, M., & Verma, A. K. (2016). Energy efficient architecture for intra and inter cluster communication for underwater wireless sensor networks. *Wireless Personal Communications*, *89*(2), 687–707. doi:10.100711277-016-3302-0

Goyal, N., Dave, M., & Verma, A. K. (2018). A novel fault detection and recovery technique for cluster-based underwater wireless sensor networks. *International Journal of Communication Systems*, *31*(4), e3485. doi:10.1002/dac.3485

Goyal, N., Dave, M., & Verma, A. K. (2019a). Protocol stack of underwater wireless sensor network: Classical approaches and new trends. *Wireless Personal Communications*, *104*(3), 995–1022. doi:10.100711277-018-6064-z

Goyal, N., & Gaba, A. (2013). A new approach of location aided routing protocol using minimum bandwidth in mobile ad-hoc network. *International Journal of Computer Technology and Applications*, *4*(4), 653.

Goyal, N., Sandhu, J. K., & Verma, L. (2019b). Machine learning based data agglomeration in underwater wireless sensor networks. International Journal of Management. *Technology And Engineering*, *9*(6), 240–245.

Guo, Y., & Liu, Y. (2013). Localization for anchor-free underwater sensor networks. *Computers & Electrical Engineering*, *39*(6), 1812–1821. doi:10.1016/j.compeleceng.2013.02.001

Guo, Z., Colombi, G., Wang, B., Cui, J. H., Maggiorini, D., & Rossi, G. P. (2008, January). Adaptive routing in underwater delay/disruption tolerant sensor networks. In *2008 Fifth Annual Conference on Wireless on Demand Network Systems and Services* (pp. 31-39). IEEE. 10.1109/WONS.2008.4459352

Heidemann, J., Ye, W., Wills, J., Syed, A., & Li, Y. (2006, April). Research challenges and applications for underwater sensor networking. In *IEEE Wireless Communications and Networking Conference. WCNC 2006* (Vol. 1, pp. 228-235). IEEE. 10.1109/WCNC.2006.1683469

Hollinger, G. A., Choudhary, S., Qarabaqi, P., Murphy, C., Mitra, U., Sukhatme, G. S., Stojanovic, M., Singh, H., & Hover, F. (2012). Underwater data collection using robotic sensor networks. *IEEE Journal on Selected Areas in Communications*, *30*(5), 899–911. doi:10.1109/JSAC.2012.120606

Huang, C. J., Wang, Y. W., Liao, H. H., Lin, C. F., Hu, K. W., & Chang, T. Y. (2011). A power-efficient routing protocol for underwater wireless sensor networks. *Applied Soft Computing*, *11*(2), 2348–2355. doi:10.1016/j.asoc.2010.08.014

Iyer, S., & Rao, D. V. (2015, Feb.). Genetic algorithm based optimization technique for underwater sensor network positioning and deployment. In *Proceedings of the IEEE Underwater Technology (UT '15)*. Chennai, India: IEEE. 10.1109/UT.2015.7108229

Jouhari, M., & Ibrahimi, K., andBenattou, M. (2015, October). Topology control through depth adjustment and transmission power control for UWSN routing protocols. In *2015 International Conference on Wireless Networks and Mobile Communications (WINCOM)* (pp. 1-5). IEEE. 10.1109/WINCOM.2015.7381310

Jun-Hong, C., Jiejun, K., Gerla, M., & Shengli, Z. (2006). The challenges of building mobile underwater wireless networks for aquatic applications. *IEEE Network*, *20*(3), 12–18. doi:10.1109/MNET.2006.1637927

Jung, L. T., & Abdullah, A. B. (2011, June). Underwater wireless network energy efficiency and optimal data packet size. In *International Conference on Electrical, Control and Computer Engineering 2011 (InECCE)* (pp. 178-182). IEEE. 10.1109/INECCE.2011.5953871

Khurana, T., Singh, S., & Goyal, N. (2012). An Evaluation of Ad-hoc Routing Protocols for Wireless Sensor Networks. *International Journal of Advanced Research in Computer Science and Electronics Engineering*, *1*(5).

Kumar, M., & Goyal, N. (2014). Reviewing underwater acoustic wireless sensing networks. *International Journal of Computer Science and Technology*, *5*(2), 95–98.

Lanbo, L., Shengli, Z., & Jun-Hong, C. (2008). Prospects and problems of wireless communication for underwater sensor networks. *Wireless Communications and Mobile Computing*, *8*(8), 977–994. doi:10.1002/wcm.654

Le, T. T., & Kim, D. S. (2013, April). Distributed cooperative transmission for underwater acoustic sensor networks. In 2013 IEEE wireless communications and networking conference workshops (WCNCW) (pp. 205-210). IEEE.

Lloret, J., Sendra, S., Ardid, M., & Rodrigues, J. J. (2012). Underwater wireless sensor communications in the 2.4 GHz ISM frequency band. *Sensors (Basel)*, *12*(4), 4237–4264. doi:10.3390120404237 PMID:22666029

Lv, C., Wang, S., Tan, M., & Chen, L. (2014). UA-MAC: An underwater acoustic channel access method for dense mobile underwater sensor networks. *International Journal of Distributed Sensor Networks*, *10*(2), 374028. doi:10.1155/2014/374028

Park, M. K., & Rodoplu, V. (2007). UWAN-MAC: An energy-efficient MAC protocol for underwater acoustic wireless sensor networks. *IEEE Journal of Oceanic Engineering*, *32*(3), 710–720. doi:10.1109/JOE.2007.899277

Prasan, U. D., & Murugappan, D. S. (2012). Underwater Sensor Networks: Architecture, Research Challenges and Potential Applications. *International Journal of Engineering Research and Applications*, *2*(2), 251–256.

Saeed, N., Celik, A., Al-Naffouri, T. Y., & Alouini, M. S. (2019). Underwater optical wireless communications, networking, and localization: A survey. *Ad Hoc Networks*, *94*, 101935. doi:10.1016/j.adhoc.2019.101935

Sarao, P., Chattu, K., & Swapna, C. (2018). *Routing issues and challenges in Underwater Wireless Sensor Networks*. Academic Press.

Syed, A. B., Andrabi, & Manoj, K. (2017). A comparative study of routing protocols in underwater wireless sensor networks. *International Journal of Computer Science and Communication*, *8*(1), 27–29.

Wahid, A., & Dongkyun, K. (2010). Analyzing routing protocols for underwater wireless sensor networks. *International Journal of Communication Networks and Information Security*, *2*(3), 253.

Xie, P., Cui, J. H., & Lao, L. (2006, May). VBF: vector-based forwarding protocol for underwater sensor networks. In *International conference on research in networking* (pp. 1216-1221). Springer. 10.1007/11753810_111

Yoon, S., Azad, A. K., Oh, H., & Kim, S. (2012). AURP: An AUV-aided underwater routing protocol for underwater acoustic sensor networks. *Sensors (Basel)*, *12*(2), 1827–1845. doi:10.3390120201827 PMID:22438740

Zhang, Z., & Blanco, L. H. (2014, May). A self adaptive MAC layer protocol for delay-tolerant underwater wireless sensor networks. In *2014 23rd Wireless and Optical Communication Conference (WOCC)* (pp. 1-8). IEEE. 10.1109/WOCC.2014.6839944

Chapter 2
Quality Enhancement in Underwater Sensor Networks Through Node Deployment Plans:
A Performance-Based Classification

Anju Sangwan
https://orcid.org/0000-0002-5433-1484
Guru Jambheshwar University of Science and Technology, India

Anupma Sangwan
Guru Jambheshwar University of Science and Technology, India

Deepti Rani
Chandigarh University, India

Rishi Pal Singh
Guru Jambheshwar University of Science and Technology, India

ABSTRACT

Due to widespread applications and an orientation towards dimensionality enhancement, the underwater sensor networks (UWSNs) are going to be in demand for today's researchers. The core idea to design a strong and reliable build under the water revolves around the master plan of sensor deployment. The quality parameters like coverage, connectivity, and network lifespan, etc. are considered as main pillars for these deployment plans. There is always a trade-off between these quality parameters. So, it is quite necessary to maintain a balance among these parameters in order to establish a successful communication system. In this chapter, the authors have categorized the node deployment plans on the basis of the above-stated quality parameters. During this categorization, the view of dimensionality enhancement from two dimensions to three dimensions is also kept in the mind.

DOI: 10.4018/978-1-7998-3640-7.ch002

INTRODUCTION

Underwater Sensor Networks (UWSNs)

UWSN is a blend of wireless technologies having tiny micro-mechanical sensor possessing communication capabilities, smart sensing and intelligent computing. UWSN is an underwater network of spatially distributed, autonomous sensor nodes (Akyildiz et al, 2005) which senses certain characteristics such as quality, pressure and temperature of water. These sensors along with the numerous vehicles are deployed in such a way so that they can collect data and perform collaborative monitoring. Sensor node transmits the data to buoyant gateway which in turn relays it to the closest control station and coastal monitoring known as remote station (Felemban, 2013).

Figure 1. Underwater wireless sensor network

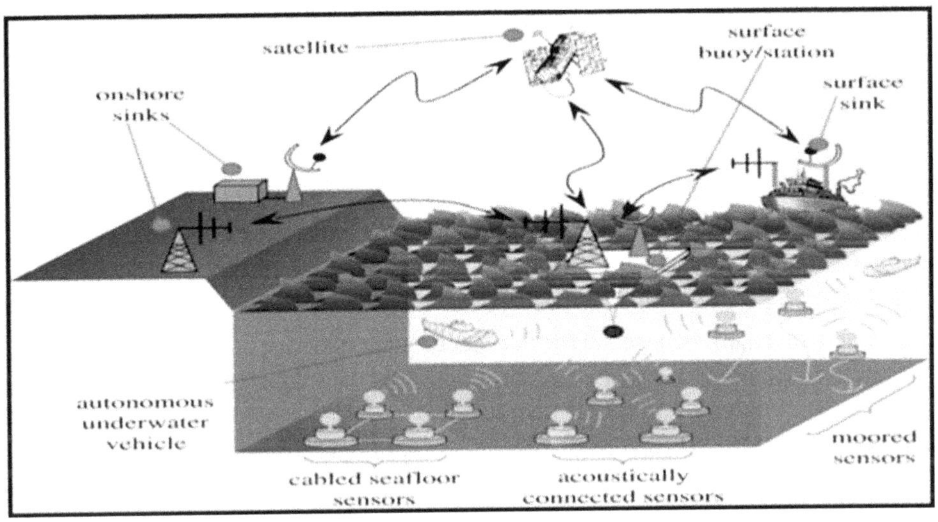

Underwater sensor networks (UWSNs) is gaining popularity because of their broad range of applications in commercial, scientific and military operation underwater. Many potential uses include surveillance of ocean tides and waves, enhanced weather forecasts, detection of climate changes, biological monitoring, estimation and understanding of human impacts on aquatic ecosystems. UWSN plays significant role in exploring precious minerals, reservoir and oilfields. It also assists in determining the route for laying the cables under the sea. Apart from this, sensors can be used to send alert to coastal areas about tsunami by measuring seismic activity from remote locations (Akyildiz et al, 2005; Soreide, 2004). UWSN help in navigation by identifying seabed threats, finding dangerous rocks or shoals in coastal seas, mooring locations, submerge wrecks and conduct profiling of bathymetries. Under-Water Acoustic Sensor Networks supports surveillance, intrusion detection systems, reconnaissance and targeting (Cayirci et al, 2004).

Communication Architecture of UWSNs

Two-Dimensional(2D) Underwater Sensor Network Architecture

Deep ocean anchors are used to collect sensor nodes in Two-Dimensions Underwater Sensor Networks. Acoustic links are used for communication among anchored underwater nodes and other underwater sinks. Underwater sinks gather information from deep ocean and transmits it to same off-shoring command stations. To accomplish this transmission, horizontal and vertical acoustic transceivers accompany these underwater sinks. Horizontal transceiver communicates with the sensor node, gather data or gives command, as received by the off-shoring command station. On the other hand, transmission of data to command station is done by vertical transceiver. The surface sink with acoustic transceivers, is able to handle parallel communication via several organized underwater sinks. The surface sink communicates with offshore sinks via extensive range of radio frequency transmitters (Awan et al, 2019; Akyildiz et al, 2004; Peach et al 2013; Akyildiz et al, 2006). Sensor nodes are organized in order to form clusters in water. There is a cluster head (known as anchor node) in each cluster. The clusters are anchored on the surface of the water to keep them stable. Each cluster node gathers and passes the underwater data to the cluster head. The cluster head gather information from member nodes and transmits it to surface beacon. Communication is performed in two dimensions in 2D underwater sensor networks and is as follows (i) each node of the cluster connects to a cluster head with a horizontal communication link. (ii) connect with buoyant node with vertical communication link.

The type of communication used depend on the application and nature of the underwater environment. In two-dimensional underwater, sensor network communication is preferred between the underwater cluster head and the surface of the buoyant node because of the large distance between them. Depending upon the requirement of the application, 2D underwater sensor network typically adopts star, mesh or ring topology. In both tolerant and time-critical applications, 2D UWSN may be used (Felemban et al, 2015).

Three-Dimensional (3D) Underwater Sensor Network Architecture

In 3Dimensional Underwater Sensor Network architecture, seabed is used for anchor sensor nodes. The depth of these nodes is managed by means of wires connected to them. Deployment of the sensors is carried out in the form of anchored clusters at varying depth under the water. This varying depth deployment of sensors leads the communication beyond two dimensions. At different depth levels in 3D- UWSNs sensor nodes are capable of inter-cluster, intra-cluster as well as anchor-buoyant node communication. Moreover, this communication can be established with the help any of acoustic, optical or RF links. Floating sensor nodes at varying heights permit to track the specific activity in the 3-Dimensional UWSN (Awan et al, 2019; Akyildiz et al, 2004; Peach et al 2013). A 3-Dimensional UWSN is engineered for a tactical monitoring device capable of identifying and classifying small delivery vehicles, submarines and divers based on sensed data from acoustic, magnetic, radiation and mechanical micro-sensors. In contrast to conventional radar systems, UWSN can achieve greater precision and enable lower signature target detection and classification by integrating measures from various sensor types (Akyildiz et al, 2005).

NODE DEPLOYMENT IN UNDERWATER SENSOR NETWORKS (UWSNS)

In underwater communication practices, network topology is one of main pillar of building an Underwater Sensor Network (UWSN) which is capable to achieve desired goals of communication. The topology of a network measures i) sensor deployment and ii) how link establishment to be done among deployed nodes. The network performance under water mainly depends on factors like- network reliability, energy consumption and network capacity etc. The deployment of sensor nodes especially under the water costs a lot of money as well as these nodes are keen to downfall. So, the deployment plan of sensor nodes should be mindfully examined as a single communication hole can affect the performance of network comprehensively (Akyildiz et al, 2005). A network can satisfy its motive as long as it is in alive state. The quality of service measures can be reduced to lifetime considerations. The parameters such as coverage and connectivity can be reduced to network lifetime also. As the deployment of sensor nodes is a decisive issue in regard to the performance of network, this study mainly focused on optimization of Quality of Service (QoS) parameters. Thus, coverage, connectivity and network lifetime, the essential parameters are the fundamental blocks of the QoS build.

Deployment of sensor node can be graded on basis of different criteria like i) method of deployment: static, self-configured and movement based, ii) quality factors: coverage, connectivity and life span and iii) potential application based: monitoring (environment and pollution, geological, underwater pipelines etc.), oil extraction, mine reconnaissance, navigation etc (Tuna et al, 2017). In the following section, a categorization of deployment plans on the basis QoS parameters, that is; deployment plans with maximized coverage, fully connected and lifetime longevity has been explained.

Coverage Maximization based Deployment Plans

In Underwater Sensor Network, coverage is one of the top quality delivery parameter in list of network performance metrics. Coverage means extent to which spotting of targets is sensible in the specified region. Efficient coverage deals with the optimization of resources in the field of interest. Today, deployments with low cost and maximize coverage is one of the demanding task in the area of UWSNs. The authors (Sangwan et al., 2015) have classified various coverage problems on the basis of network deployments, sensing models, target characteristics, application attributes and monitored areas etc. A number of strategies to cope up with these coverage problems have been discussed.

Grid based Deployment Plans for Two-Dimensional and Three-Dimensional UWSNs

The authors (Pompili et al, 2009) have proposed sensor deployment plans for both two-dimensional as well as three-dimensional UWSNs communication architectures. In case of two-dimensional architecture, the authors have mainly followed equilateral triangular grid deployment approach as shown in Figure 3. After this, a mathematical analysis in order to attain the desired level of coverage has been worked out in terms of '*d*' - the distance among sensor nodes and '*r*' – the sensing range of communicating node. It is derived that when $d = \sqrt{3}\,r$, the overlapped sensing area among nodes can be reduced to zero extent. But doing so, it triggers need of higher number of sensor nodes and vice-versa. Also, authors have worked out some theorems with aim to find out minimum number of sensor nodes to cover the trusted region. In next step, authors have studied about the trajectory of sinking object. Using geometric properties, a formula has been derived to find out the deploy area at sea bottom at which sensors place-

Figure 2. Categorization of node deployment plans in UWSNs

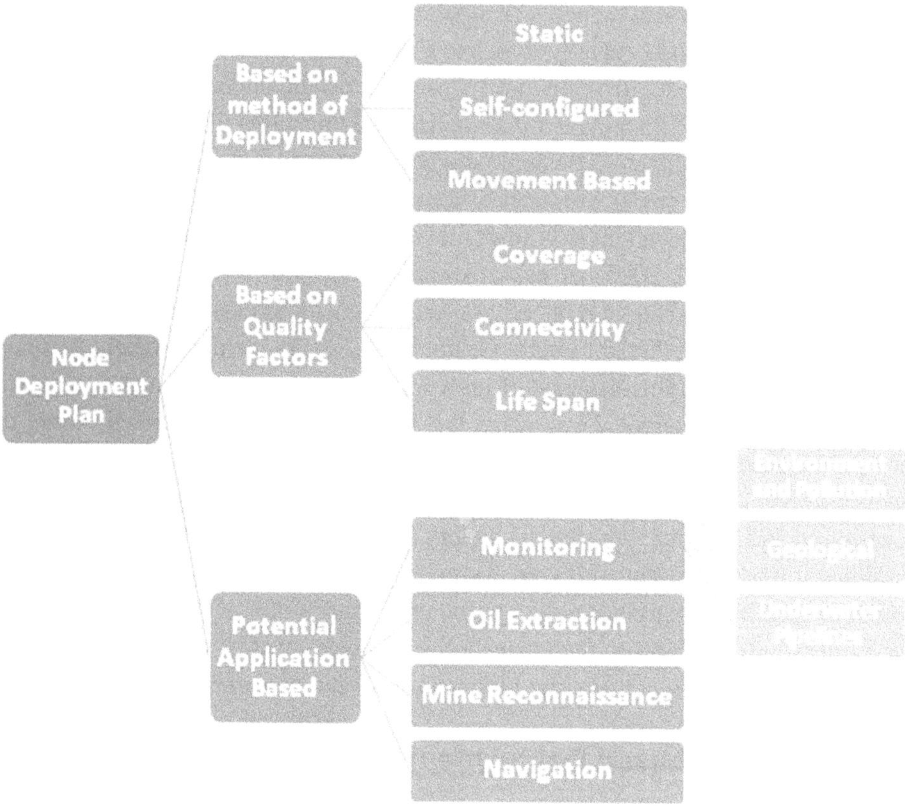

ment has to be done to target the interested point in two-dimensional plane. Further, the deployment in three-dimensional area is characterized by three techniques, namely; 3D-random, bottom-random and bottom-grid. 3D-random technique is one of the unembellished techniques in which randomly placed sensors choose their depth point and communicate their final locations to surface station. In contrast to previous technique, the depth of each node is determined by the surface station to reach at desired coverage ratio in bottom-random technique. In third bottom-grid technique, each deployed node is aided with an autonomous underwater vehicle to gain a depth in accordance to predefined locations for grid deployment. With this study, the authors have provided work directions to optimize the number of sensor nodes for coverage and surface area for deployment. Authors have also given estimation in regard to the fault tolerance capacity of the network.

Self-Deployment of Sensors for Maximized Coverage in UWSNs

The authors (Akkaya et al, 2009) have shown their interest in the feature of self-deployment of sensors and presented a distributed topological adjustment approach which not only promises to maximize coverage but also gives the assurance of network connectivity. In topological arrangement, first of all, clusters are formed for randomly distributed sensor nodes using their node IDs. The highest ID determines the node

Figure 3. Grid based node deployment in UWSNs
Source: (Pompili et al, 2009)

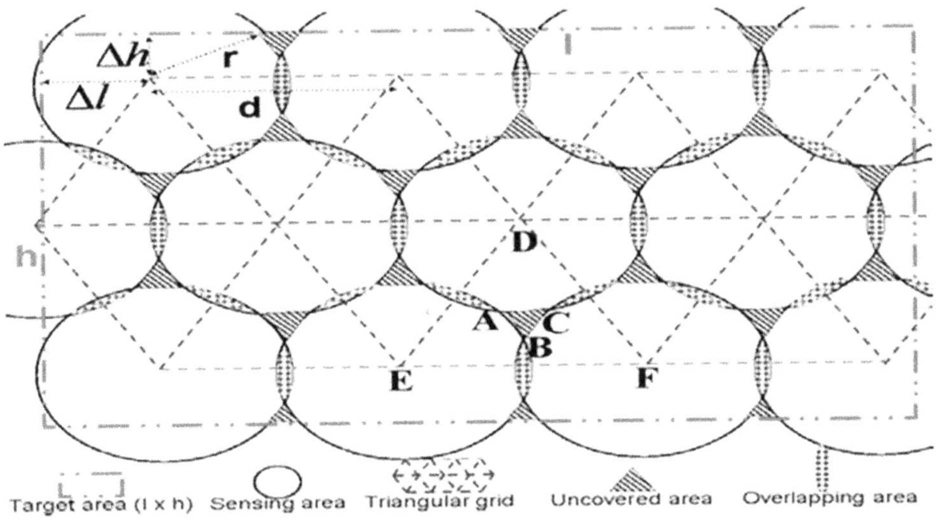

as cluster head. In next step, these heads point outs the coverage overlap present within their clusters. The nodes possessing overlapping with their neighbours are grouped together. Then, group IDs are assigned to different groups. To proceed further, nodes depth adjustments are to be done on the basis of formula given as $D/(G+1)$, where '*D*' represents the depth of water and '*G*' defines the total number of groups constituted in all. Again, the nodes find out its neighbours by measuring distance and coverage overlap for further attempt. The authors also restricted the number of attempts up to two as there is no significant enhancement seen in the network coverage further. The authors have done a comparison of this distributed approach with optimal centralized approaches on basis of performance factors such as network coverage and connectivity. The observations noted simply show that proposed one downfalls a bit under mobility model but it overtakes the random approaches.

Coverage-Aware Connectivity-Constrained Unattended Sensor Deployment in UWSNs

Senel (2016) has designed an algorithm attributed with the facts of coverage enhancement and topological expansion, that is, from two-dimensional topological to three-dimensional topological structure. The main concern of structured algorithm is to augment coverage percentage at maximum limits in finally constituted three-dimensional structure. The algorithm is shaped into three steps, these are; initialization, optimization of depth level, depth distribution among sensor nodes. In accordance to the algorithmic analysis, a tree-structured formation is build-up by the sensor nodes and each node completes the process of whole communication with the help of four messages. In its first step, three messages are required to transmit information from leaf to parent or root node. In second step, only computational procedures are there. Thus, there is no need of message transmission. In the last third step, one message is required to circulate the depth information. The performance assessment of discussed approach has been done on the basis of coverage percentage, connectivity, average node degree and expected path length. In

Figure 4. (a) – Grouping of nodes based on coverage overlaps and (b) – Grouping based depth adjustment for nodes
Source: (Akkaya et al, 2009)

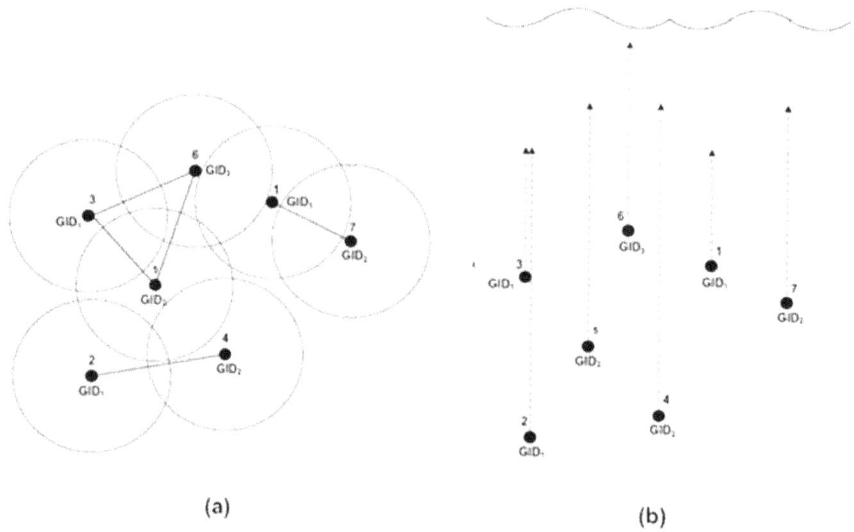

simulation, it is clearly seen that proposed approach enhances the coverage percentage at least by 10% in comparison to its contemporary schemes.

Connected Dominating Set (CDS) based Depth Computation Approach for UWSNs

The authors (Senel et al, 2014) have modelled a fully distributed random deployment approach to cope with trade-off between minimization of coverage overlapped area while maintaining the network connectivity. In this direction, authors have explored topological changes caused by currents, winds, chan-

Figure 5. Assumed three-dimensional system model
Source: (Senel, 2016)

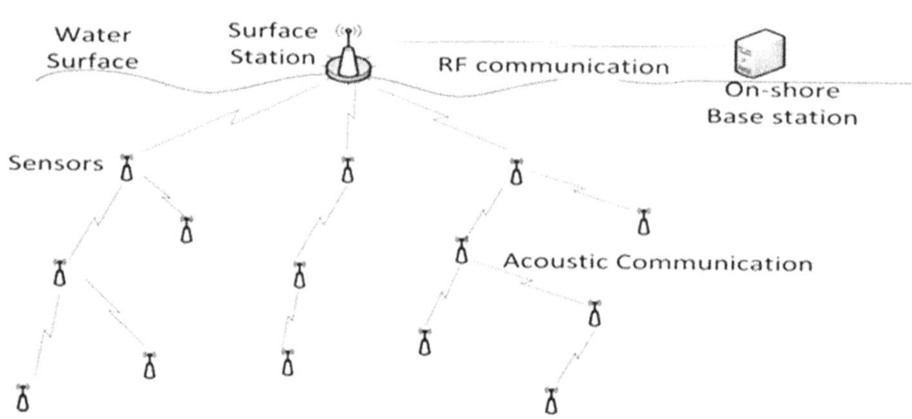

nel randomness etc. The authors have worked out with an orientation towards network expansion from two-dimension to three-dimension characterized by node mobility. In any algorithm, the deployment time effects significantly to the performance factors like coverage, connectivity, lifetime etc. A formula to calculate best deployment time was also suggested. Authors have investigated best-suited time before the network got disconnected. The mainstay of given approach is based on two aspects; i) formation of Connected Dominating Sets, ii) depth calculation for nodes. The nodes which are part of CDS are known as dominators and rest are known as dominate. Here, CDS is responsible for connected roadmap for the system. The process of depth computation is responsible to maintain the position of nearby nodes as far as possible without any kind of disconnection to its dominator. The results of discussed approach provides promised connectivity with coverage extent as in coverage-aware deployment scheme discussed previously in section 2.1.3.

Figure 6. (a)- Initially two-dimensional connected UWSN and (b)- Black nodes as CDS
Source: (Senel et al, 2014)

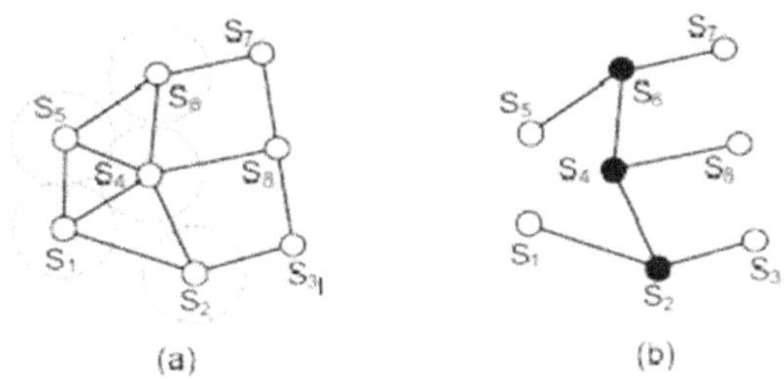

Multiple Autonomous Underwater Vehicles (AUVs) and Surface Gateways (SGs) based Deployment Scheme for UWSNs

The authors (Bharamagoudra et al, 2016) have offered a node placement strategy with the content of coverage and connectivity reinforcement. Inauspicious channel and harsh environmental conditions can be major reasons for the destruction of network system. In order to obtain better coverage and connectivity, the authors have mainly focused on attenuation factor, integration of multiple autonomous underwater vehicle and route planning for AUVs & SGs to improve coverage and connectivity respectively. The authors have worked out a mathematical model for different three phases of stated approach, these are, i) sensor node deployment, ii) route planning for AUVs corresponding to distinct mobility patterns and iii) placement strategy for SGs. The considered mobility patterns are as follows: i) single AUV each layer, ii) two AUVs between two consecutive layer of nodes, iii) single AUV two consecutive node layer and iv) multiple AUVs. A comparison is also given between distinct mobility patterns on account of number of AUVs, AUV movement, delay and applications. The performance evaluation of the stated approach has been done on basis of covered area percentage, connectivity, number of sensor nodes required, messages communicated by sensor nodes to each other and distance moved by sensor nodes to cover the specified area. And, it is observed that the proposed one dominates over CDS-based purely distributed node placement strategy as explained in section 2.1.4 in terms of all considered performance factors.

Connectivity Maximization based Deployment Plans

Network connectivity is on the next step of quality ladder due to its significant role in network establishment.Network connectivity calculates different ways so that there exist at least one link between each and every node to communicate.Today, k – connected topological establishment and fully connected networks are the focus points of deployment plans in order to formulate a good network. A network is said to be k – connected if there exist at least 'k' independent links for every node pair to communicate with each other (Akram et al, 2016). Resultantly, there is always possibility to cope with the problem of network disconnection. Recently, connectivity problem has been explored by many researchers on priority basis in order to design fault tolerant networks.

Prediction Assisted Dynamic Surface Gateway Placement (PADP) for Mobile UWSNs

The authors (Han et al, 2012) have given a placement strategy for dynamic SGs in mobile UWSNs. This approach is associated with a tracking algorithm which identifies the node's future positions due to its movements in coming time rounds. This strategy is mainly targeted to meet four goals of network performance. These are- i) to enhance the coverage at top priority, ii) to optimize the energy consumption for the case of redeployment, iii) to preserve good connectivity and iv) to bound budget limits. In order to meet the desired requirements, the tracking algorithm is incorporated with an Interacting Multiple Model (IMM) and one of the operational research techniques, branch-and-cut for optimization. Further, the selection of Disjoint set benefits the proposed scheme in terms of good network connectivity. The mentioned IMM is an estimation method which matches all possible node movement patterns to a specified pattern of node movement. The proposed PDAP has considered two node movement patterns defined as –

i) uniform motion given by Kalman filter in which movement is considered along with a straight forward line with constant speed only and
ii) maneuver motion given by an extended Kalman filter in order to coordinate turn with constant turn rate and speed. The result analysis shows PADP, the dynamic approach, has more weightage over the static deployment approach on the account of performance factors as stated earlier.

Optimal Node Placement in UWSNs with Shadow Zones

The author (Domingo, 2009) has introduced an optimal placement strategy for sensor nodes in three-dimensional UWSNs with the presence of shadow zones. Presence of shadow zones reveals negative influence on the network. These zones may cause connectivity holes resulting in a partitioned network. To resolve this problem the authors have adopted topological reorganization as an option to obtain optimal placement of sensor nodes. Doing so, the transmission loss can be minimized while maintaining the network connectivity. Different propagation phenomena as explained by authors in (Domingo, 2008) based on different channel models were also considered. Author has proposed one mathematical model after completing theoretical and numerical evaluations. These investigations point out that the stepped procedure under all phenomena for all water depths improves the quality of three-dimensional networks with minimum transmission loss while keeping the whole network connected in continuous manner.

Figure 7. Block Diagram for moving patterns
Source: (Han et al, 2012)

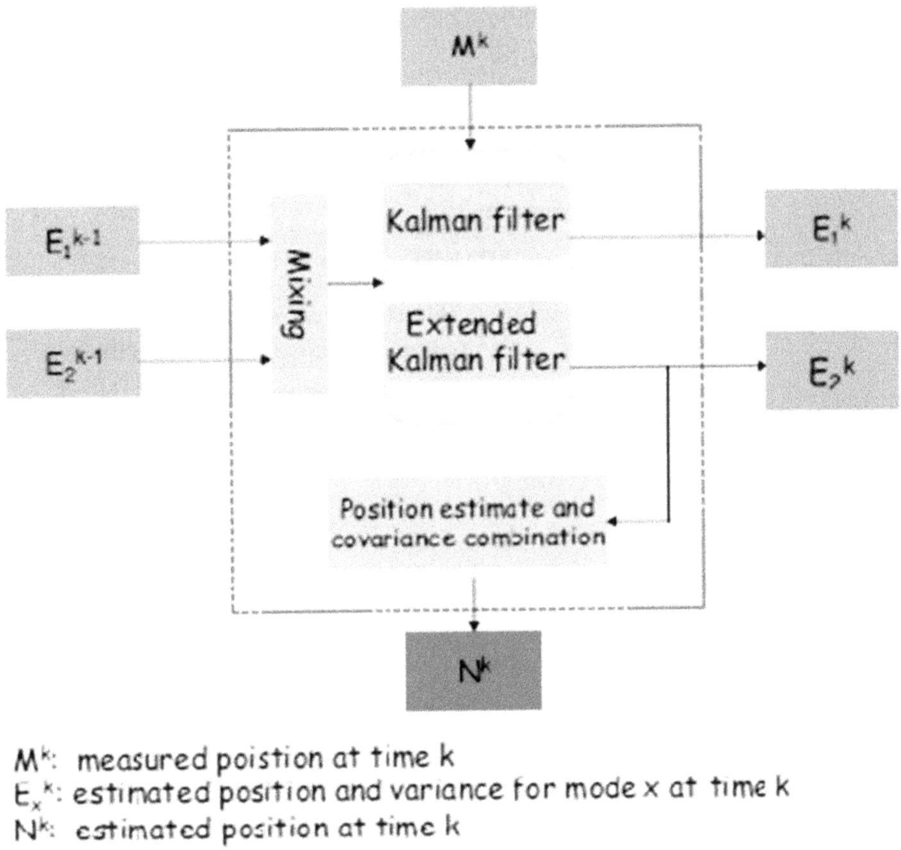

Multiple Underwater Unmanned Vehicles (UUVs) Connectivity Enhancement Approach for UWSNs

The authors (Winston et al, 2005) have proposed a solution to the problem of communication or link disconnection by introducing multiple underwater unmanned vehicles (UUVs) to maximize the connectivity in the network. Because of unfavourable channel and environmental conditions, there is always possibilities of connectivity holes in the communication network. The UUVs are aimed to monitor these communication gaps which arises due to some type of interference or noise etc. Also, an additional number of sensor nodes can be deployed for repairing purpose. The strategic policy for the deployment of these UUVs lies on the fraction taken as number of subnetworks resulted from link loss to the total number of UUVs available in the network. The working methodology of proposed scheme is divided into two task phases, these are; i) to find out communication gaps and ii) to search out critical areas for to obtain complete coverage. The authors have analysed the performance of network on the basis of connectivity, average hop count and *k*-connectivity. In this network, *k*-connectivity is attributed due to deployment of nodes to heal communication or link holes. Thus, given approach also adds characteristic of faulty tolerance to the deployed sensor network.

Figure 8. Three-dimensional UWSN model with shadow zones
Source: (Domingo, 2009)

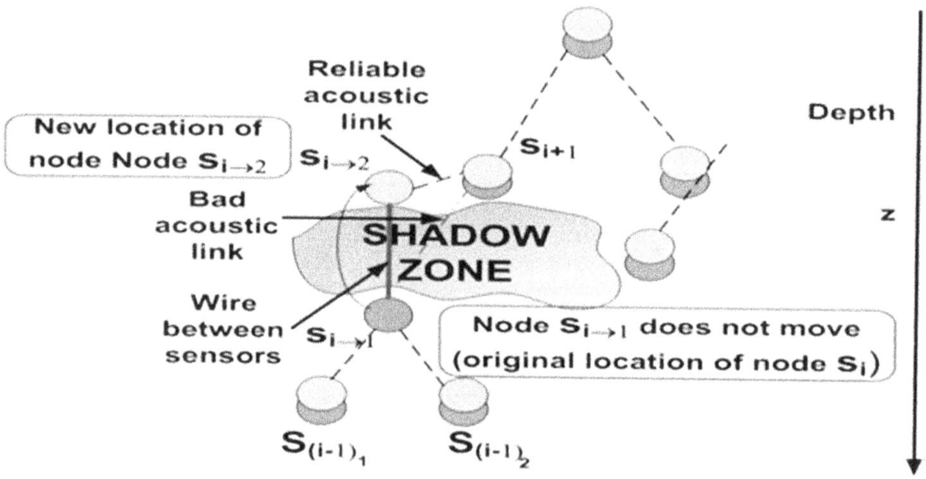

Node Sinking Algorithm for Coverage and Connectivity in UWSNs

The authors (Wang et al., 2016) have studied a node sinking algorithm with an enhancement to the network dimensions, i.e. from two to three- dimensional network. Initially, randomly scattered nodes on the water surface constitutes a two-dimensional network. Then, nodes are moved along z – axis to expand the network in third dimension. The whole process for the proposed algorithm goes through three steps, namely, i) pattern node selection, ii) network connectivity restoration, and iii) coverage hole repair. In the algorithm's first step, the 3D-sphere packing pattern is considered. In next step, a check is imposed on the number of nodes deployed randomly; M and number of sank nodes, N. If $M>N$, then the node

Figure 9. Identifying and bridging gaps
Source: (Winston et al, 2005)

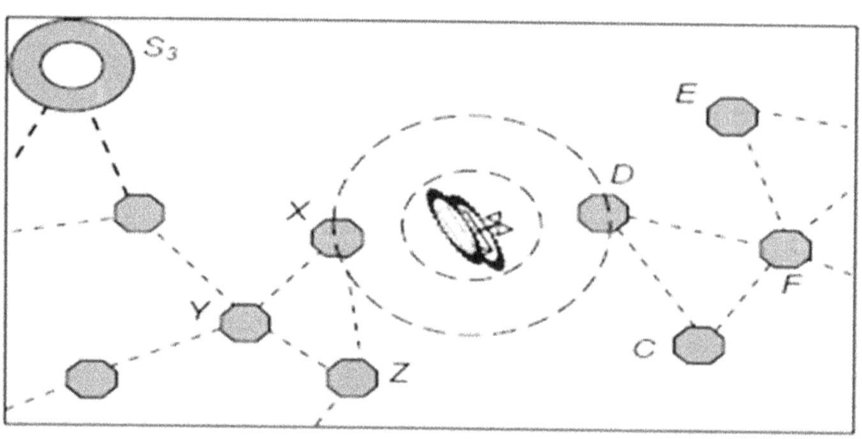

restoration is done through an Inserting and Adjusting Algorithm, otherwise, it is done through Adjusting Algorithm (Wang *et al.*, 2016). Further, 3D-voronoi diagram and restricted *K*- means clustering were used to repair the coverage holes. In the given solution, coverage ratio, giant connectivity ratio, connected components and time complexity have been taken as simulation parameters. The authors have also compared the proposed solution to the existing algorithms. These comparisons clearly show that offered solution outperforms on account of coverage ensuring a connected underwater network.

Connectivity and Localization Analysis for Underwater Optical Wireless Sensor Networks

The authors (Saeed et al., 2018) have prescribed an idea of multi-hop communication to improve the quality of network connectivity. Due to poor channel impairments and imbalance optical transceivers, there is always chances of intermittent connectivity in UWSNs. In the present paper, the authors have studied about connectivity of underwater optical wireless sensor networks and investigated how it influences localization and overall performance of the network. The authors have considered parameters

Figure 10. (a)- Network partitions, (b)- Network connectivity restoration adjusting and (c)- Network connectivity restoration inserting and adjusting
Source: (Wang et al., 2016)

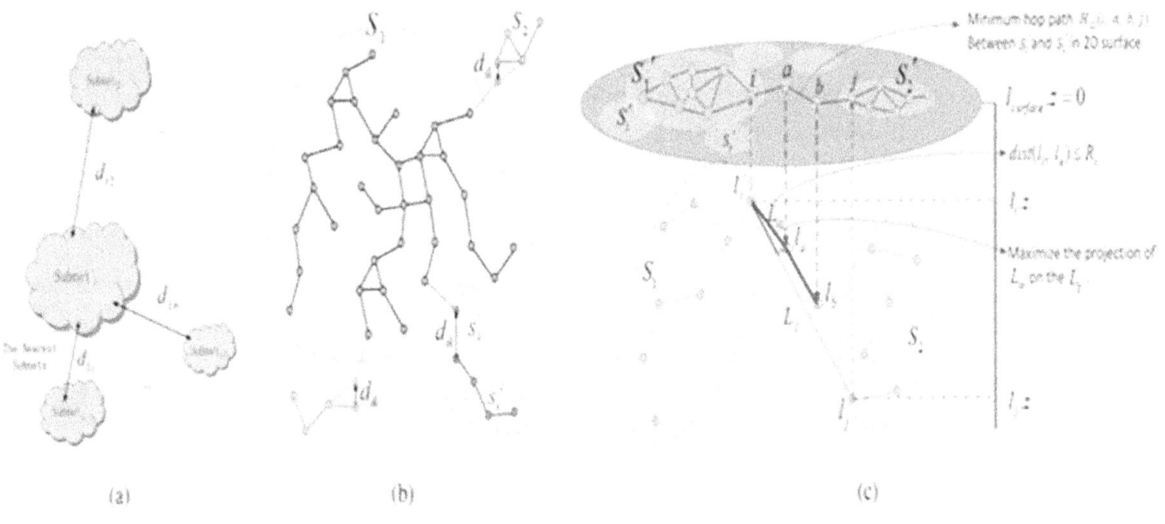

such as number of sensor nodes, transmission range and divergence angle etc. for performance analysis in multi-hop underwater optical wireless sensor networks. The authors have defined the system in term of uni-directional random graphs with angular sector coverage region for each optical transmitter node. A stochastic analysis for network connectivity of proposed system is also performed. And, the authors have worked out *k*- connectivity for underwater optical wireless sensor networks. The descendant and antecedent neighbours have played a significant role in order to derive closed expression for *k*- connectivity. The results obtained clearly show that performance parameters influence network connectivity. The authors have presented a relationship between connectivity and root mean square error. They found that there is a proportional relation between connectivity and root mean square error which knocks the localization performance in positive direction.

Figure 11. Demonstration of descendant and antecedent neighbours
Source: (Saeed et al.; 2018)

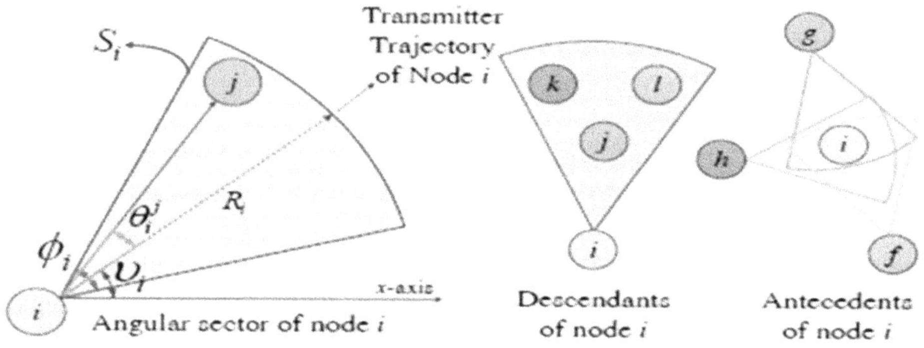

Lifespan Maximization based Deployment Plans

The efficient network lifespan can enhance the network performance to the next level of quality ladder. The network lifespan or network lifetime can be defined as continuation span in which the network nodes can communicate or can be connected to each other through base station. Network lifespan is directly related to the energy consumption during the communication process and the battery life of sensor nodes. Most of the authors have worked out in this direction by imposing constraints on the number of nodes so that some nodes can be kept as spare nodes without any increase in the deployment cost of the network. Cluster based algorithms are another solution in this direction. In clusters, nodes can go in sleep modes or be inactive and switches their states as per requirement to preserve energy. The lifetime of a network mainly depends on the energy consumption by nodes over the time and the residual energy of the sensor nodes. The energy management among the sensor nodes should be done with very careful and tactful manner. So, the network lifetime is one of the key characteristic related to any sensor network as the parameters like coverage and connectivity can also be reduced to lifetime considerations.

Delay Tolerant Placement and Routing (DTPR) in UWSNs

Authors (Alsalih et al., 2008a) have proposed a distributed algorithm with the objective of maximization of network lifetime without any latency constraint. This means, the length of the route from the sensor node to the data collector is not constrained. This scheme also tends to resolve the problem of three-dimensional placement, where data collectors can be positioned under the water at any place until there is at least one sensor node within the acoustic transmission range. Under latency constraints, a nearest-optimal route and location of mobile data collector is identified. The network lifespan is divided in fixed size round such as day or week. At the outset of every round, the data collectors move to a new location. A new route will be required to transmit data from sensor node to data collector due to this repositioning of the data collector. Integer Linear Program (ILP) solver, is used to identify optimal multi-hop routing path along with the optimal positioning of the data collector. The current residual energy and potential energy consumption of every sensor node is kept into consideration by the objective function of ILP. Experimental results verified the ability to significantly extend the lifespan of UWSNs, especially in comparison with other random and static scheme.

Delay Constrained Placement and Routing (DCPR) in UWSNs

Another scheme named delay constrained placement and routing (DCPR) in UWSN for routing and placement of mobile data collectors was discussed (Alsalih et al., 2008a; 2008b). However, DTPR (described in previous section) and DCPR are striving to achieve the same goal in maximizing the network duration; DCPR has a constraint on how long a data unit can take its way to data collector. Just like DTPR, the network lifespan is divided in fixed size cycle. At the start of each cycle, optimal position of data collector and routing path to transmit sensed data to data collector need to be identified keeping in consideration the maximum data delay constraint. Integer Linear Program (ILP) solver with constant time limit, is used identify optimal multi-hop routing path along with the optimal positioning of the data collector. The current residual energy and potential energy consumption of every sensor node is kept into consideration by the objective function of ILP. Experimental results show that although the life of the DCPR is slightly shorter than that of DTPR, but it is able to fulfil all QoS commitments based on delays.

Multipath Virtual Sink Architecture for UWSNs

A centralized multipath virtual sink architecture was designed in order to boost the energy efficiency and robustness of the network, under adverse network conditions (Seah et al., 2007). Multipath data delivery is used in such environments to conquer adverse link conditions and long propagation delays. Spatially diverse physical sinks are grouped together to form a cluster. A local aggregator, generally known as virtual link, is associated with each cluster. These virtual sync designs avoid the contention near the sink while it is commonly seen in conventional multipath routing. High-speed mesh network is used for interconnection between the local sinks and the virtual sinks. Although multipath data delivery scheme balances the load of local sink but tend to enhance data redundancy.

Surface Gateway Placement Strategy for UWSNs

A strategy for maximization of the network lifespan under a specific requirement has been given (Ibrahim et al., 2010). Deployment of the underwater sensor network is accomplished by integrating a set of surface-level gateway nodes. Surface gateways are equipped with underwater acoustic communication interface and radio interface located above the sea level. Surface gateways are used to relay data either between UWN and central control station or to remote areas of the underwater network more efficiently. Authors have formulated a multi-objective function to establish the necessary balance between the lifespan and performance. The use of gateway node reduces the total distance that data need to travel under the water. This in turn, increases the durability and reliability, and decreases delay as well as power consumption. The authors have used this gateway placement strategy in UWSN deployment scenarios at multiple instances. Experiments reveal that if the lifespan of the network is taken as the goal of implementation optimization, it gives 20 percent better results when other metrics are the primary optimization aim.

Underwater Deployment Algorithm (UDA) for UWSNs

Liu (2011) has worked on an underwater deployment algorithm for enhancing the lifetime of 3D sensor network when sensing and communication radius are given randomly. Bottom nodes and anchored nodes are linked to each other and responsible for data transmission to surface sinks. This phenomenon is based on the scenario of using number of static sensors in a specific 3D ocean environment. Each node has limited-energy batteries. UDA identifies and selects the best cluster shape by mathematical analysis. 3D space is divided into multiple layers made-up of such shape clusters while keeping full coverage and connectivity. A cluster-head elected from each cluster, act as backbone of network. Cluster-head keeps other sensor nodes asleep in order to minimize energy consumption. Furthermore, nodes close to sinks can carry a heavy data-relaying task. In reaction to potential relay mismatch, UDA sets varying node deployment density at each layer. OMNeT++ is used for simulation. Simulation results shows that if number of nodes is large (more than 10500), Truncated octahedron cluster model can maximize the UWSN lifespan. However, when communication radius is large (greater than 40 meters), Rhombic dodecahedron obtains the longest UWSN lifespan and if sensing radius is large enough (greater than 20 meters), Truncated octahedron is the best cluster shape.

CONCLUSION

This chapter presents a collaboration between the network quality parameters and node deployment strategies. Authors have tried to combine various deployment schemes in this study to strengthen the base for new researchers. Most of the node deployment strategies presented in this chapter shows how a two-dimensional underwater sensor network can be expanded to three-dimensional while maintaining the quality parameters.This study provides a platform to new researchers to invent new node deployments in order to build stronger and reliable builds than the present ones. Some of future directions for research expansion in the field of underwater sensor networks are as follows (Han et al., 2015; Sandeep et al., 2017; Nasri et al., 2019):

[1] The real scenario based working model can be helpful to optimize the coverage overlaps.
[2] The security issues related with deployment algorithms in UWSNs have not been addressed properly. As any raider can falsify data gathering which can lead to demolishment of the network.
[3] For dynamic environment, algorithms to re-connect the network partition can be designed for the availability of some standby options in case of emergency.
[4] The concept of *k*-coverage and *k*-connectivity have not gained so much attention by the researchers in case of UWSNs. However, now a days, due to widespread applications of UWSNs, some efficient criteria can be explored in this direction.
[5] The development of algorithms to optimize the delay constraint can be used as one of best option to explore research in the field of UWSNs.

REFERENCES

Akkaya, K., & Newell, A. (2009). Self-deployment of sensors for maximized coverage in underwater acoustic sensor networks. *Computer Communications*, *32*(7), 1233–1244. doi:10.1016/j.comcom.2009.04.002

Akram, V. K., & Dagdeviren, O. (2016). On k-connectivity problems in distributed systems. In N. Meghanathan (Ed.), *Advanced Methods for Complex Network Analysis* (pp. 30–57). IGI Global. doi:10.4018/978-1-4666-9964-9.ch002

Akyildiz, I. F., Pompili, D., & Melodia, T. (2004). Challenges for efficient communication in underwater acoustic sensor networks. *SIGBED Review*, *1*(2), 3–8. doi:10.1145/1121776.1121779

Akyildiz, I. F., Pompili, D., & Melodia, T. (2005). Underwater acoustic sensor networks: Research challenges. *Ad Hoc Networks*, *3*(3), 257–279. doi:10.1016/j.adhoc.2005.01.004

Akyildiz, I. F., Pompili, D., & Melodia, T. (2006). State-of-the-art in protocol research for underwater acoustic sensor networks. *Proceedings of the 1st ACM International Workshop on Underwater Networks*, 7–16. 10.1145/1161039.1161043

Alsalih, W., Hassanein, H., & Akl, S. (2008a). Placement of multiple mobile data collectors in underwater acoustic sensor networks. *Wireless Communications and Mobile Computing*, *8*(8), 1011–1022. doi:10.1002/wcm.656

Alsalih, W., Hassanein, H., & Akl, S. (2008b). Delay constrained placement of mobile data collectors in underwater acoustic sensor networks. *Proceedings of 33rd IEEE Conference on Local Computer Networks*, 91-97. 10.1109/LCN.2008.4664156

Awan, K. M., Shah, P. A., Iqbal, K., Gillani, S., Ahmad, W., & Nam, Y. (2019). Underwater wireless sensor networks: A review of recent issues and challenges. *Wireless Communications and Mobile Computing*, *2019*, 1–20. doi:10.1155/2019/6470359

Bharamagoudra, M. R., & Manvi, S. K. S. (2016). Deployment scheme for enhancing coverage and connectivity in underwater acoustic sensor networks. *Wireless Personal Communications*, *89*(4), 1265–1293. Advance online publication. doi:10.100711277-016-3315-8

Cayirci, E., Tezcan, H., Dogan, Y., & Coskun, V. (2006). Wireless sensor networks for underwater surveillance systems. *Ad Hoc Networks*, *4*(4), 431–446. doi:10.1016/j.adhoc.2004.10.008

Domingo, M. C. (2008). Overview of channel models for underwater wireless communication networks. *Elsevier Physical Communications Journal*, *1*(3), 163–182. doi:10.1016/j.phycom.2008.09.001

Domingo, M. C. (2009). Optimal placement of wireless nodes in underwater wireless sensor networks with shadow zones. Proceedings of IEEE 2nd IFIP Wireless Days (WD). doi:10.1109/WD.2009.5449707

Felemban, E. (2013). Advanced border intrusion detection and surveillance using wireless sensor network technology, *International Journal of Communications. Network and System Sciences*, *6*(5), 251–259. doi:10.4236/ijcns.2013.65028

Felemban, E., Shaikh, F. K., Qureshi, U. M., Sheikh, A. A., & Qaisar, S. B. (2015). Underwater sensor network applications: A comprehensive survey. *International Journal of Distributed Sensor Networks, 2015*(11), 896832. doi:10.1155/2015/896832

Han, G., Zhang, C., Shu, L., & Rodrigues, J. J. P. C. (2015). Impacts of deployment strategies on localization performance in underwater acoustic sensor networks. *IEEE Transactions on Industrial Electronics, 62*(3), 1725–1733. doi:10.1109/TIE.2014.2362731

Han, J., Al-Bzoor, M., Zuba, M., Cui, J. H., Ammar, R. A., & Rajasekaran, S. (2012). PADP: Prediction assisted dynamic surface gateway placement for mobile underwater networks. *IEEE Symposium on Computers and Communications (ISCC)*, 139-144.

Ibrahim, S., Ammar, R. & Cui, J-H. (2010). Surface gateway placement strategy for maximizing underwater sensor network lifetime. *The IEEE symposium on computers and communications.* DOI: . doi:10.1109/ISCC.2010.5546562

Liu, L. (2011). A deployment algorithm for underwater sensor networks in ocean environment. *Journal of Circuits, Systems, and Computers, 20*(6), 1051–1066. doi:10.1142/S021812661100775X

Nasri, N., Mnasri, S., & Val, T. (2019). 3D node deployment strategies prediction in wireless sensors network. *International Journal of Electronics*, 1–19.

Peach, C., & Yarali, A. (2013). An overview of underwater sensor networks. *Proceedings of the routing techniques regarding network layer*, 31–36.

Pompili, D., Melodia, T., & Akyildiz, I. F. (2009). Three-dimensional and two-dimensional deployment analysis for underwater acoustic sensor networks. *Ad Hoc Networks, 7*(4), 778–790. doi:10.1016/j.adhoc.2008.07.010

Sandeep, D. N., & Kumar, V. (2017). Review on clustering, coverage and connectivity in underwater wireless sensor networks: A communication techniques perspective. *IEEE Access: Practical Innovations, Open Solutions, 5*, 11176–11199. doi:10.1109/ACCESS.2017.2713640

Sangwan, A., & Singh, R. P. (2015). Survey on coverage problems in wireless sensor networks. *Wireless Personal Communications, 80*(4), 1475–1500. doi:10.100711277-014-2094-3

Seah, W. K. G., & Tan, H.-X. (2007). Multipath virtual sink architecture for underwater sensor networks. *Proceedings of the OCEANS—Asia Pacific*, 1–6.

Senel, F. (2016). Coverage-aware connectivity-constrained unattended sensor deployment in underwater acoustic sensor networks. *Wireless Communications and Mobile Computing, 16*(14), 2052–2064. doi:10.1002/wcm.2667

Senel, F., Melike, K., Kantarci, E., & Yilmaz, T. (2014). Self-deployment of mobile underwater acoustic sensor networks for maximized coverage and guaranteed connectivity. *Ad Hoc Networks*, 1–25.

Soreide, N. N., Woody, C. E., & Holt, S. M. (2004). Overview of ocean based buoys and drifters: Present applications and future needs, *Proceedings of 16th International Conference on Interactive Information and Processing Systems (IIPS) for Meteorology, Oceanography and Hydrology.*

Tuna, G., & Gungor, V. C. (2017). A survey on deployment techniques, localization algorithms, and research challenges for underwater acoustic sensor networks. *International Journal of Communication Systems*, *30*(17), 1–21. doi:10.1002/dac.3350

Winston, K. G. S., Xian, T. H., Zheng, L., & Marcelo, H. (2005). Multiple-UUV approach for enhancing connectivity in underwater ad-hoc sensor networks. *Proceedings of MTS/IEEE on Oceans*. DOI: 10.1109/OCEANS.2005.1640102

Chapter 3
Recent and Future Node Deployment Strategies in the Underwater Sensor Network (UWSN)

Krishna Pandey
University Institute of Engineering and Technology, Kurukshetra University, India

Manish Kumar
University Institute of Engineering and Technology, Kurukshetra University, India

ABSTRACT

The chapter focuses on the recent development in the field of the sensor node deployment in the UWSN (under water wireless sensor network). In the chapter, the technical challenges during the node deployment of the sensor nodes in the UWSN (under water wireless sensor network) are represented with prefacing the background. The chapter focuses on the different methods of node deployment and presents a generalized model for ensure the reliability. A view of analyzing the deployment of sensor nodes is also shown in the example by following the recent researches in the domain. Finally, the future scope and conclusion is represented with the idea of new paradigms in the deployment of sensor nodes in the UWSN.

INTRODUCTION

First, under the water communication scenario was introduced after the II-World War, the underwater phone was designed by the US Navy for communication (Almalkawi et al. 2010). Electromagnetic waves, optical waves, and acoustic waves can have more potential than radio frequency waves because of the high attenuation factor in underwater communication. Transmission power always plays a crucial role when uses the RF waves for the underwater communication and applications of the Underwater Wireless Sensor Network (UWSN). Many researchers use the optical waves to ensure more reliability in case UWSN and others use a large number of modems.

DOI: 10.4018/978-1-7998-3640-7.ch003

Nowadays, underwater WSN plays a vital role in the field of oceanographic observations, marine surveillance, and water quality monitoring. The advancement in the UWSN technologies results in the bright scope for the researchers. Due to the requirement of low transmission power and sensor autonomy nature of UWSN, it becomes a master key for the oceanographic observations. UWSN has several advantages when used for underwater solutions. It provides large area monitoring by using GPRS/GSM technologies and can be designed with low cost having low power consumption abilities (Albaladejo et al. 2010). UWSN provides an innovative and efficient way to the deployment of sensor nodes having low cost and proposed the idea of the network using logical and physical topologies of the sensor node to form a network and communicate the information with the base station (Akyildiz et al. 2002).

UWSN has several applications in the field of environmental monitoring such as water quality monitoring (Postolache et al. 2014). In recent days, UWSN technologies are also utilized for monitoring natural areas such as lakes, rivers, ponds, and small marine areas (Seders et al. 2007). As the underwater sensors are expensive, the implementation cost of the UWSN monitoring system may be high. Consequently, there are several factors like failure and location of the sensor node must be taken as primitive of the implementations. To avoid failure during underwater monitoring, a good network topological design must be implemented to reduce energy consumption by increasing reliability and for ensuring the security of the nodes in the underwater condition (UWSN) with safe communication, the secure authentication and secure data aggregation methods can be utilized (Goyal et al. 2020).

Figure 1 depicts a UWSN model having sensor nodes with gateway node and AUV (automated underwater vehicle). The surface sink does monitoring of UWSN and sends the desired data or information to the earth station via satellite for communication.

Figure 1. Underwater wireless sensor network (Nadeem et al. 2015)

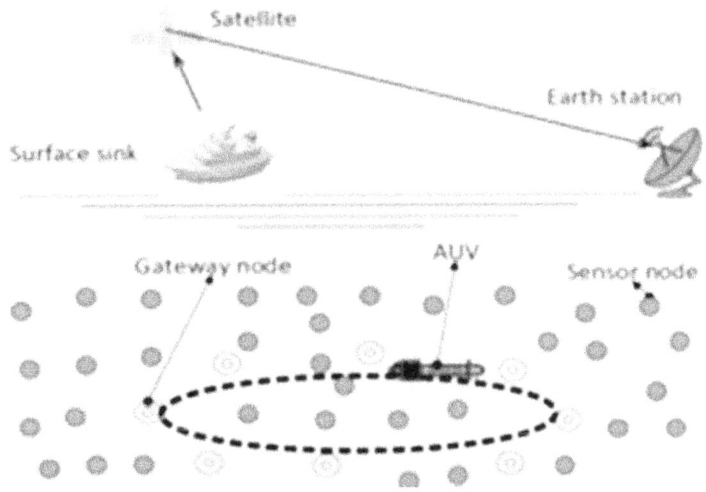

TECHNICAL CHALLENGES IN THE DEPLOYMENT OF SENSOR NODES IN UWSN

UWSNs are amazingly valuable in military and adversary observation applications. Communication environment, Network topologies, power challenges, secure data, and safe localization are the basic constraints for the deployment of sensor nodes in UWSN. The challenges are discussed below:

Communication Demands

Electromagnetic waves are not effective within the transmission signal in the underwater environment since over the long-distance signal confront damping impacts. There are a few modes of communication other than EM waves to serve the reason for communication like acoustic communication and optical communication that can perform well beneath water communication due to the less signal diminishment overtime during propagation. Acoustic communication is broadly utilized within underwater communication particularly for profound seawater communication.

Power Consumption

Xie. et al. (2005), Tilak et al. (2002) proposed the assessed information for the optimization strategies and battery life in shallow water, the bandwidth, the shallow water environment, and the network topology are the constraints that affect the life of the battery of a particular node.

There is a great challenge for the node deployed in the underwater environment compare with terrestrial WSN in the term available energy resources. In the underwater, there is no supply of sunlight, so here no provision of solar energy whereas the terrestrial WSN has always a permanent renewable resource in the form of solar energy to short the energy problem for the node battery. Ovaliadis et al. (2010) represented the possible battery technologies for the underwater wireless sensor nodes. The work showed that the nodes using the Li-ion (Lithium) batteries provide the best results in the UWSN rather than Nickel-cadmium and Nickel-metal hydride systems by following the given benefits (Gitzendanner et al. 2004).

1. Less maintenance required in comparison to the Nickel-cadmium and Nickel-metal Hydride batteries.
2. Have a good system survivability mechanism due to the no thermal and magnetic impacts and low discretion rate.
3. Less memory effect.
4. (d)Provides the higher power density
5. Available higher levels of energy
6. Flexible in design
7. Secure and independent battery technology
8. Ease of supports for the communicative devices and readable battery status.
9. Overall cost-effective and power-saving life cycle of the battery.

Network Topology

Network topology defines the geometrical arrangement of the sensor nodes. In the WSN, the position of the sensor nodes is almost constant. In the underwater environment, the sensor node movement represents critical issues as the network topology changes in every instance of time. One, the behavior of the movement of the sensor nodes in the shallow water where the position of the sensor nodes continuously changes with the time due to the wave activities in the water. Second, the situation is sensor nodes in the deep water. In the deep-water, sensor nodes almost settled down and represent the constant topology. The challenging issue is in the shallow water and the nodes deployed close to the water surface. Cui et al. (2006) represented that the object might move in the underwater environment with a speed of 3-6 Km/hr. So, node mobility is a crucial factor, which suggests the concept of an intelligent node with self-learning capabilities. In UWSN, the system should have the ability to adapt to the dynamic changes and decide the network topology accordingly by the movement of the sensor nodes. The network control strategies can only provide the best network topologies having effective channel distribution with optimal results. Liu (2010) had provided some network topology control techniques for UWSN.

SECURITY

Cong et al. (2010) presented the different aspects of security issues of the UWSN. In the UWSN, the security algorithm can be designed by taking the constraint of the energy consumption and computational power consumption of a secure algorithm with ensuring the strong cryptographic mechanism for encryption and decryption. There are several challenges in the UWSN to maintain the secure localization and synchronization with maintaining the Confidentiality, Integrity, and Authentication (CIA) of nodes for secure communication.

Confidentiality and Integrity

In military applications of UWSN, it is expected that the data should be shared between the nodes which are actually in the communication. In addition to the military application, some other types of applications like earthquake and disaster management and target monitoring systems required to ensure data integrity to maintain the purpose of the services.

Authentication

To avoid the attacks by the foe or actually in the term of security "to avoid the man in the middle" type of attack several authentication algorithms will be required to ensure the authentication of the user node. An energy and power constraint with the authentication algorithm should be developed to maintain secure UWSN.

Synchronization and Secure Localization

In the UWSN, all the real-time data analysis required application must be incorporate with synchronization techniques in every sensor node. In the deployment of the sensor nodes in the UWSN, the secure localization plays a crucial role, which is the first placement of the nodes in the underwater, an intruder may change the energy or power level of the nodes to disturb the localization. (Salva-Garau et al. 2003) suggested the GPS (Global positioning system) to determine the position of the randomly deployed nodes in the UWSN to maintain and perform the secure localization. Although due to the high power requirement of the GPS this method is not advisable (Domingo et al. 2007). In the UWSN, energy, and power-based secure localization is required to ensure the node deployment in the UWSN.

RECENT THEORIES

As time elapsed, so many researchers are contributing their research work in the area of the Node deployment of UWSN. A glimpse of recent years of publication is presented here to give a structural idea of the research in this domain.

Wong et al. (2006) proposed a wake scheme of ocean monitoring UWSN. In the work a two-stage (AWSF) adaptive wakeup schedule function scheme proposed with (WSF) wakeup schedule function pruning followed by WSF reconstruction, where parametric details and various properties of the AWSF scheme with delay and energy overhead analysis is defined. Ibrahim et al. (2007) used Integer Linear Programming (ILP) for solving the problem and further do the optimization for the deployment. The considered problem was focused on the cost function of surface gateway deployment (not focus on SFG) with their pre-defined number of nodes and locations of deployment based on resource optimization such as a minimum number of nodes, and design objectives with several constraints like delay, energy consumption, and fault tolerance to maintain the high reliability.

Ibrahim et al. (2010) proposed a surface deployment strategy as a function of UWSN lifetime with deployment, flow, and, Interference constraints. This work focused on the maximization of a lifetime with trade-off end to end delay was shown in the simulation results. Vieria et al. (2011) proposed the analytical approach for decreasing the energy consumption and increasing the life cycle of the sensor nodes by providing the proper coverage area, link probability, and node degree for the UWSN. Yingying et al. (2011) introduced the OPNET simulator-based results for minimizing the end to end delay, and energy consumption with adequate methods to maintain scalability, connectivity with proper coverage area.

Felamban et al. (2013) provided a nonlinear programming model to minimize the transmission loss under a predefined number of nodes and their volume with different methods to increase the channel throughput under the depth of UWSN. Hui et al. (2015) proposed the Voronoi diagram based depth adjustment deployment method with high coverage, connectivity, and network lifetime. Iyer et al. (2015) proposed the genetic-based optimization algorithm for position deployment in the UWSN with a focus on the coverage area issue. Rahman et al. (2016) proposed an energy-efficient protocol for enhancing the network lifetime with maximum throughput. In the scheme, two sinks and four-node were considered to perform the depth-based routing in the underwater environment.

Ge et al. (2016) proposed the CFD method for the deployment and recovery process of the sensor nodes in the UWSN. For analysis, minimum depth and minimum time of the sensor nodes were considered with their inclined angle and velocity under the water environment. Alhumyani et al. (2016) utilized the heuristic approach and proposed the graph and ILP formula based mathematical model for the deployment of the nodes to attain the optimum solution. The work represented the three algorithms to ensure the best-suited method of deployment using a layer-based approach by following the greedy path approach and path-based approach. To avoid the energy consumption problem Ding et al. (2017)

proposed the double coverage algorithm for sensor deployment in underwater wireless sensor networks. The algorithm reduced the energy consumption with equal distribution of sensor nodes in the network to ensure the minimized cost by decreasing the number of sensor nodes during the deployment. To ensure the minimum consumption of energy and complete the coverage area, FST (Four-Sensor-Triangle) was proposed to cover the water area using grid strategy in the underwater environment.

Radha et al. (2017) proposed the flooding protocol in which the node can transmit the data to nearby all nodes by setting the transmission power of all the nodes so that the minimum energy consumption will be achieved with the optimal deployment of sensor nodes in the underwater environment. Yuan et al. (2019) proposed the EELA (Energy-Efficient Localization Algorithm) for the localization problem as a Single-Leader-Multi-Follower Stackelberg game to ensure the optimized the trade-off between the high localization capabilities with the low energy consumption by using the sensor nodes and anchor nodes as a leader-follower relationship. Wang et al. (2019) proposed the optimal relay deployment algorithm by using the K-mean algorithm for dividing the network space, with minimizing the energy consumption to ensure the optimal deployment of sensor nodes in the UWSN.

Figure 2. Operating framework (Wang et al. 2019)

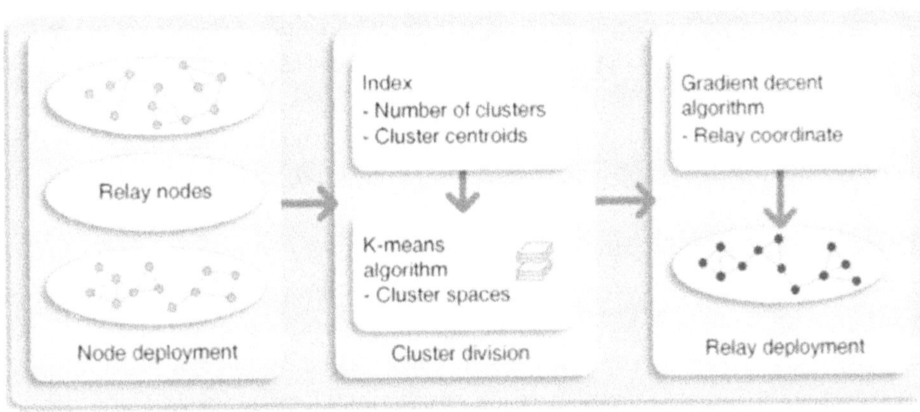

Figure 2 represents the operating framework of UWSN having three major parts Node deployment for find out the location of nodes, cluster division for the organizing the sensor deployment space and relay deployment to achieve optimal deployment scheme for the UWSN (underwater wireless sensor Networks) with low transmission delay and less multipath fading effect.

Deployment Methods

In the deployment of sensor nodes underwater wireless sensor networks, deployment methods focus on the node movement and node path correction (Cui et al. 2005) and several challenges for the node deployment based on node placement and localization because of the node mobility in the underwater environment. The sensor node deployment is already explained by many researchers. (Li et al. 2009) defined the several deployment schemes of sensor nodes.

The UWSN underwater wireless sensor network node deployment scheme can be done by the following approach.

1. **Virtual Force (Vector-Based) Approach:** In this technique, the movement of the sensors of the UWSN follows the movement of the already computed vector using the neighbor's relative positions. In this uniform deployment can be achieved by placing the sensor nodes using a threshold distance value according to the summation of vector forces.
2. **Voronoi-Based Approach:** In this approach, by using the Voronoi polygon the underwater sensors try to reduce the uncovered local area many times for attaining an optimized result.
3. **Load-Balancing Approach:** In this approach, the sensor region will be partitioned and gone through the multiple iterations until the balance is achieved.
4. **Stochastic Approach:** In this approach, the nodes are deployed randomly. Random deployment is mostly used in the scenario where no proper data of the geographic environment is available as the military applications. In many scenarios, the random deployment method is very cost and time effective but not ensures the uniform distribution of sensor nodes. Due to its several benefits, it serves as an initial phase of the movement-based deployment schemes.
5. **Point-Coverage Approach:** In this approach, the area coverage explored as a point coverage problem to ensure the deployment over the certain geographic environment of the underwater.
6. **Incremental Approach:** It is a centralized approach, where the sensors are deployed in the underwater environment based on the deployment data of previous sensor node. Each sensor is deployed using one by one approach.
7. **Maximum-flow Approach:** In this approach, the sensor nodes underwater are deployed as a function of the minimum cost of maximum flow to ensure the coverage of the desired region of deployment.
8. **Genetic Algorithm Approach:** In this approach, the sensor deployment is done based on natural and reproduction simulating genes by following a multi-round selection method.
9. **Optimization Approach with ILP (Integrated Linear Programming) & Non-ILP(Integrated Linear Programming):** In this approach, by optimizing the formulated problem with constraint and non-constraint value with the help of the optimizing algorithms one can achieve the perfect deployment of nodes in the UWSN scenario.

GENERALIZED PROCEDURE TO THE DEPLOYMENT OF SENSOR NODES IN THE UWSN

1. Identify the problem after previewing the technical challenges
2. Select X= t whereas t belongs to (increase coverage area, increase the lifetime, reduce energy consumption, High throughput, Better connectivity, Distance vector method, etc.).
3. Identify the power and energy constraint or any other constraints.
4. Formulate the problem X with defined constraints or no-constraints.
5. Use any of the above-given deployment methods according to the appropriate stated problem.
6. Validate the results with simulation tools or with mathematical theorems to attain the desired output from the deployment strategy.

Example: Steps for the node deployment in UWSN (underwater wireless sensor network) using the Optimization approach.

1. Firstly find out the problems to increase coverage area, increase the lifetime, reduce energy consumption, High throughput, Better connectivity, Distance vector method, etc.
2. Representation of the Problem ILP or Non-ILP must be defined.
3. Find out the constraints, Constraints always plays a vital role in binding the problems. Although, the problem doesn't require to have constraints.
4. Use the appropriate optimization techniques for the deployment of sensor nodes underwater environment which helps to solve the complexity of the problems. Optimization of the problem with given constraints or unconstrained has to be done for providing the optimal solution.

FUTURE OF THE DEPLOYMENT OF THE UWSN:

In the future, several types of research will be possible for the materials which reduce the cost of underwater sensors, and the TWDP (two way with diffuse power) or other ray theory produces an exact model for analyzing the channel prediction of the Underwater. In addition to coverage and the connection, several routing, multicast, load balancing, and data securities issues should be considered to ensure more reliability in the communication as the formulation of the problem in the deployment of the nodes, Convex optimization techniques and Nature-inspired optimization techniques will be balanced the optimization problems to ensure the QoS (Quality of Services). Artificial Intelligence and machine learning-based algorithm will resolve the challenges which come in the deployment of the sensor nodes in the underwater environment. Secure deployment of the sensor nodes is also a major concern for the efficient and safe communication between the nodes or node to base station.

CONCLUSION

In the chapter, the discussion of the deployment of the sensor nodes in the UWSN begins with the past background and technical challenges. The main focus is on the deployment methods of sensor node underwater environment scenarios with recent researches and examples. Description of the different approaches of deployment of sensor nodes in UWSN is presented. The optimization approach is a new way to represent the deployment strategy. The chapter provides a generalized method of deployment of nodes in the underwater environment. The chapter concludes with the future of node deployment strategy mentioning the deployment of a sensor node in the UWSN as an optimization, artificial intelligence problem, and solving approach.

REFERENCES

Akyildiz, I. F., Su, W., Sankarasubramaniam, Y., & Cayirci, E. (2002). subramaniamSankara, Y., and Cayirci, E. (2002). Wireless sensor networks: A survey. *Computer Networks*, *38*(4), 393–422. doi:10.1016/S1389-1286(01)00302-4

Albaladejo, C., Sanchez, P., Iborra, A., Soto, F., Lopez, J. A., & Torres, R. (2010). Wireless sensor networks for oceanographic monitoring: A systematic review. *Sensors (Basel), 10*(7), 6948–6968. doi:10.3390100706948 PMID:22163583

Alhumyani, H., Ammar, R., Albarakati, H., & Alharbi, A. (2016). Deployment strategies for underwater sensing and processing networks. *IEEE Symposium on Computers and Communication (ISCC)*, 358-363, 10.1109/ISCC.2016.7543766

Almalkawi, I. T., Guerrero Zapata, M., Al-Karaki, J. N., & Morillo-Pozo, J. (2010). Wireless multimedia sensor networks: Current trends and future directions. *Sensors (Basel), 10*(7), 6662–6717. doi:10.3390100706662 PMID:22163571

Cong, Y., Yang, G., Wei, Z., & Zhou, W. (2010). Security in underwater sensor network. *Communications and Mobile Computing, International Conference on, 1*, 162–168.

Cui hong, J., Kong, J., Gerla, M., & Zhou, S. (2005). *Challenges building scalable and distributed underwater UWSN*. Academic Press.

Cui hong, J., Kong, J., Gerla, M., & Zhou, S. (2006). Challenges: Building scalable mobile underwater wireless sensor networks for aquatic applications. *IEEE Network, 20*.

Ding, Y., Li, N., Song, B., & Yang, Y. (2017). The mobile node deployment algorithm for underwater wireless sensor networks. *Chinese Automation Congress (CAC)*, 456-460. doi: 10.1109/CAC.2017.8242811

Domingo, M. C., & Prior, R. (2007). Design and analysis of a gps-free routing protocol for underwater wireless sensor networks in deep water. *International Conference on Sensor Technologies and Applications, SENSORCOMM '07*, 215–220. 10.1109/SENSORCOMM.2007.4394923

Felamban, M., Shihada, B., & Jamshaid, K. (2013). Optimal Node Placement in Underwater Wireless Sensor Networks. *IEEE 27th International Conference on Advanced Information Networking and Applications (AINA)*, 492-499. doi: 10.1109/AINA.2013.40

Ge, W. (2016). Numerical simulation on the deploying and recovering process for a node of underwater wireless sensor network. *OCEANS 2016*, 1-4. doi: . doi:10.1109/OCEANSAP.2016.7485602

Gitzendanner, R., Puglia, F., Martin, C., Carmen, D., Jones, E., & Eaves, S. (2004). High power and high energy lithium-ion batteries for under-water applications. *Journal of Power Sources, 136*(2), 416–418. doi:10.1016/j.jpowsour.2004.03.032

Goyal, N., Dave, M.,, & Verma, A. K. (2018). Adaptive Error Control Technique for Cluster-Based Underwater Wireless Sensor Networks. In *International Conference on Wireless Intelligent and Distributed Environment for Communication* (pp. 269-280). Springer. 10.1007/978-3-319-75626-4_20

Goyal, N., Sandhu, J. K., & Verma, L. (2019). Machine Learning based Agglomeration in Underwater Wireless Sensor Network. Vol. 9, Issue 6, pp 240-245 DOI:16.10089.IJMTE.2019.V9I6.19.28495

Goyal, N., Dave, M. & Verma, A. K. (2020). SAPDA: Secure Authentication with Protected Data Aggregation Scheme for Improving QoS in Scalable and Survivable UWSNs. *Wireless Personal Communication*, Vol 113, Issue 1, pp 1-15. Doi:10.100711277-020-07175-8

Hui, W., Liu, M., & Lin, Z.S. (2015). An Efficient Depth adjustment Deployment Scheme for Underwater Wireless Sensor Network. *Proceedings of the 34th Chinese Control Conference*, 7771-7776.

Ibrahim, S., Ammar, R., & Cui, J. (2010) Surface gateway placement strategy for maximizing underwater sensor network lifetime. *The IEEE symposium on Computers and Communications, Riccione*, 342-346. doi: . doi:10.1109/ISCC.2010.5546562

Ibrahim, S., Jun-Hong, C., & Reda, A. (2007). Surface-Level Gateway Deployment for Underwater Sensor Networks. *Proceedings of IEEE Military Communications Conference*, 1 - 7. 10.1109/MILCOM.2007.4454996

Iyer, S., & Rao, D. V. (2015). Genetic algorithm based optimization technique for underwater sensor network positioning and deployment. *2015 IEEE Underwater Technology (UT),* 1-6. doi:10.1109/UT.2015.7108229

Lanbo, L., Shengli, Z., & Jun-Hong, C. (2008). Prospects and problems of wireless communication for underwater sensor networks. *Wireless Commununication Mobile Comput.*, *8*(8), 977–994. doi:10.1002/wcm.654

Li, X., Nayak, A., Simplot-Ryl, D., & Stojmenovic, I. (2009). *Sensor placement in sensor and actuator networks in wireless and actuator network: algorithm and protocol*. Academic Press.

Liu, L. (2010). A qos-based topology control algorithm for underwater wireless sensor networks. *International Journal of Distributed Sensor Networks.*

Nadeem, J., Naveed, I., Ashfaq, A., Nabil, A., & Umar, Q. (2015). An Efficient Data-Gathering Routing Protocol for Underwater Wireless Sensor Networks. Sensors. *Sensors (Basel), 15*(11), 29149–29181. doi:10.3390151129149 PMID:26593924

Ovaliadis, K., Savage, N., & Kanakaris, V. (2010). Energy efficiency in underwater sensor networks: A research review. *Journal of Engineering Science and Technology Review*, *3*(1), 151–156. doi:10.25103/jestr.031.25

Postolache, O., Pereira, J. D., & Girao, P. S. (2014). Wireless sensor network based solution for environmental monitoring: Water quality assessment case study,'' IET Sci. *Measurement Techniques*, *8*(6), 610–616.

Radha, S., Josemin, G., & Bala, J. (2017). *Flooding in static and free deployment for underwater wireless sensor network. International Conference on Signal Processing and Communication (ICSPC'17)*, 28-32.

Rahman, Z. U. (2016). On Utilizing Static Courier Nodes to Achieve Energy Efficiency with Depth Based Routing for Underwater Wireless Sensor Networks. *IEEE 30th International Conference on Advanced Information Networking and Applications (AINA),* 1184-1191. doi: 10.1109/AINA.2016.170

Salva-Garau, F., & Stojanovic, M. (2003). Multi-cluster protocol for adhoc mobile underwater acoustic networks. In *OCEANS 2003* (Vol. 1, pp. 91–98). Academic Press.

Seders, L. A., Shea, C. A., Lemmon, M. D., Maurice, P. A., & Talley, J. W. (2007). LakeNet: An integrated sensor network for environmental sensing in lakes. *Environmental Engineering Science*, *24*(2), 183–191. doi:10.1089/ees.2006.0044

Tilak, S. N., Abu-Ghazaleh, B., & Heinzelman, W. (2002). Infrastructure tradeoffs for sensor networks. In *Proceedings of the 1st ACM international workshop on Wireless sensor networks and applications, WSNA '02*, (pp. 49–58). ACM. 10.1145/570738.570746

Vieira, L. F. M., Almiron, M. G., & Loureiro, A. A. F. (2011). 3D MANETs: Link Probability, Node Degree, Network Coverage and applications. *IEEE Wireless Communications and Networking Conference*, 2042-2047, 10.1109/WCNC.2011.5779443

Wang, S., & Shin, Y. (2019). 3D-Deployment of Magnetic Induction Relays in Underwater Sensor Networks. *International Conference on Information Networking (ICOIN)*, 222-226. 10.1109/ICOIN.2019.8718105

Wong, Y. F., Ngoh, L. H., Wong, W. C., & Winston, K. G. (2006). Wakeup Scheme for Ocean Monitoring Underwater Sensor Networks (UWSN). *OCEANS 2006 - Asia Pacific:ScholarBank@NUS Repository*. doi:10.1109/OCEANSAP.2006.4393874

Xie, P., Cui hong, J., & Lao, L. (2005). Vbf: Vector-based forwarding protocol for underwater sensor networks. *Proceeding of IFIP Networking*, 1216–1221.

Yingying, Z., Xia, L., & Shiliang, F. (2011). Deployment analysis in two-dimensional Underwater Acoustic Wireless Sensor Networks. *IEEE International Conference on Signal Processing, Communications and Computing (ICSPCC)*, 1-5. 10.1109/ICSPCC.2011.6061819

Yuan, Y., Liang, C., Kaneko, M., Chen, X., & Hogrefe, D. (2019). Topology Control for Energy-Efficient Localization in Mobile Underwater Sensor Networks Using Stackelberg Game. *IEEE Transactions on Vehicular Technology, 68*(2), 1487-1500. doi:10.1109/TVT.2018.2886904

Zhou, B., Li, S., Wang, W., Wang, J., Cheng, Y., & Wu, J. (2019). An Efficient Authentication Scheme Based on Deployment Knowledge Against Mobile Sink Replication Attack in UWSNs. *IEEE Internet of Things Journal, 6*(6), 9738-9747. doi:10.1109/JIOT.2019.2930823

Chapter 4
Underwater Localization Techniques

Manisha Bharti
National Institute of Technology, Delhi, India

Poonam Rani Verma
National Institute of Technology, Delhi, India

ABSTRACT

Underwater acoustic communication uses sound waves to trans-receive information, diving deep inside water, environment scanning, undersea explorations, disaster prevention, etc. In this chapter, an attempt has been made to cover stationary and mobile localization algorithm. They are further subdivided into distributed and centralized. Each one is further subcategorized into estimation-based and prediction-based schemes. The category therefore extends on the basis of ranging method, communication, and synchronization, some of which are area localization, sensor-based localization, forming a sensor array, motion-aware self-localization, silent localization. Each one will be discussed in detail in this chapter. At last, hybrid technique is also discussed, which combines stationary and mobile techniques. The discussion includes various nodes including anchor node, unknown node, sink node, and reference node. Various methods to follow the techniques are also discussed, which include anchor-based method, ranging method, and message communication.

INTRODUCTION

The occurrence of UWSNs introduces us with the latest opportunities to study and understand the communication through ocean.

Underwater sensor nodes are the one that are capable to communicate through acoustic signals and are also less expensive.

Broadly, anchor nodes, reference node and unknown nodes are used. The unknown nodes usually sense the surrounding signals, if any, and anchor nodes are used to find the location of the unknown nodes. They pre obtain their location information with the help of GPS systems. Anchor nodes and unknown

DOI: 10.4018/978-1-7998-3640-7.ch004

nodes are embedded in the reference node. Acoustic signal produces less attenuation and higher speed. In distributed localization algorithms, localization information is collected and the algorithm is made to run. In centralized localization algorithms, base station or a sink node is used to detect. Further, the previous or present data is used to find the location of next node and therefore the classification is based on estimation and prediction.

Basically, three types of nodes do exist:

Anchor node, unknown node and reference node. Since the environment data is to be sensed, so the unknown node will be used. Therefore we should know the position of the unknown node, for which anchor nodes are responsible . There is third type of node that is, reference node, which contains the localized unknown node, and initial anchor node. Estimation based technique uses current information to predict the next location. Whereas, prediction based, uses the previous data to find next node location.

LITERATURE REVIEW OF UNDERWATER WIRELESS SENSOR NETWORK

Before the advent of Underwater wireless communication, Sound navigation and Ranging, came into effect. SONAR operators then predicted underwater acoustics.

Then the question comes, why underwater communication via acoustics? It is because, the earlier used approach of ocean monitoring, have limitations, including online real time prediction, unable to detect the failure of location, and requirement of large storage capacity of information. The information, contains the location and task performed on the nodes .The applications of underwater thereby includes,ocean networks,undersea security,environmental detection,catastrophic prevention,etc. Since to perform the above applications, what is needed is desired location, with efficient power and energy consumption,the chapter discusses the major techniques of finding the respective node location.Localization applications include tracking of a desired node or underwater vehicle,detecting the targets, and improvement of medium access and network protocols.

The technique could have used adhoc,mobile networks,but the later ones' protocols are not suitable for the wireless sensor network applications.Wireless sensor networks are not only used in underwater, but also used in sensing environment, patient health control,traffic control monitoring.

ISSUES OF WIRELESS SENSOR NETWORKS

1. Hardware and operating system environment, requiring large power consumption.
2. Medium access methods control the node radio functionality.
3. Proper synchronization of clock issues, in case of non stationary nodes.
4. Deployment problems, with varying terrain.

Since the sensors are very small in size, we do need them in large number,in order to increase the processing capabilities.

Motivation and Objective

1. 1.To minimize the node size,in order to reduce the operating area.

2. To minimize the energy requirement of the node.
3. 3.To reduce the overhead cost.

To find the exact data,location and time of the node is, mainly to be considered,because with the changing time,the location of the node also changes, along with the moving water current.

In this chapter, we discuss three types of schemes, Static, Movable, and Hybrid.

If we use Static algorithm, the regular update is required, because the nodes location changes continuously.The periodic update, therefore requires larger power, and increased communication overhead cost. So we prefer designing algorithms for the movable sensor nodes. Furthermore the aim of the paper is to study the hybrid algorithm,which reduces the communication cost, as well as increases the efficiency.

Motivation to find the localization comes from the fact that the small number of sensor nodes are very large in size, and the base station is not capable enough to find all these positions.The technique should be so chosen, in order to avoid the obstacles of the interfering signals coming from the nearby nodes.

Challenges and Issues

Since electromagnetic waves are not able to go farther away under water, we make use of acoustic communication, which further leads to a set of issues

1. 1.Large latency, bandwidth limitation, multipath propagation with respect to time.
2. For continuous tracking of movable nodes, with static algorithm, continuous updation is required,which increases the power requirement.
3. 3.The design of hybrid localization is based on the condition given below
4. Self localization period=data observed period

Methods to Follow the Techniques

Anchor Type

Anchor node position is determined by using GPS technique, now since the GPS is not suitable for underwater communication, sub-sea environment, so,we use below mentioned methods, Global positioning system, uses accurate clock timings, and in underwater, the nodes, practically are movable, they are not stationary, so they are unable to provide, synchronized clock timings. Another reason for GPS, to not operate underwater is that radio waves do not propagate within the water surface. This can be better explained from the example of losing connectivity, when we go deep into the water surface.

Ranging method:

Involves time of arrival and time difference arrival. Since the refractive index of water is more than that of air, so the speed of sound in air is more than that in water, and ranging uses it as an advantage.

The nodal estimation is done based on distance and angle calculation.It uses time of arrival(ToA),Time difference of arrival(TDoA),Angle of arrival Arrival (AoA) or Received Signal Strength Indicator (RSSI) to find the distances to other nodes .In the case of UWSN,the classification is further into infrastructure based, distributed and mobile beacons.

Message Communication

It is divided into Silent and Active types

In silent type, a vessel whose magnetic dipole is already known, is used along with the magnetometers. The magnetometers and the dipoles create the ferromagnetic field. The vessel so used can be traced down, using a Kalman filter, further discussed in the chapter.

In active type, the targets are actively illuminated by the transmitters, and the position is further detected by the use of their locations. Example may include Multiple input multiple output RADAR.

Time Synchronization

Perfect time synchronization of the sensor nodes with one another.

Computational Complexity

Complexity should be less. It should be noted that the localization estimation time should be very less.

Because in larger time duration, the estimated position of the sensor nodes will move to the new location. But this implies mainly for mobile localization algorithm, unlike stationary,in M. Stojanovic et al.(1996).

Energy Consumption

Energy consumption should be less for the algorithm to function.

TECHNIQUES

Stationary Localization Algorithms

Depends on the position of the nodes,when the nodes are stationary.The example of stationary localization can be obtained, from the mobile communication, in which the sender and receivers are stationary. There occurs no Doppler shift in case of stationary localization, and therefore it is prone to less interference as in P. Zorzi et al. (2012).

Centralized

The centralized scheme means the process of finding the respective location, is thereby based on a single authority, of one central node .

Area Localization Scheme (ALS)

It was first designed for the terrestrial wireless sensor network. The area localization finds out the position of that node which is not known around in a given area instead of its exact location. The signals are sent, to localize unknown nodes, but they all should be of different power levels, so as to differentiate easily between them and their location as in FengWang et al. (2012). The signals come with the ID so anchor nodes, and the unknown nodes listen and record these IDs. All of this collected data is received by the sink node. Sink node contains all the data of the power levels of the anchor node as well as their position. Now comparing the recorded data and the data received by the unknown node, the sink node draws a map or a graph of the respective location.

Advantages of area localization is that it do not require any synchronization. Also, the calculations are to be done by the anchor nodes, and undertaken by the sink nodes . Therefore the power requirement of the anchor nodes reduces to a great extent.

The basic disadvantage of the ALS is that it requires high communication overhead and also it is not suitable for finding the accurate data about the position. Once it is installed, their positions should not change in future, otherwise the throughput, and efficiency is highly reduced.

Localization Scheme Based on Hyperbola

As shown in figure.1, the signals detected by the unknown node is acknowledged back to the anchor node. And, t1 and t2 are the respective times of reception of the nodes A1, and A2.The difference between both time duration does not change, and so the difference between the distance r1 and r2 do not change and the unknown node is localized as per hyperbola properties.The intersection of any two curves helps to detect the position of unknown nodes .It might be possible, that the various hyperbolas so obtained, may not be able to converge the points accurately, and in precise form, so here least square algorithms are adopted. If we want to find the position of larger number of unknown nodes, hyperbola technique, as compared to circle based technique is used. Also the intersection of circle takes place at two or more points and that of the hyperbola at a single point, therefore using hyperbola based algorithm detects the location of the unknown nodes more accurately than that of the least squares algorithm. As compared to area based localization technique, hyperbola based one is more suitable for accurate position detection. Provided that, in hyperbola, the long range signals are sent, and consumes high energy. The same disadvantage as that of area localization, hold here also, that is, they should be stationary not movable.

Wired connection between nodes is made, creating array. Hence each target radiates an acoustic signal,as can be seen in figure1.

One more type of Energy based algorithm is used in the localization, but it is not used in large distance communication, because then, noise effect is also added.

A Probabilistic Localization Method (PLM)

It performs the operation based on two steps, thereby increasing the accuracy.

In the first step, two nearby anchor nodes uses circle or hyperbola technique to find the unknown position .Then final location is determined, by error measurement and its probability. As the number of anchor nodes is increased, the coverage area also increases and so does the accuracy. As the nodes increases, the complexity and the power thereby required also increases.

Asymmetrical Round Trip Based Localization (ARTL) Algorithm

The above method is based on the prediction that anchor nodes assumes their packets which are not known but other nodes cannot do so, as can be seen from figure2.

Figure 1. Nodes and their location from the source

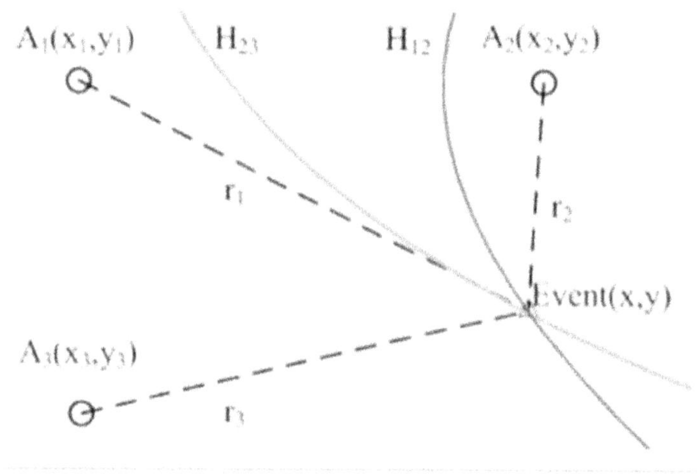

We can easily extend ARTL, to larger networks. By simply increasing new anchor nodes. The coverage area can be extended as per the desired amount by increase in the addition of new anchor nodes. Also the advantage is that by addition of the additional node, neither program modifications nor the extra complex calculations are required. Higher overheads do in occur with extending area and anchor nodes .Also the disadvantage of ARTL is that, it takes long localization time, and more energy

Figure 2. Acknowledgement sent and received

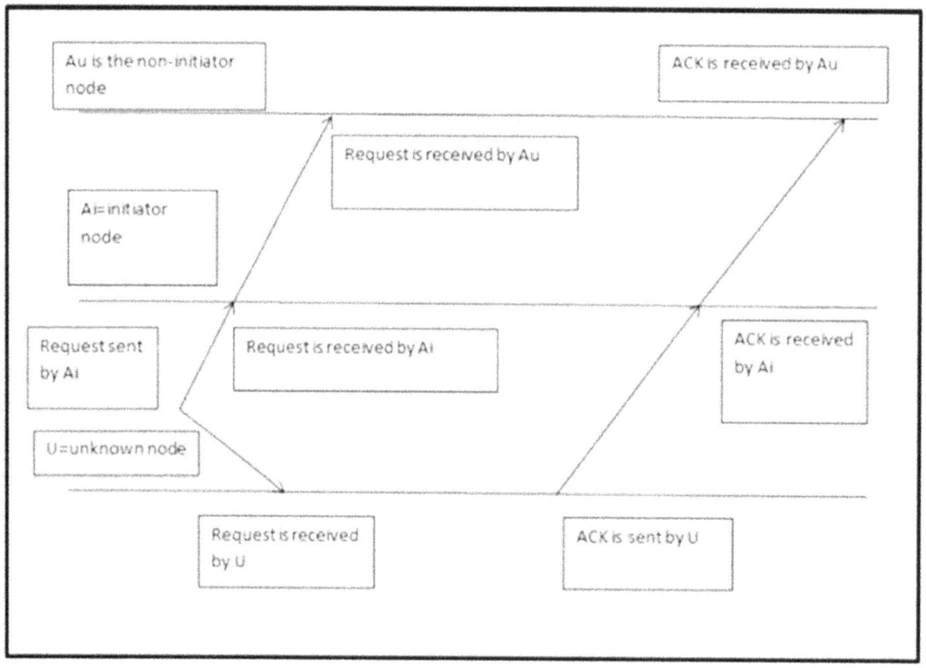

Distributed Localization Algorithms

Distributed location is based on the scheme that there are more than one central node, and if a single node fails to perform its task of localization, then another nodes, responsible for the control, thereby takes the authority, of position finding.

Node Discovery and Localization Protocol (NDLP)

NDLP has one of the nodes whose position is known. There is one primary seed node to be taken. That is S1.Then we choose second seed node say s2, but the distance of s2 should be the largest, in order to extend the range of the communication .Then a third seed node is so selected that it is at a maximum distance from S1 and S2.As can be seen from Figure3.

Figure 3. Common area selection

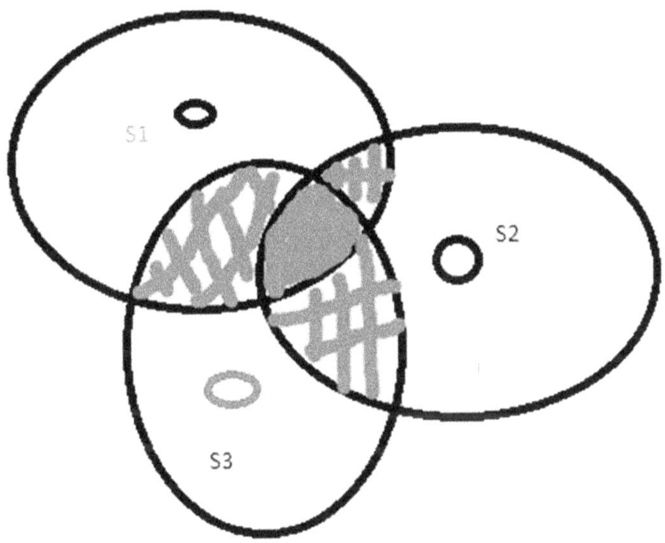

Each node uses a triangulation method, and the overlapping region, gives out the position. If we have to detect the node position in the cross hatched region, as shown in the figure 3, we need to select a next seed node s4,as per the rules we have used for the above three nodes.

Large-Scale Hierarchical Localization (LSHL) Approach

This type basically contains,anchor node localization and unknown node localization.The unknown node localization is therefore further categorized into two types,that is, reference node and non-localised node. Non-localised node is the one whose position is not known, and is to be determined by the below given techniques, mentioned.

Now here we use reference and non-localized nodes to find the position of unknown nodes. Each node sends a signal or a message with its ID. All the nearby nodes can thereby detect the threshold set,

it is labeled as new reference node. The position, based on different measuring techniques is thereby detected. Non localized nodes contain information about the reference node in the variable say 'm'. If m<4,message for localization is sent. Euclidian distance estimation is performed, to find its distances to more nearby reference nodes. And therefore the value of 'm' is updated. Until m>=4,the localization node selects 4 reference nodes, with high probability to localize. Now this process helps to label non localized node as a reference node. After we localize the non reference node, a value 'n', is found. LSHL has drawback of confidence threshold .

Reactive Localization Algorithm (RLA)

RLA localizes only that node which finds an event, instead of localizing all of the nodes. Oncea sensor node detects an event, RLA starts .Initially it finds the anchor nodes . The sensor node in the first step sends an introductory message containing ID and power level to its nearby nodes. The K-Node Coverage Algorithm, finds at least 4 non-coplanar anchor nodes. The second step involves the sensor node reactive localization.

Underwater Positioning Scheme (UPS)

It is performed in two steps. The signals that are arriving is used to detect the difference in time, in the very first step. The time difference is thereby used to find out the difference in the distance, provided the speed of the medium is given, and since the medium used here is water,so distance estimation prediction becomes easy.it should be noted that the Salinity of the water may change its refractive index, thereby

Figure 4. UPS location with source node

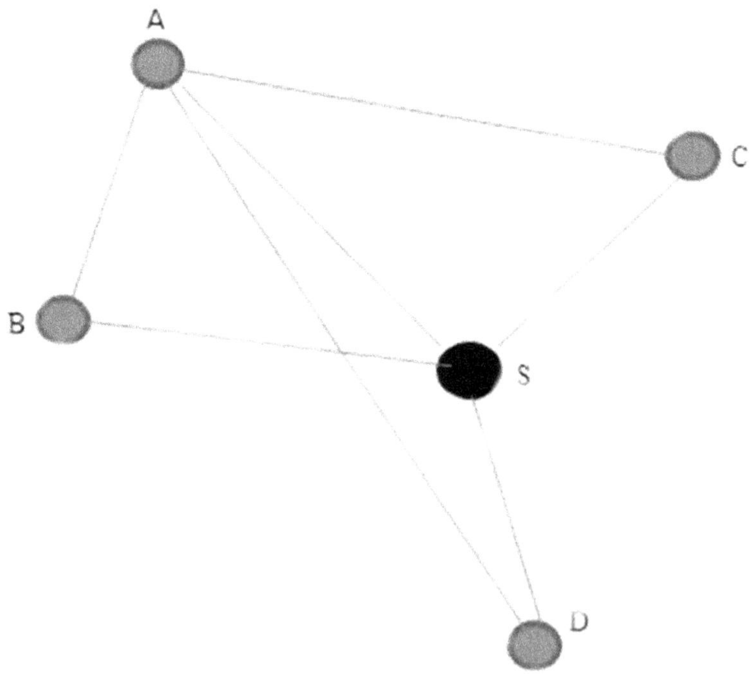

changing its speed, and hence the difference may vary, but the factor of salinity of the ocean water is highly neglected, or ignored, but it could be the future motivation for research field. The transmission anchor node, sends the signals after a fixed duration of time interval say T seconds, B also transmits, but only after receiving the signal from A. And to localize unknown nodes, trilateration is performed. C transmits, only after receiving signal from A and B. Similar is the case for D, and other nodes. D transmits only after receiving signals from A B and C. Then all the respected signals are stored and distance estimation difference after time difference prediction is thereby calculated.As can be seen from Figure4

Ray Bending Based Localization (RBL)

We assume the line trajectory is straight and the velocity is constant, and we have spherical range duration. For the model which traces the rays, constant range interval surface is not spherical, but it retains axial symmetry about the longitudinal axis containing anchor node, since velocity varies only with depth. According to the studies on the above models discussed.

MOBILE LOCALIZATION ALGORITHMS

It is advantageous, if we use stationary algorithms, but since the nodes move along with the water current, the stationary localization techniques, become quite impractical, and thereby we have to deal with the movable sensor nodes.

Centralized Localization Algorithms

Similar to the classification of stationary algorithms, mobile localization is also classified into centralized and distributed.

Centralized means having a single control node .

Absolute Positioning Scheme (APS)

As in the figure5, the autonomous underwater vehicle, transmits a signal, after every fixed duration. AUV sends the signals to the surface as well as to the seafloor. In the next step, the signals are again received to the surface, but this time, they are received from the transponders lying on the sea floor. The time difference is thereby used to find the position of the vehicle. There is limited coverage due to "acoustic interrogation pulse".

There comes a little error in the exact time difference of signal arrival, because of respective movement of both the AUCV as well as the ship,as in D. Green(2007)

GPS positioning is also a drawback, as it takes greater energy and the hardware installation cost is also high.As can be seen from Figure5

Energy-Efðcient Ranging Scheme (EERS)

Sufficient distance map estimation is used to detect the range, by two way process of synchronization and distance estimation step. Distance estimation selects a subset node which will broadcast the localization

Figure 5. GPS positioning and AUV

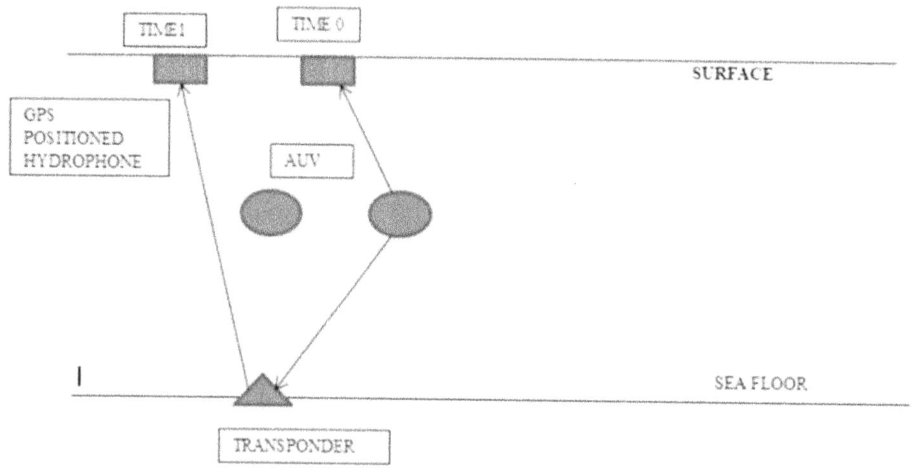

message, irrespective of all the nodes sending the message. Time synchronization is required, so full localization scheme, taking messages from all the nodes is not used. Only distant estimate is collected. So the energy is also saved,in Mushtaq Khan et al. (2019).

Motion-Aware Self-Localization (MASL) Scheme

This technique do not uses anchor nodes, reduces the signals coming to the unknown node and hence the computational burden. It is used for large coverage area,by using iterative algorithm. Since the algorithm is iterative so, it do not provides location real time.For the real time location to find out,we have to perform online monitoring.

Collaborative Localization Scheme (CLS)

Here the nodes are autonomous and do not use long range transponders on surface.Sensor nodes here, use buoyancy control, and thereby can travel, deep into the sea, upto a certain depth and then back again to the surface of the ocean.The depth of the nodes are known by the pressure sensor used in them .The position continuously changes because of the continuous varying motion currents.

The classification of the nodes descending deep into the ocean is basically of two types:

Profilers and followers

Initially both the profilers and the followers are on the surface.

In the next process the profiles travel, and then the followers, as the name suggests, follow the path traced by the profilers. The nodes at the surface trace their position by the GPS.

The speed of travel of all the nodes is same, and the future location of the follower is the present profiler location. As can be seen from Figure6.

Figure 6. Follower and Profiler

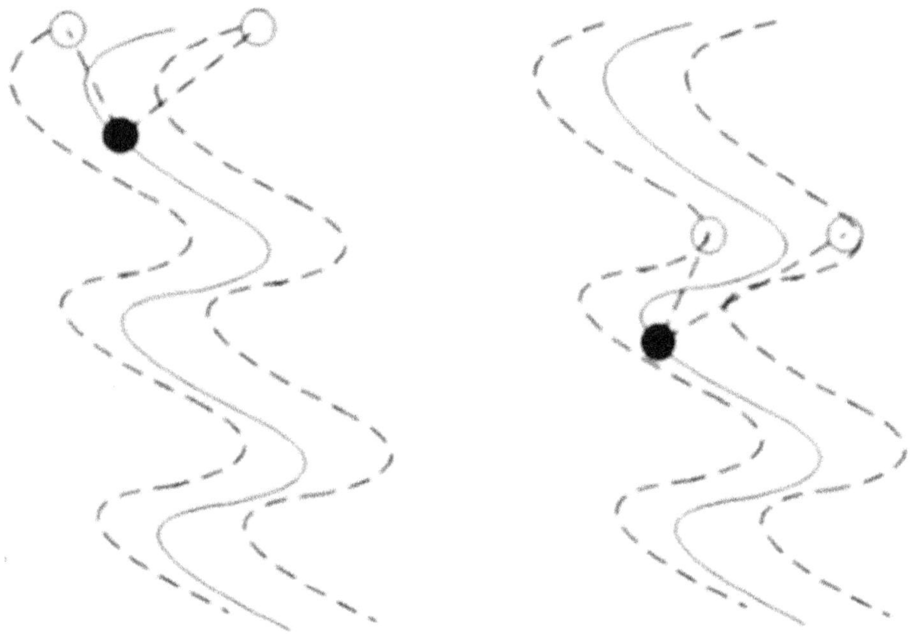

Three-Dimensional Underwater Target Tracking (3DUT) Scheme

It involves two types of listening Active and passive.Figure7,explains 3DUT Scheme.

Figure 7. Anchor node,projector and target representation

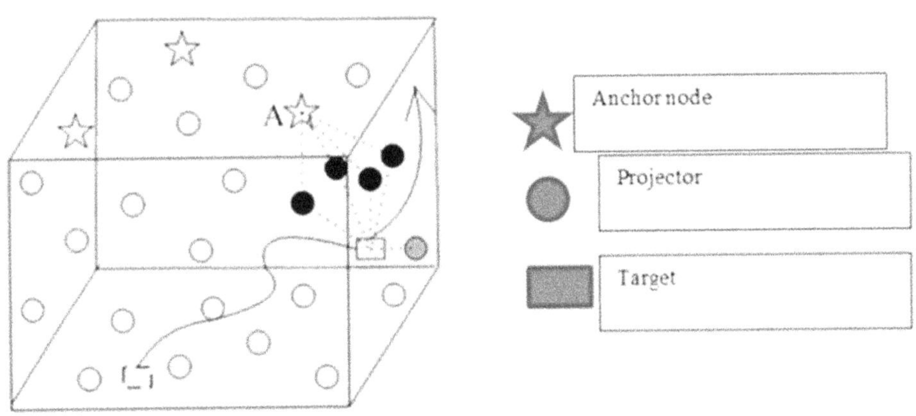

Firstly passive listening is done, where the nodes find out and detect the signals from the potential targets. Secondly active ranging is performed. Now since the 3d targeting scheme is used so the echoes are transmitted, because to undergo the estimation based on radial distance, and axial angle.As the signals are received, periodic acknowledge is sent, by selecting a projector node.Isotropic radiation, equal in all directions is radiated, because the target on which the acknowledgement is sent is a point target. Now after receiving all the projector finds the target point distance and forwards it to the sink node,as stated in J. Catipovic (1991).

Distributed Localization Algorithms

If centralized algorithm has a single node to control and find the respective positions of all other nodes, it has a basic disadvantage that if the single authorized node fails to do so, the whole system collapses, therefore, to overcome its limitation, we use distributed localization algorithm, in which the control is under more than one node, in NA Alraje (2013).

Multi-Frequency Active Localization Method Base on TDoA

It uses ordinary nodes, relay and buoy nodes. Bouy nodes are the first to localize themselves using GPS. They then send the acoustic signals with very low frequency,and broadcast it to the other nodes.Relay nodes uses that low frequency sound signal to communicate with each other.

To calculate the amount of maximum hops for each domain.

The receiving devices are low frequency and the sending devices operate on high frequency.

Message contain the information about the ordinary node.This information needs to be sent,so for broadcasting,signals should be operated with high frequency,or if they are not, they can be modulated for the same reason, in order to be broadcasted.The information so broadcasted, is essential, because it contains the location of a particular node.The package do not broadcast when the value of the maximum hops is zero.

The package is disposed, by sending the received package to buoy nodes, by the relay nodes,for further disposal.Once the package is disposed, the buoy nodes will provide ACK signals to ordinary nodes, which go on power off mode, until the new event or signal.

Dive and Rise (DNR) Positioning Scheme

Here the sinking and rising process is used .All the anchor nodes here are embedded to GPS. Anchor node sink and rise and in the duration, they also broadcast signals to the unknown nodes. The target is thereby determined.

Multi-Stage DNR (MS-DNR) Positioning Scheme

The localized unknown nodes acts as reference to other nodes. Since it uses iterative schemes, so it suffers from communication overhead cost and high power consumption. Compared with DNR, MS-DNR implies a more realistic underwater mobility model, which is termed as Meandering Current Mobility (MCM) model. The model considers sensor nodes' movement. The range in MS-DNR represents a large coastal space covering several kilometers. It should not be assumed that the sensor nodes would be evenly and uniformly distributed over the surface. The technique here is expensive because in 1*1*1 cubic km,25,down and rise anchor nodes are found, and each such anchor node would therefore require the same number of GPS, and the same number of moving equipments.

Autonomous Underwater Vehicle (AUV)-Aided Localization Technique

Here the vehicle move inside the ocean. B finds out the location of the vehicle at time t1,t2.

C in the diagram below is the node whose position we have to find. As shown in Figure8. If the difference between the position of unknown node and A is y, then, the x coordinate of unknown node C is bounded by y to the left. The y coordinates of C are bounded by ya-y and ya +y

Figure 8. AUV tracking

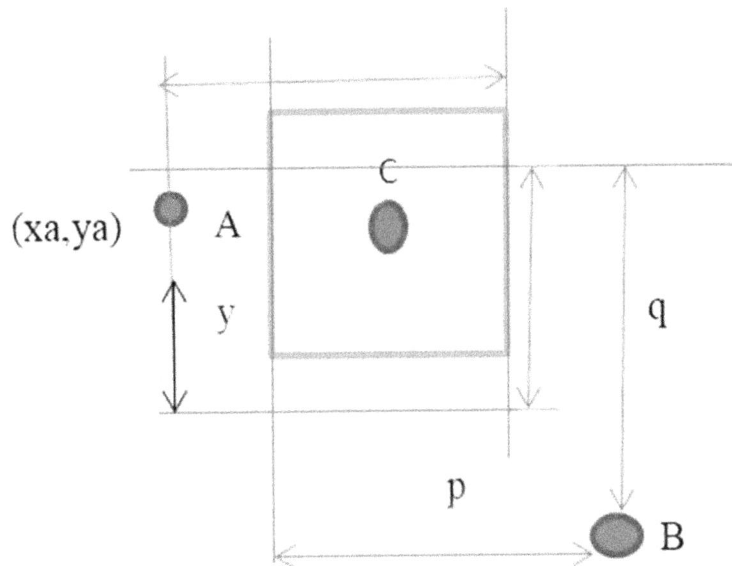

The position of unknown node is determined by the respective coordinates, which is provided by the intersection of diagonals. The position of the node is more accurate when the direction of signals so send is opposite to one another .

Multi-Stage vehicle-Aided Localization (MS-AUV) Scheme

It utilizes the advantages of both, vehicle aided, and silent localization. The efficiency is so highly increased that the node position is detected within 10 min, with a network span of 95%.

Mobility Prediction Scalable Localization Scheme

It involves anchor node localization and unknown node localization.The future nodes are predicted with the help of past available set of patterns. The future location of every node is based on its mobility pattern so predicted. After a fixed interval of time, the sensor nodes have to send its location . By sending their information anchor nodes communicate with buoys on the surface and thereby measure their location.The current estimated localization period is predicted, based on the previous measured one. Then a comparison is made between the measured and the predicted location,and the Euclidian distance is found. The so found Euclidian distance is compared with the threshold value, the algorithm runs, if the Euclidian distance is greater than the threshold value. Then a fresh localization message containing current location and new mobility pattern is sent. If the node has not received any signal for long period of time, it is labeled as un-localised.When the propagation period is short, a large cost for the communication overhead is then required.

HYBRID LOCALIZATION ALGORITHMS

Hybrid localization algorithm, takes the advantages of both mobile and stationary localization scheme, therefore is more efficient way of underwater location, as compared with the other two, as PS.Tsai(2017).

Centralized Localization Algorithms

3D Multi-Power Area Localization Scheme (3D-MALS)

In hybrid technique, the surface of the water is covered by a surface buoys, which are less in number. Each buoy has a multipower acoustic transreceiver and an elevator.

The work of the elevator as the name suggests, is to rise and dive into the water,and the transreceiver sends signals, in order to transmit and receive information with unspecified nodes.

Silent Localization Using Magnetometers (SLUM)

SLUM secretly localizes underwater unknown nodes embedded with magnetometers by a equipment whose magnetic dipoles is known. Messages of the dipole helps in localization. The ferromagnetic field localizes unknown nodes.Pressure sensor so used here is used for the estimation of depth and accelerometer helps in finding the sensor orientation. The path of the vessel and the positions are thereby found.

Distributed Localization Algorithms

A Range-Free Scheme Based on Mobile Beacons (RSMB)

As shown in the figure 9, the unlocalised node can thereby receive five beacons. from the mobile anchor node.It is to be noted that the unlocalised nodes may not know there position, but they are aware of the information about their depth from the water surface .The beacons or the signals so received are made to project on a plane to help find the position of the respective nodes .Initially three becons are selected,out of all the beacons, which are received .The location of unknown node is thereby estimated. Here independent node estimation takes place.

Figure 9. Sensor nodes and mobile beacon

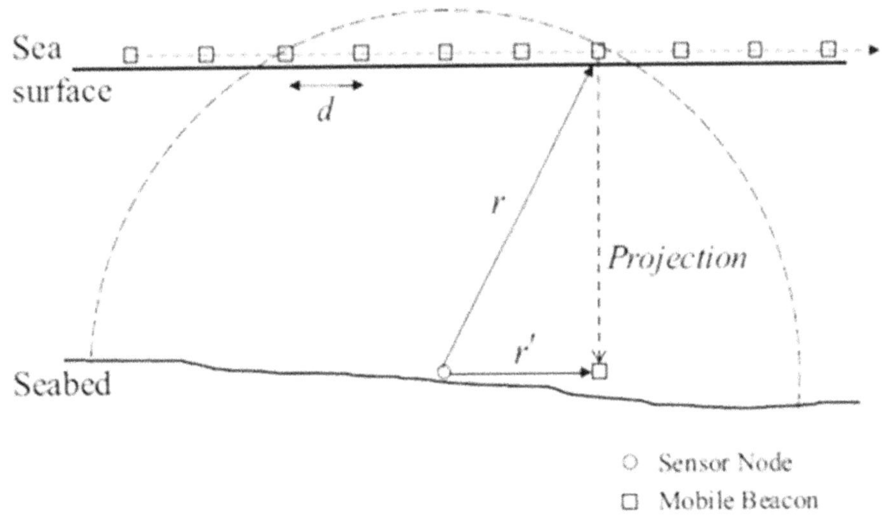

"Three-Dimensional Underwater Localization (3DUL)"

Buoys as we know, are kept on the water surface.so the first step in 3DUL is that the unlocalised nodes has to find the position of the adjacent buoys. For this, they make use of bidirectional message technique to transfer. The extraction of the depth information is thereby called ranging. The second step then involves, projecting the signals and dynamic trilateration, where this technique do not make use of time synchronization and thereby adds to its advantage. But its drawback is that it requires very large localization time.

" Underwater Localization Based on Directional Beacons (UDB)"

As in figure.10,Ra,basically defines the sensing range of the vehicle,and the respective angle is controlled by the directional antenna. It is to be noted that the vehicle is moving with a stable speed. Ru is the sensing range of the node whose position is to be noted.If the time interval taken by the vehicle to send the signals is high, then there will be less number of nodes,localized and so will be the accuracy, FengWang(2012).

Localization scheme with Directional Beacons (LDB)

During time interval T1 to T6,the AUV sends beacons at regular intervals,which is heard by unknown node S, but all these six signals are not stored by the unknown nodes.Only the first –heard and the last heard beacons are stored, and they only helps to find the respective position. It should be noted that the signals sent at time T1 and the signal sent at time T6 are the first and the last heard beacon, respectively. AUV sends beacons with time intervals. The ðrst-heard beacon point and last-heard beacon point are shared by the nodes in the same area.

Figure 10. Projection of beacons

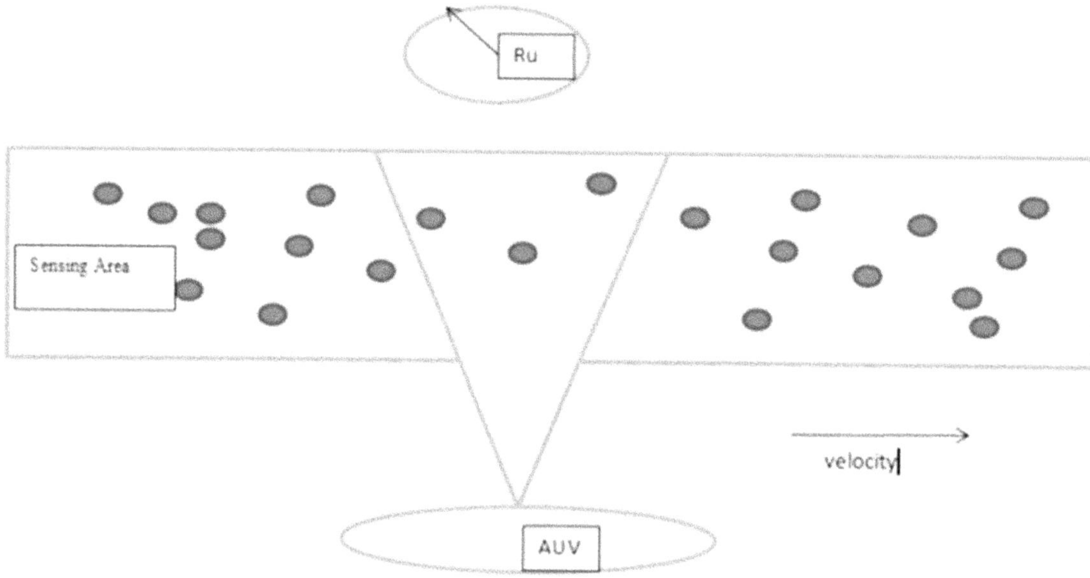

As shown in the figure 11and Figure12, The centre of the four circles are T1,T2,T8,and T9,these thereby forms a small area. Generally there is only one unlocalized node in it,but if there is more than one,then the node with unknown position share the first heard and last heard signal point with S,which are at T2 and T8 respectively.The centroid of the area is so computed,to find the maximum localization error and position

Figure 11. Continuous case

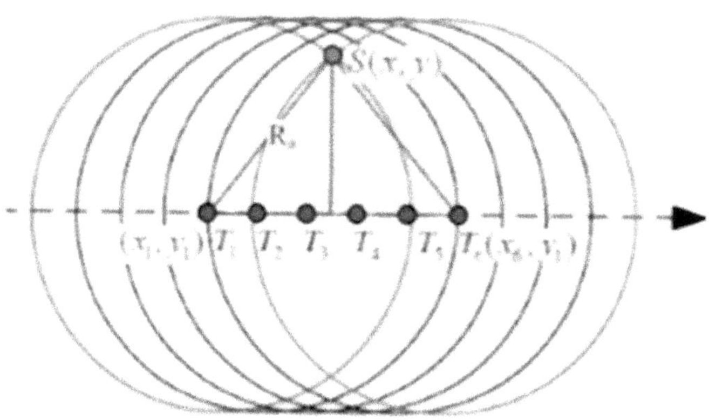

Figure 12. Discrete case

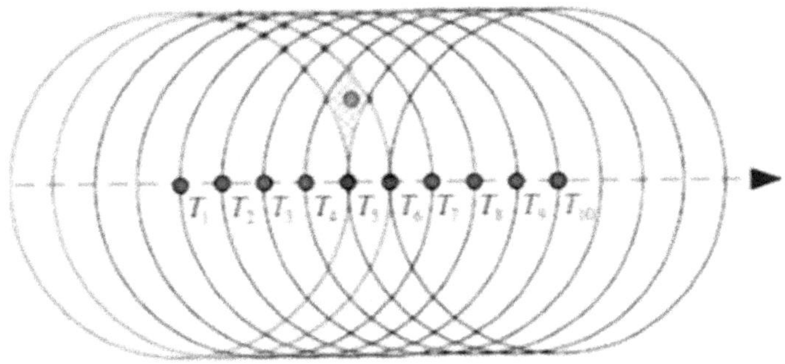

OTHER METHODS

One way travel time estimation:

The setup is shown inFigure13

Figure 13. Setup

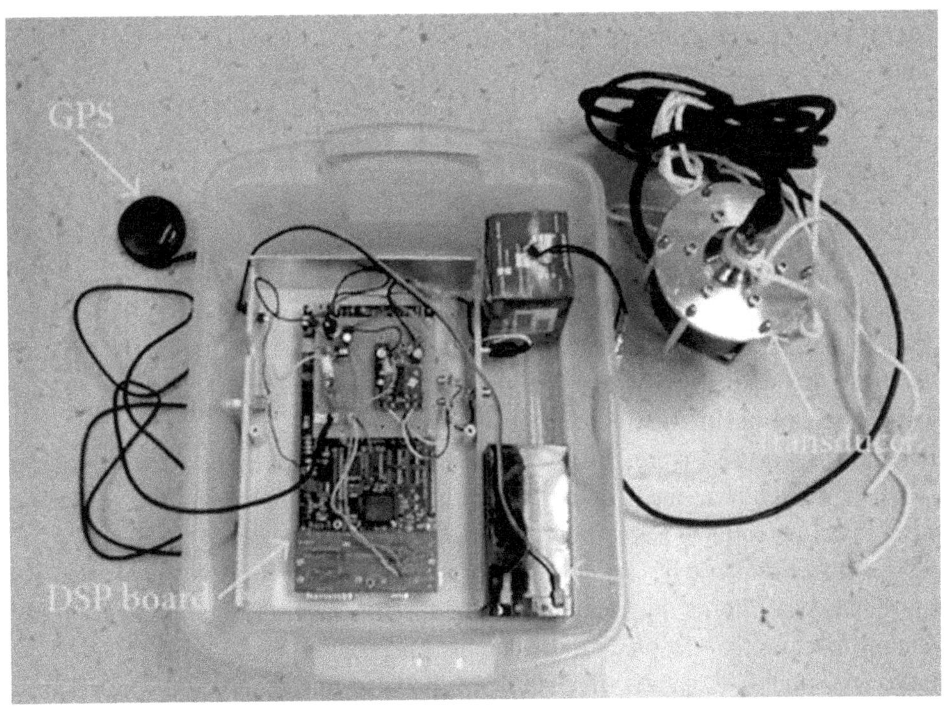

Least square solution:

Shown in Figure.14

Figure 14. Least square solution (green line represents the exhaustive search and blue line represents the least square search)

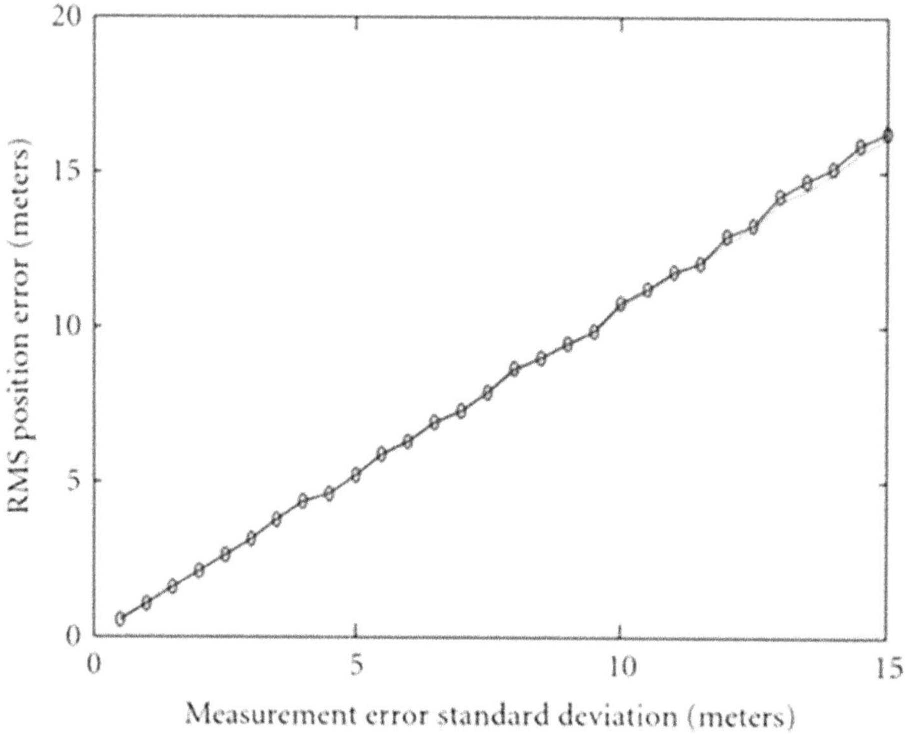

SIMULATION RESULTS

Simulation is carried out by using a simple noise model. Z is known, and x,y is to be calculated. Four transmitters are kept on a square grid with coordinates (0,0), (100,0), (0,100), and (100,100). One receiver is also kept at the (0,50) point and moves at a predefined constant rate of 0.125m/s parallel to the x-axis.

After every 16 sec,the position of the respective node is updated.

Kalman filtering,as shown in Figure15 or linear quadratic estimation,is a type of tracking algorithm,which has long periods of motion.Here the node movement is set as white noise varying velocity model.

Kalman filter is also good and advantageous for time series analysis.

It is a two step process,first step is known as prediction, where we find the state variable .The advantage of using the kalman filter is this that it do not require the past information, it is highly recursive. The disadvantage includes that the prediction is highly uncertain.

As in figure 16, Simulation path is the path of the node so traced, denoted with respect to x and y coordinate.

Figure 15. Kalman Filter(green line and red line represents the point estimate and kalman filter output respectively)

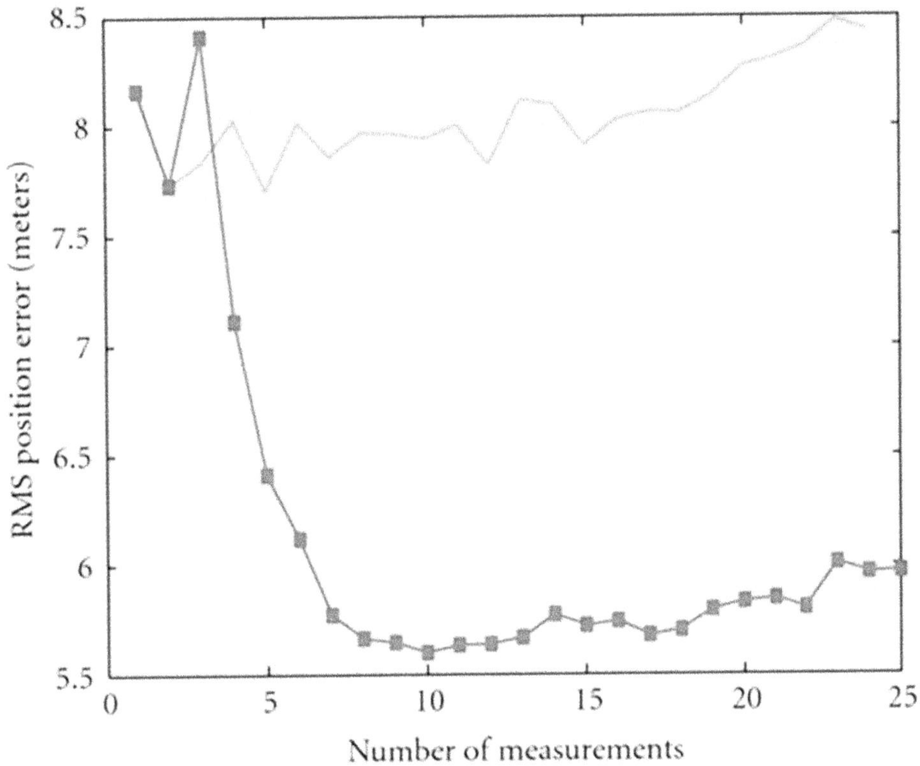

Figure 16. Simulation path

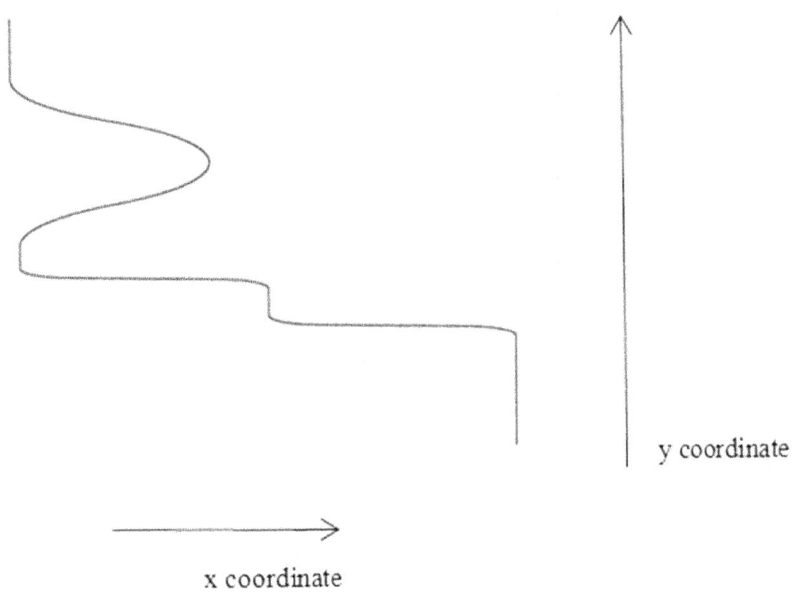

CONCLUSION

The basic disadvantage of GPS is that using sensor nodes along with it is quite expensive so, anchor nodes are used to localize the unknown nodes. Also we use hierarchical approaches. Some parameters which we need to take care are bandwidth constraint, sensor nodes mobility, channel behavior varying conditions. ToA and TDoA methods are used because they do not need synchronization of clock. Multi step localization methods is used, to tackle the undergoing problems of sparse deployment and mobility. The nodes we locate, uses range based PSO algorithm, which are used to calculate the velocities. And then using spatial correlation, unknown nodes are predicted.

Practical Implementation Challenges

Power Source

The sensors get power from the batteries, so hybrid scheme is highly energy efficient. The low power states, in hybrid technique, activates the MODEM, only after a fixed time span, otherwise they go to sleep. During deep sleep, the current consumption is in microamperes, increasing the lifetime of the battery, stated in How Voon Ee et al. (2006).

Modulation Methods

The modulation method used in the case of underwater acoustic communication is DSSS(direct sequence spread spectrum).Due to water currents, noise, wind or other factors,the DSSS is used to resist the effects of reflections.

Message Transit Times

The speed of sound is less than the speed of light, so the message takes larger time to get transmitted .In the meanwhile another packet or signal from different sensor nodes may cause the overlap.

Routing Protocols

Architecture of the wireless sensor network should be such that the level of energy, so consumed is low.

Scope of Enhancement

Optical –Acoustic Hybrid Network

Because the speed of light is more than that of speed of sound, we can overcome the limitation of the bandwidth by the use of acoustic –assisted alignment between optical transmitters and receivers.

Energy Harvesting Based Routing

Amplify and forward technique is used, at the transmitting node, and fixed combining ratio method at the receiver node, in Catalin David (2011).

Improving the Routing Algorithm Based on Less Power Consumption Requirement

By the proper alignment of sensor nodes, we can minimize the use of power consumption.

Weakness and Strengths of the Chapter

The chapter helps the reader to understand how to detect a location of a node, efficiently, and effectively, with less resource consumption. It however leaves a scope for future development, if the researcher wants to increase the speed of the communication and further minimize the resources, including the constraints of energy, overhead cost, and area utilization.

Over-Emphasis and Under –Emphasis of the Problems In Wireless Sensor Network

The main problem concerning the deployment of the sensor nodes is that the non-stationary nodes, need continuous updation of their location. The regular movement of the sensor nodes, due to water currents, therefore, results the above problem, Thereby, increasing the requirement of power .In order to amplify the signals from the sensor nodes, they need to be operated under high frequency. The process can be termed as modulation.Since modulation technique so used is DSSS,the described problem is under-emphasized.

REFERENCES

Alraje. (2013). *Localization Techniques in Wireless Sensor Networks*. Academic Press.

Catipovic, J. (1991). Underwater acoustic local area network for ROV and instrument communications. *Proc. AUVS'91*, 446-460.

Chandrasekhar, V., & Winston, K. G. (2006). Localization in underwater sensor networks-survey and challenges. *Proceedings of the ACM International Conference on Mobile Computing and Networking*, 33–40.

David. (2011). *Energy consumption analysis of underwater acoustic sensor networks*. Academic Press.

Green, D. (2007). *Underwater Acoustic Communication and Modem-Based Navigation Aids*. Springer. doi:10.1007/978-3-540-77090-9_43

Han, Jiang, Shu, Xu, & Wang. (2012). Localization Algorithms of Underwater Wireless Sensor Networks: a survey. *MDPI*.

Khan, A., Khan, M., Ahmed, S., Mohd, A. A. R., & Khan, M. (2019). Energy harvesting based routing protocol for underwater sensor networks. *PLoS One*, *14*(7). doi:10.1371/journal.pone.0219459 PMID:31314772

Otnes, R., Asterjadhi, A., Casari, P., Goetz, M., Husøy, T., Nissen, I., … Zorzi, P. (2012). Underwater Acoustic Networking Techniques. Springer.

Stojanovic, M. (1996). Recent advances in high-speed underwater acoustic communications. *Oceanic Engineering IEEE Journal of*, *21*(2), 125–136. doi:10.1109/48.486787

Tsai. (2017). *Hybrid localization approach for underwater sensor network*. Academic Press.

KEY TERMS AND DEFINITIONS

Anchor Node: Nodes with predetermined location.

APS: Access points.

AUV: Unmanned, autonomous underwater vehicle.

CLS: Collaborative localization scheme, which determines the position, without the need of transponders.

Dive and Rise: Continuous repetition of steep descent and moving to the surface.

Projection: Display of location.

Reference Node: Used to find the position of unknown node.

Sink Node: They are the nodes, where the information so collected, of a particular node is sent.

Target: On which the projection is made.

Unknown Node: Node whose location is to be predicted.

Chapter 5
Data Fusion in Underwater Wireless Sensor Networks and Open Research Challenges

Kalpna Guleria
Chitkara University Institute of Engineering and Technology, Chitkara University, Punjab, India

Saira Banu Atham
https://orcid.org/0000-0002-4469-5153
HKBK College of Engineering, Bangalore, India

Ashok Kumar
https://orcid.org/0000-0003-3279-5111
Chitkara University Institute of Engineering and Technology, Chitkara University, Punjab, India

ABSTRACT

Underwater wireless sensor networks consist of numerous devices such as vehicles and sensor nodes, which are positioned underwater, and these devices collaboratively perform data collection, monitoring, and control task. The most critical issues related to underwater wireless sensor networks include limited bandwidth, data aggregation, data fusion, data collection, routing, media access control, higher propagation delay, and higher power consumption. Underwater data fusion is one of the critical issues for underwater military surveillance applications. It is also considered an equally important and challenging aspect for underwater intelligent traffic control and underwater vehicle navigation systems. This chapter provides an insight into basic concepts about multi-sensor data fusion and detailed review of most popular data fusion architectural models available for underwater wireless sensor networks. Finally, the chapter also provides open research directions to researchers for multi-sensor data fusion.

DOI: 10.4018/978-1-7998-3640-7.ch005

INTRODUCTION

Recent technological advancements have led to numerous opportunities for underwater explorations using underwater wireless sensor networks (UWSNs) (Goyal et al., 2016, Goyal et al., 2017, Goyal et al., 2020). Major difference between WSN (Guleria and Verma 2018, Guleria and Verma A.K.2019a, Guleria and Verma 2019b) and UWSN is the communication method. The WSN uses radio wave communication whereas UWSN utilizes acoustic signals for communication (Goyal et al., 2018). The most critical issues related to underwater wireless sensor networks include limited bandwidth, data aggregation, data fusion, data collection, routing, media access control, higher propagation delay and higher power consumption.

Ocean, rivers or canals covers almost 75% of earth's surface and various invaluable resources lie underwater which are still unexplored. Due to the significance of underwater target tracking, sonar array based underwater target tracking methods have been proposed (Dalberg et al., 2006; Georgy et al., 2012). Sonar arrays are pulled by a submarine or ship. These sonar array-based algorithms are not adaptive which do not match the highly dynamic nature of underwater environment. Further, if the whole target tracking system may lead to a failure if the submarine or ship is attacked. Underwater Wireless Sensor Networks (UWSNs) provides a very reliable solution in comparison to sonar array. UWSNs offer a better solution with higher reliability, precision and persistence for target tracking because it comprises a varied number of small sized, inexpensive sensor nodes which exhibits self-organized behavior and are deployed in a large geographical area (Lloret, 2013). UWSNs consists of numerous tiny sized sensor nodes spatially deployed underwater in an acoustic medium and collaboratively perform various tasks such as sensing (temperature or pressure), processing, monitoring and control. Underwater communication is possible through acoustic signals since optical communication and radio wave communication is not possible in UWSNs. Acoustic transceivers are utilized for UWSNs communication, which collects the data from underwater sensor nodes and transfers it to their nearest buoyant gateway. Further, the data is transmitted from gateway to the nearest remote station for coastal monitoring and control. Underwater communication is possible through acoustic waves. The acoustic waves exhibit important properties which defines that acoustic waves possess low bandwidth and long wavelength. Hence, acoustic waves can communicate over long distances up to several kilometers (Berger et al., 2008). Major application areas in UWSNs include disaster prevention through underwater monitoring, scientific exploration, coastline surveillance and monitoring, water pollution monitoring and water sports facilitation.

In recent past mobile robots have been used in numerous applications and this concept has got a lot of attention. "Navigation" relates to its multifold aspects which include various tasks related to sensing, mapping, localization and control. Navigation is considered as one of the most challenging tasks for exploratory missions because there is no prior knowledge about the underwater environment about the location of the mission. This relates to the scenario about Unmanned Underwater Vehicles (UUVs) are deployed to carry out missions in a large region however the challenging aspects include highly dynamic and unstructured environment. Hence, it requires an intelligent and reliable navigation and control (liu et al., 2016). In past few years, underwater navigation has got a lot of stimulus since varied variety of numerous sensors have been deployed in underwater for underwater exploration and navigation. Various varieties of underwater navigation sensors include: GPS sensor, acoustic sensor, dead-reckoning sensor, vision and range finder sensor. GPS sensors provide perfect positioning as well as better navigation.

Sensor Fusion

"Sensor fusion" or "information fusion" or "data fusion" or "multi-sensor integration" or "multi-sensor data fusion" are varied number of terms which are used to represent sensor fusion systems. Sensor data fusion is an important phenomenon in WSN (Guleria and Verma, 2018a, Guleria and Verma, 2018b, Guleria et al., 2019c) and it is also one of the most important aspect which is responsible for reliable navigation in UWSNs (Bakr et al., 2017).

Data fusion refers to a process called data integration where data from multiple sensors is integrated into one uniform format. Multiple Sensor Data Fusion combine the information received from complementary, redundant and single sensor after a predefined interval. The data fusion leads to various advantages in navigation which include noise removal, removal of inconsistency and incompleteness of data, improved coverage as well as increased resolution. The uncertainty of data can also be dealt very well using multiple sensor data fusion.

Multi sensor data fusion is a process of collaborating and merging the data received from various sensor and to draw the inference results which cannot be attained from a single data source. Multi sensor data fusion has varied number of applications which include battlefield monitoring analytics or threat analysis automatic target tracking in military surveillance operation. Multi-sensor data fusion has been widely used in the area of remote sensing which can determine location information about mineral resources, compositional characteristics of vegetation. Industrial applications include controlling nuclear plants processes, automated production processes and to handle complex machinery where human intervention cannot be made. The technologies which deals with multi sensor data fusion are signal processing statistical estimation, machine learning, pattern characterization, deep learning and theory of computational intelligence (Mitchell, 2007, Azar and Vaidyanathan, 2015). The process of data fusion is like the inferences human's make based on integrated information collected by their senses.

Significance of Sensor Fusion

Multi-sensor fusion methods have numerous advantages in comparison to single sensor because the observations attained by single sensor systems have the following discrepancies (Elmenreich, 2002, Bakr and Lee, 2017).

- **Sensory Deprivation:** If the deployed sensor fails down then the system will have no other way to get the desired information.
- **Limitation of Spatial Coverage:** Individual sensor systems suffers from limited area coverage.
- **Limitation of Temporal Coverage:** Individual sensor systems suffers from limited temporal coverage means it will have limited frequency of sensor measurements.
- **Imprecision:** Individual sensor system measurements has limited accuracy or may be imprecise.
- **Uncertainty:** Individual sensor system measurements may lead to uncertainty due to ambiguous (Murphy, 1996, Conde et al., 2013). The reason for uncertainty is due to limited view provided by single sensor information (Foote and Huebner, 1995, Li et al., 2016).

These are the following benefits achieved from multi-sensor data fusion (Bosse et al., 1996; Grossmann, 1998, Khaleghi et al., 2013):

- **Fault Tolerance and Enhanced Reliability:** Multi-sensor structure provides fault tolerance which leads to reliable behavior of the system.
- **Enhanced Precision:** Multi sensor observations leads to a better observed precision of a measured event.
- **Less Ambiguous and Uncertain Behavior:** Multi sensor data fusion system reduces the ambiguity and uncertainty of observed behavior.
- **More Robust Behavior Against Interference:** Multi – sensor data fusion systems exhibits more robust behavior against interference caused in system

APPLICATIONS OF MULTI SENSOR DATA FUSION

Multi sensor data fusion applications related to military and non-military domain have been classified as given in table 1.1 and table 1.2 respectively (Blackman and Broida, 1990). Ocean monitoring and surveillance, military intelligence, target tracking and air-air, air-surface monitoring and defense are the military based applications. Non- military based applications cater to monitoring and controlling robotic process-based operations, remote sensing and control, automated process monitoring, law enforcement and medical diagnosis (Liggins et al.,2017, Song and Deng, 2019) .

Table 1.Summary of multi-sensor data fusion application in military domain

Application Domain	Observed Data Inference	Types of observed Data	Volume of monitoring	Required Platforms
Ocean monitoring control and surveillance	To detect, track or identify events or targets	Presence of radiation (nuclear or electromagnetic), or presence of acoustics	Air, Surface Several (hundreds) miles	Ocean based, Submarine, ships, Aircrafts
Air-Air Surveillance monitoring and defense	To detect track and identify an aircraft	Electromagnetic radiation signals	Several hundred miles	Aircraft (Ground-based)
Surface-Air Surveillance monitoring and defense	To detect track and identify an aircraft	Electromagnetic radiation signals	Several hundred miles	Aircraft (Ground-based)
Military Intelligence and monitoring	To detect and identify targets	Electromagnetic radiation signals	Several miles (hundred or tens) in a military battle field	Aircraft (Ground-based)
Strategic defense and warning	To detect and identify strategy-based actions, detect and track Missiles	Nuclear particle-based signals and electromagnetic radiation-based signals	Very large and global monitoring and surveillance	Aircraft (Satellite based)

SPECTRUM OF SENSOR'S FOR DATA FUSION

This section provides the details about spectral range of sensors along with its detectable attribute's characteristics and types of sensor systems used for data fusion as illustrated by Waltz. Table 1.3 Summary of spectral range of sensor for Data Fusion (Waltz and Buede, 1986, Zeng et al., 2017). Sensors

Table 2. Summary of multi-sensor data fusion application in non-military domain

Application Domain	Observed Data Inference	Types of observed Data	Volume of monitoring	Required Platforms
Robotic based automated process	To identify the location of obstructions	Acoustics, Electromagnetic radiations-based signals	Nearest Locational based information statistics of robot	Robot
Medical Diagnosis	To detect diseases, physical, sentimental (Physiological) analysis	Electromagnetic radiations-based signals, Nuclear, biological data, physiological data	Respective volume of human body and observations	Position and placement of sensor nodes in/ on human body
Remote sensing	To identify the location of minerals in the ocean, crop or forest monitoring and control	Electromagnetic radiations-based signals	Several hundreds of miles	Satellite or ground based Aircraft systems
Law enforcement	To detect the location-based attributes of drug shipments	Acoustics, Electromagnetic radiations-based signals	State borders, country borders	Ground based aircraft systems
Automated process monitoring and automated equipment monitoring and control	To detect and identify health and fault status of an equipment	Electromagnetic radiations-based signals, pressure or temperature-based observations	Volume of machines or volume of factory	Sensors associated with machine or factory

Table 3. Summary of spectral range of sensor for data fusion

Types of Sensor Systems	Spectral Range	Detectable attributes 'characteristics
Acoustic detectors, Active and Passive sonar Seismometers	1Hz-10KHz	Acoustic Frequency
Infrared camera Focal plane arrays Scanning in search track	300nm-1micron	Infrared (IR) radiation
Magnetometers; passive electronic support measure (ESM) receivers	1Hz-1MHz (low Frequency)	Electromagnetics
High-frequency (HF), very high frequency (VHF); Ultra – High Frequency (UHF) radars	10MHz-100MHz (HF/VHF/UHF)	
Surveillance and fire control radars	1-10GHz	
Super high frequency (SHF); extreme high frequency (EHF)	10-50 GHz (SHF/EHF)	
Millimeter wave (MMW) radar and radiometers	30-300GHz (MMW)	
Nuclear particles	$3x\ 10^{2} – 3 \times 10^{-4}$ Angstroms	X-ray detectors Gamma ray detectors
UV spectrometer	200-380 nanometers	Ultraviolet light
Multispectral arrays Laser radars Electro optic sensors (TV)	780-380 nanometers	Visible light

uses either active or passive phenomenon throughout the energy spectrum. Radars, X-rays machines, lasers, devices which emits nuclear magnetic resonance and electromagnetic waves are called as active sensors and they keep on emitting energy so that a phenomenon of detection of an event or target can be carried out. However, passive sensors detect natural emissions. IR or visual camera gamma ray or x-ray detectors, acoustic sensors are examples of passive sensors.

CLASSIFICATION OF SENSOR FUSION

Durrant-Whyte's Classification of Sensor Data Fusion Based on the Relationship Among Data Sources

Author(Durrant-Whyte, 1990, Li and Wang, 2018) has categorized the sensor fusion based on the following:

- **Complementary Sensor Fusion:** More complete and global information can be obtained about a system by complementary fusion, where sensors nodes represent information about a area which can give the better insight since the presented information represents different area of the system being monitored. In a surveillance monitoring system, the complementary fusion relates to a situation where two camera focus on the same target however they perform monitoring in different field of view.
- **Redundant Data Fusion:** Redundant fusion takes place when multiple sensors gives the information of similar target which provides better resolution /precision and enhances confidence of the obtained observation.
- **Cooperative Data Fusion:** Cooperative data fusion relates to the situation where the observed information provides a new information which is more complex in nature in comparison to original obtained observation. Figure 1 depicts Whyte's classification based on the relations between the data sources.

Figure 1. Whyte's classification based on the relations between the data sources

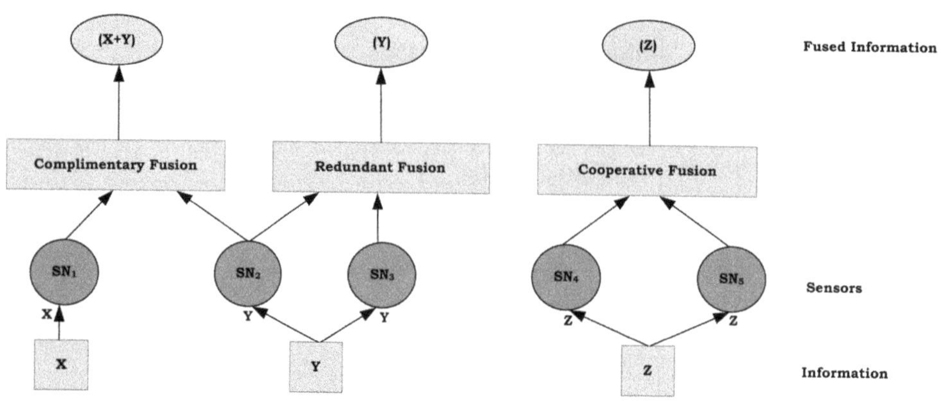

Dasarathy's Classification of Sensor Data Fusion

Author (Dasarathy, 1997, Dasarathy and Merli, 2016) has provided sensor data classification in the following five categories:

- Multi sensor fusion based on Data In -Data Out (DAI-DAO): In this category the multi sensor fusion, takes place directly on inputs and provides output which is more precise. Signal processing or image processing algorithmic methods are utilized to carry out fusion at this level. Fusion performed on multi- spectral data is an example of DAI-DAO.
- Multi sensor fusion based on Data In-Feature Out (DAI-FEO): In this type of sensor fusion process takes the raw data and provide the extracted features or attributes to define an observation about the environment. To extract the shape DAI-FEO fusion is carried out.
- Multi sensor fusion based on Feature In-Feature Out (FEI-FEO): In this type of fusion, the input observation and the result of fusion process, both are features. Hence, the fusion process performs enhanced features and called as symbolic, feature or mid-level fusion. Fusion performed on image or non-image is an example of FEI-FEO.
- Multi sensor fusion based on feature in-decision out (FEI-DEO): This type of fusion process takes input observations as features and the result provides decision based on the input features. FEI-DEO fusion is carried out to recognize the objects.
- Multi sensor fusion based on Decision In-Decision Out (DEI-DEO): This fusion process is called as decision fusion because input decisions are fused to get much better or refined decisions. DEI-DEO fusion is carried out when there is incompatibility among sensors. DEI-DEO fusion is carried out to perform pattern recognition task.

Luo et al. Classification Based on Levels of Abstraction

The following classification based on level of abstractions is provided (Luo et al.,2002 Cardarelli et al., 2014)

- **Signal Level:** Signal level fusion process directly perform fusion on the signals obtained from sensors.
- **Pixel Level:** This type of fusion process works at pixel level for tasks related to processing of images.
- **Feature or Characteristic Level:** This type of fusion process performs fusion on features obtained after feature extraction process.
- **Symbol:** In this type of fusion, the observations represent symbols and it is called as decision level fusion.

The information fusion also exhibits another category of abstraction based on observations, attributes and decisions. These categories can be defined as:

- **Level-0 or Low Level:** In this category the input to a fusion process is raw data and the output of fusion process is more precise data in comparison to independent sensors.

- **Level- 1 or Mid-Level Fusion:** In this fusion process the input is attributes and these attributes are fused together to obtain attributes to perform another task. This is also known as characteristic or feature level fusion.
- **Level-2 or High-Level fusion:** The input to high level fusion is symbols which are fused together to get a refined decision.

MULTI-SENSOR DATA FUSION DATA FUSION MODEL ARCHITECTURES

There are various advantages of multiple sensor systems which include enhanced fault tolerance and reliability in case of failure, enhanced and complete coverage, higher confidence, minimal uncertainty (Thomopoulos, 1991, Nicosevic et al., 2004, Majumder et al.,2001). However, a trade-off must be considered between numbers of sensors to be deployed to minimize the time for the measurement of the observed parameters.

There are some important factors which must be taken into consideration in order to define the fusion algorithm which include the architectural characteristics of sensor distribution in the network, type of sensors used, computational characteristics of sensors, homogeneity or heterogeneity of sensor nodes. If homogeneous sensors are deployed in the network, then the complexity of the system will reduce as well as it will also take minimal time for the computations and data analysis.

It is extremely important to understand the data fusion models which have been developed in research and commercial domain in order to know the most perfect approach or algorithm which will provide the solution for a given problem.

JDL Data Fusion Model

The Joint Directors of Laboratories (JDL) data fusion model is the most popular model among all the data fusion systems available. It was proposed by Joint Directors of Laboratories (JDL) and the American Department of Defense (DoD) (Llinas et al., 2004, Hall et al.,2015). Authors (Hall and Llinas,1997) have defined various levels where the fusion process can take place. This model was basically designed for military applications. Figure 2 depicts the JDL data fusion model architecture.

Figure 2. JDL data fusion model architecture

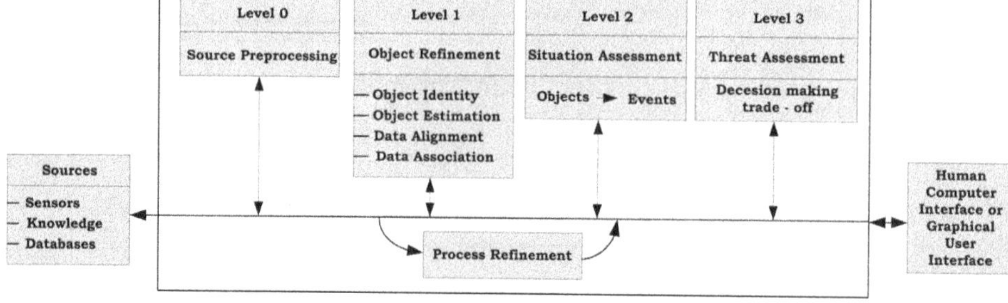

Level 1 relates to object refinement: Object refinement deals with multi sensor data fusion to get a broader view of the fused features of an object from different sources. It focusses on identification of an object as well as to provide location information of the same. It performs alignment of data, attribute prediction, data or behavioral associations among entities and identity refinement.

Level 2 relates to situation assessment: This level enhances the incomplete view provided by level 1 by assessing the situation through an observed event.

Level 3 relates to threat analysis which is the interpretation of results from previous level 2 for the initiation of a possible operation thereof.

Level 4 relates to process refinement: This level repetitively works towards monitoring the performance of the system, identification of sources to improvise the data and to find the optimal arrangement or positioning of sensors. Preprocessed data retrieval, data storage and graphical user interface are the other support systems of the presented model. The JDL model presents a hierarchical architecture which allows division of levels into further sub levels. Centralized or decentralized process can be utilized for data fusion. Centralized process carries out fusion on raw data whereas decentralized process carries out fusion on preprocessed data. Hybrid data fusion can be carried out on raw as well as preprocessed data. Authors (Paradis et al., 1998) proposed JDL blackboard data structure which include fusion agent-based process enhancements known as fusion center.

Thomopoulos Data Fusion Model

A data fusion model which consists of three major modules which performs integration on data at different levels was presented (Thomopoulos, 1991, Bevilacqua et al.,2015, Zhang et al.,2011). Signal level data fusion performs correlation on observed data through learning because it lacks mathematical model for measurement of observed pattern. Evidence Level fusion utilizes statistics-based inferences and user's decision-based inferences. Dynamic level fusion utilizes mathematical model for performing multi sensor data fusion. Figure 3 presents the Thomopoulos fusion model. Thomopoulos has stated that major factors which a fusion model must consider include monotonicity of the fused information

Figure 3. Thomopoulos Data Fusion Model based Architecture

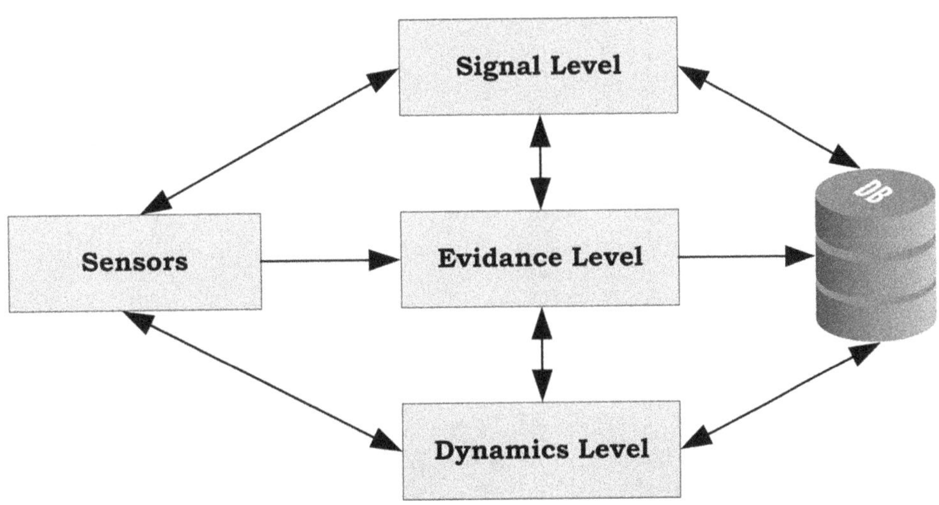

and monotonicity related to cost and robustness related to uncertainty to a prior information inference. Further, this model emphasizes that transmission delay, communication errors and various other aspects must be considered during multi sensor data fusion.

Luo and Kay Multi-Sensor Integration and Fusion Model

Authors (Luo and Kay, 1988) have provided a multi-sensor data fusion model performs the multi sensor data integration. This model follows a hierarchical approach where data from multiple sources is fused at embedded data fusion and integration centers. The authors clearly stated the distinction between multi-sensor fusion and integration. The inference drawn from of multiple sensor information is called as multi sensor fusion whereas multi sensor integration is one of the states during multi- sensor data fusion. During Multi- sensor data integration step the actual combination of data is presented. Figure 4 shows Luo and Kay multi-sensor data fusion model which depicts multi-senor data fusion and integration as well. It depicts that the data is received at sensor level and sent further so that fusion can take place at fusion center. Data fusion process utilizes either hierarchical or sequential methodology as depicted in figure 4. To support fusion process, appropriate database as well as libraries are included. After performing data fusion, at various fusion centers higher abstraction level is achieved for data representation. Table 4 exhibits comparison among various fusion levels based on data representation.

Pau's Behavioral Knowledge-based Data Fusion Process

Behavioral knowledge was utilized to build up framework for data fusion model (Pau, 1988). At first stage attribute vector / feature vector extraction is performed on raw data. Thereafter, vector alignment

Figure 4. Luo and kay multi-sensor integration and fusion model architecture

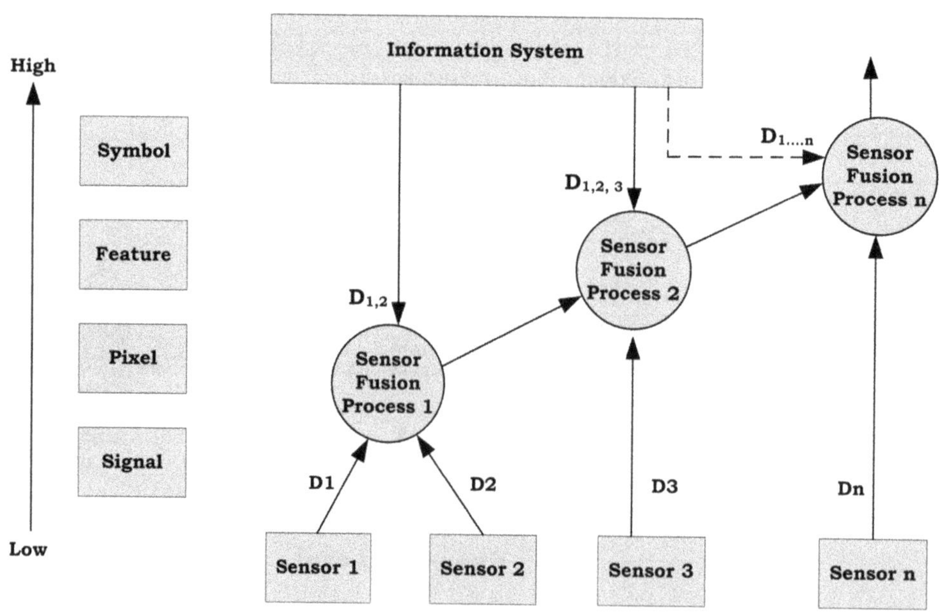

Table 4. Summary of Characteristics/ Features of data representation among various fusion levels (Luo and Kay, 1988)

Features of Fusion Information Representations	Information representation level	Categories of Sensory Information	Model based on Sensory Information
Signal level	Low Level	Multi-dimensional Signal Information	Random Variable with noise
Pixel Level	Low Level	Multiple number of images	Random process across the pixel
Feature Level	Medium Level	Attribute/feature extraction from the image/signal	Non-invariant form of attributes/ features
Symbol Level	Higher Level	Decision inference obtained based on signal/ image	Symbolic representation with degree of uncertainty

is performed on extracted features. Fusion is carried out on sensor attributes and at data analysis level as well. Finally, the behavioral attribute-based inference rules are applied to obtain fused data output. Figure 5 shows Pau's Behavioral Knowledge-based data fusion Process Model

This data fusion process model architecture utilizes hierarchical methodology instead of blackboard architecture which consists of three levels of data representation. The most basic and lower level for every sensor comprises of vector space which uses coordinate system dimensions and values. The second level pulls out appropriate characteristics from these vectors and allocate labels. The top level comprises of set of formal inferences which connect vectors to events. Examples include smart home/ building, surveillance and control.

Harris Waterfall Fusion Model

Authors (Harris et al., 1998) have provided a new hierarchy based fusion architecture which is known as waterfall data fusion model. Figure 6 shows waterfall model which depicts that information flow in model works from data level to decision level. A continuous update is performed on sensor system based on feedback received from decision module. The feedback suggests system re-calibration/ system reconfiguration. Waterfall model also exhibits three levels of information representation. At most basic level raw data provides information about environment which requires sensor models based on experimental analytics and measured phenomena. Level 2 performs feature/attribute extraction and fusion implementation. Thereafter, symbolic inferences are drawn, and output is represented as probabilistic estimates. The third level performs object and event relevance

RESEARCH CHALLENGES AND OPEN RESEARCH DIRECTIONS FOR MULTI-SENSOR DATA FUSION PROCESS

The challenges which may arise during application of Multi-sensor fusion process for UWSNs can be described as mentioned:

Figure 5. Pau's Behavioral Knowledge-based data fusion Process Model

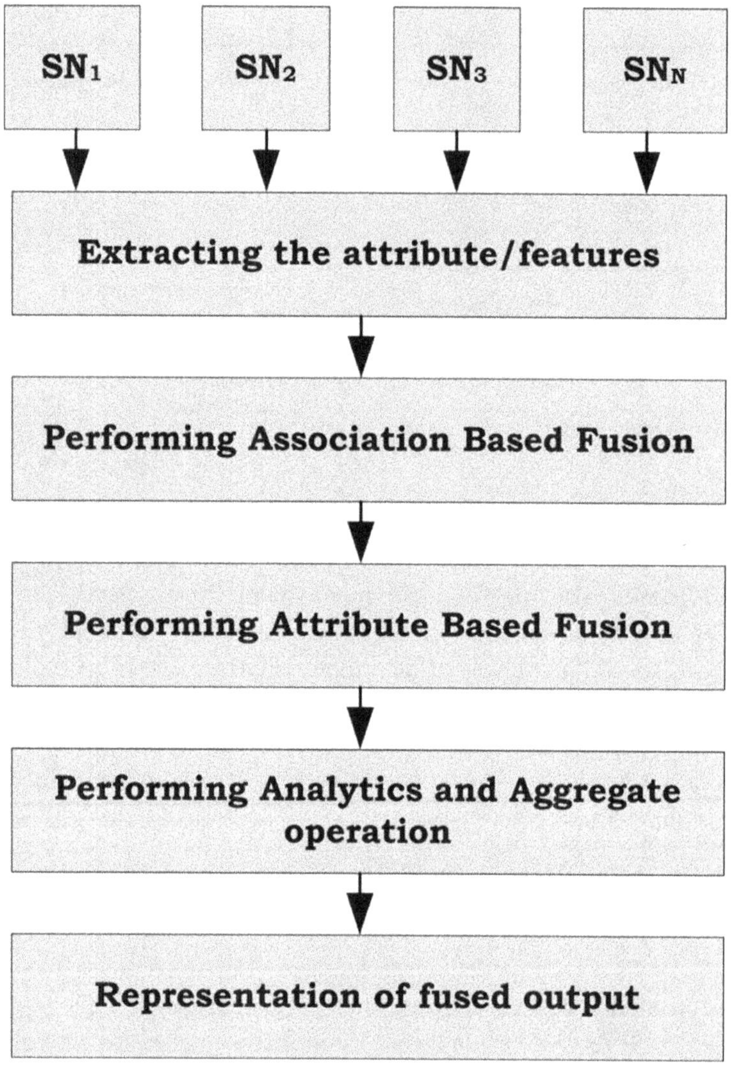

- The multi-sensor data fusion process may involve heterogeneous sensors. The level of heterogeneity varies with respect to various features such as location attribute, synchronized behavior or type of output.
- The heterogeneity among different types of data deals with text date, image or statistical data.
- The consistency of sensed data is another major issue which must be tackled to eliminate malicious data.
- Limitation of interoperability among sensors and limited algorithm development for data fusion domain is another major issue in UWSNs which must be addressed.

The challenges related to architectural aspects can be summarized as:

Figure 6. Harris waterfall fusion model

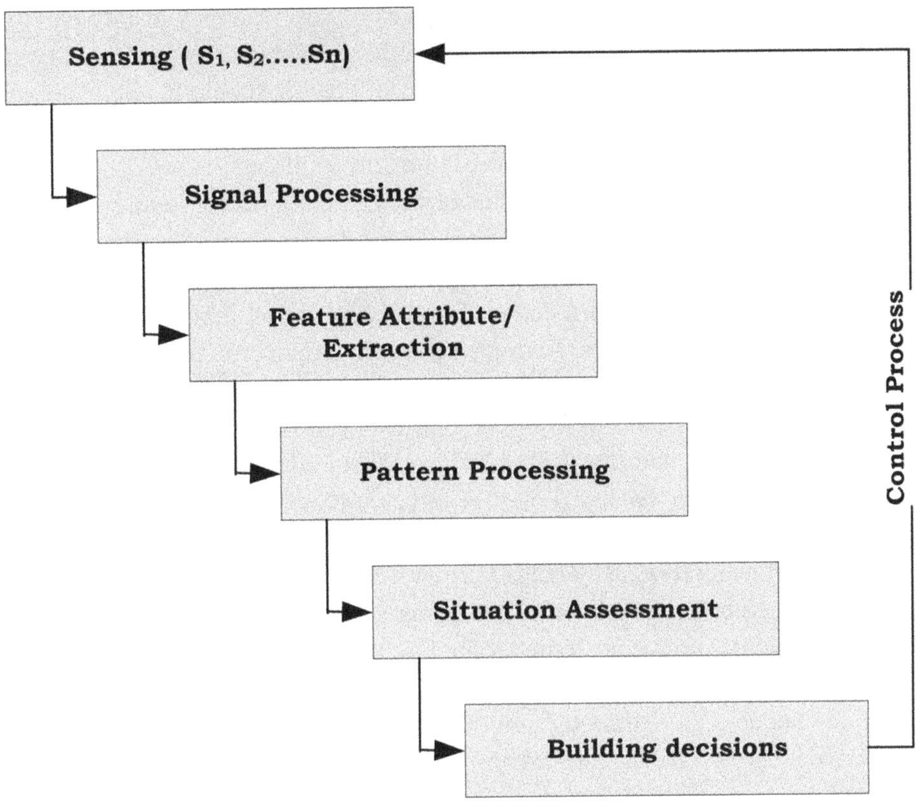

- The architectural aspects deal with how the sensors are configured whether they follow parallel or serial configuration, however in UWSN it becomes difficult to deal and maintain the proper sensor configuration.
- To choose among appropriate levels to represent the information is another issue which arises from architectural point of view. However, 3-level system for information representation is most widely. UWSNs can follow the various architectural model depending upon application domain.
- To deal with sensor node failures and data corruption is another major aspect which must be taken care so that the UWSNs can function properly. Appropriate algorithms to deals with the dynamicity of nodes is very important aspect which must be dealt properly.
- The selection of proper data mining algorithm is also one of the important aspects in multi-sensor data fusion (Dalberg et al.,2006) Clustering algorithms and neural network techniques are important machine learning techniques which enhances the positional level fusion and process related to identity recognition of an object.
- Analyze the performance and quality of a data fusion process is another challenging aspect which must be explored especially in UWSNs because it has higher uncertainty level (Oxenham et al.,1996, Shchurov, 1991).

Open Research Directions in UWSNs can be described as mentioned below:

- Quality of Data in fusion: The quality of data fusion in UWSNs basically deals with sensing coverage which is most challenging aspect in UWSNs. Poor coverage may lead to poor data quality or poor data representation. Hence, it requires the higher sensor density in the area of deployment in order to ensure proper coverage. Further, the sensor density should not be overemphasized in order to avoid higher complexity of operations. Therefore, the algorithm development for proper coverage in UWSNs to ensure the quality of data must be explored.
- Representation of data: There is a tremendous increase in data sources and formats for data representation are not compatible with each other. The data ontology deals with this problem to provide interoperability among different data formats. However, there is still a gap which shows that ontology-based languages still focus on limited application domain however UWSNs has wide variety of applications domain. Therefore, to explore the ontology of data is still an area to be explored (Ubaldi, 2013).
- Privacy and Security: Collection of UWSNs data for security and surveillance-based applications is utmost challenging task due to its highly sensitive nature. However, these issues related to privacy of data have been discussed by various researchers (Chan et al.,2008, Li et al.,2015). The data security deals in two major aspects which involves infrastructure level security and data security. Infrastructure level security deals with system infrastructure as well as services. Data security deals with storage of data, communication and data generation. The researchers have provided various security protocols, however there is still a need to explore this domain to ensure higher performance analytics for security related applications in UWSNs.
- Data fusion Techniques: To extract the information for various UWSNs based applications require the involvement of various data mining operations. The research in this domain involves machine learning based approaches such as Deep Neural Networks (DNN) (Zhang et al., 2018), unsupervised data fusion using machine learning (Li and Zhou, 2014). Therefore, the data fusion techniques based on machine learning is an emerging area to be explored for various application domains of UWSNs.

CONCLUSION

The chapter provides an overview about UWSNs, its applications, sensor fusion, its significance and advantages. It provides a detailed insight into the applications of multi-sensor data fusion for ocean monitoring and surveillance, military intelligence, target tracking and air-air, air-surface monitoring and defense with respect to various parameters which include: types of observed data, volume of monitoring and required platform as well. The spectral range of sensors for data fusion has also been presented. Sensor fusion classifications have been discussed in detail. Further, this chapter provides a detailed review of most popular data fusion architectural models available for UWSNs which include JDL data fusion model, Thomopoulos Data Fusion, Multi-sensor integration and fusion model, Behavioral Knowledge-based data fusion process model and Waterfall Fusion Model. Finally, the chapter highlights research challenges and open research directions for multi-sensor data fusion process. The recent trends in UWSNs data fusion applications emphasize the need of new evaluation methods as well.

REFERENCES

Azar, A. T., & Vaidyanathan, S. (Eds.). (2015). *Computational intelligence applications in modeling and control*. Springer International Publishing. doi:10.1007/978-3-319-11017-2

Bakr, M. A., & Lee, S. (2017). Distributed multisensor data fusion under unknown correlation and data inconsistency. *Sensors (Basel)*, *17*(11), 2472. doi:10.339017112472 PMID:29077035

Berger, C. R., Zhou, S., Willett, P., & Liu, L. (2008). Stratification effect compensation for improved underwater acoustic ranging. *IEEE Transactions on Signal Processing*, *56*(8), 3779–3783. doi:10.1109/TSP.2008.924801

Bevilacqua, M., Tsourdos, A., Starr, A., & Durazo-Cardenas, I. (2015, February). Data fusion strategy for precise vehicle location for intelligent self-aware maintenance systems. In *2015 6th International Conference on Intelligent Systems, Modelling and Simulation* (pp. 76-81). IEEE. 10.1109/ISMS.2015.37

Blackman, S. S., & Broida, T. J. (1990). Multiple sensor data association and fusion in aerospace applications. *Journal of Robotic Systems*, *7*(3), 445–485. doi:10.1002/rob.4620070309

Cardarelli, E., Sabattini, L., Secchi, C., & Fantuzzi, C. (2014, September). Multisensor data fusion for obstacle detection in automated factory logistics. In *2014 IEEE 10th International Conference on intelligent computer communication and processing (ICCP)* (pp. 221-226). IEEE. 10.1109/ICCP.2014.6937000

Chan, A. B., Liang, Z. S. J., & Vasconcelos, N. (2008, June). Privacy preserving crowd monitoring: Counting people without people models or tracking. In *2008 IEEE Conference on Computer Vision and Pattern Recognition* (pp. 1-7). IEEE. 10.1109/CVPR.2008.4587569

Conde, M. E., Cruz, S., Munoz, D. M., Llanos, C. H., & Fortaleza, E. L. (2013, February). An efficient data fusion architecture for infrared and ultrasonic sensors, using FPGA. In *2013 IEEE 4th Latin American Symposium on Circuits and Systems (LASCAS)* (pp. 1-4). IEEE. 10.1109/LASCAS.2013.6519059

Dalberg, E., Lauberts, A., Lennartsson, R. K., Levonen, M. J., & Persson, L. (2006, July). Underwater target tracking by means of acoustic and electromagnetic data fusion. In *2006 9th International Conference on Information Fusion* (pp. 1-7). IEEE 10.1109/ICIF.2006.301613

Dasarathy, B. V. (1997). Sensor fusion potential exploitation-innovative architectures and illustrative applications. *Proceedings of the IEEE*, *85*(1), 24–38. doi:10.1109/5.554206

Dasarathy, S., & Merli, M. (2016). Sarcopenia from mechanism to diagnosis and treatment in liver disease. *Journal of Hepatology*, *65*(6), 1232–1244. doi:10.1016/j.jhep.2016.07.040 PMID:27515775

Durrant-Whyte, H. F. (1990). Sensor models and multisensor integration. In *Autonomous robot vehicles* (pp. 73–89). Springer. doi:10.1007/978-1-4613-8997-2_7

Elmenreich, W. (2002). *An introduction to sensor fusion*. Vienna University of Technology.

Fan, X., Zhang, F. B., Zhang, Y. Q., & Wang, G. (2011). Information Fusion Technology and its Application to Multi-sensor Integrated Navigation System for Autonomous Underwater Vehicle. *Huoli yu Zhihui Kongzhi*, *36*(3), 78-81

Foote, K. E., & Huebner, D. J. (1995). *The Geographer's Craft Project*. Department of Geography, University of Texas.

Galan, B., & Grossmann, I. E. (1998). Optimal design of distributed wastewater treatment networks. *Industrial & Engineering Chemistry Research*, *37*(10), 4036–4048. doi:10.1021/ie980133h

Georgy, J., Noureldin, A., & Mellema, G. R. (2011). Clustered mixture particle filter for underwater multitarget tracking in multistatic active sonobuoy systems. *IEEE Transactions on Systems, Man and Cybernetics. Part C, Applications and Reviews*, *42*(4), 547–560. doi:10.1109/TSMCC.2011.2154329

Goyal, N., Dave, M., & Verma, A. K. (2016). Energy efficient architecture for intra and inter cluster communication for underwater wireless sensor networks. *Wireless Personal Communications*, *89*(2), 687–707. doi:10.100711277-016-3302-0

Goyal, N., Dave, M., & Verma, A. K. (2017). Improved data aggregation for cluster based underwater wireless sensor networks. *Proceedings of the National Academy of Sciences, India Section A: Physical Sciences*, *87*(2), 235-245. 10.100740010-017-0344-y

Goyal, N., Dave, M., & Verma, A. K. (2018). A novel fault detection and recovery technique for cluster-based underwater wireless sensor networks. *International Journal of Communication Systems*, *31*(4), e3485. doi:10.1002/dac.3485

Goyal, N., Dave, M., & Verma, A. K. (2020). SAPDA: Secure Authentication with Protected Data Aggregation Scheme for Improving QoS in Scalable and Survivable UWSNs. *Wireless Personal Communications*, *113*(1), 1–15. doi:10.100711277-020-07175-8

Guleria, K., Kumar, S., & Verma, A. K. (2019 c). Energy Aware Location Based Routing Protocols in Wireless Sensor Networks. *World Scientific News*, *124*(2), 326–333.

Guleria, K., & Verma, A. K. (2018). An energy efficient load balanced cluster-based routing using ant colony optimization for WSN. *International Journal of Pervasive Computing and Communications*, *14*(3/4), 233–246. doi:10.1108/IJPCC-D-18-00013

Guleria, K., & Verma, A. K. (2018a, December). Ant Colony Optimization Techniques–An Enhancement to Energy Efficiency in WSNs. In *International conference on Computer Networks, Big data and IoT* (pp. 670-677). Springer.

Guleria, K., & Verma, A. K. (2018b, December). Unequal Clustering-An Enhancement to Network Longevity in WSNs. In *2018 4th International Conference on Computing Communication and Automation (ICCCA)* (pp. 1-4). IEEE. 10.1109/CCAA.2018.8777605

Guleria, K., & Verma, A. K. (2019a). Comprehensive review for energy efficient hierarchical routing protocols on wireless sensor networks. *Wireless Networks*, *25*(3), 1159–1183. doi:10.100711276-018-1696-1

Guleria, K., & Verma, A. K. (2019b). Meta-heuristic Ant Colony Optimization Based Unequal Clustering for Wireless Sensor Network. *Wireless Personal Communications*, *105*(3), 891–911. doi:10.100711277-019-06127-1

Hall, D. L., & Llinas, J. (1997). An introduction to multisensor data fusion. *Proceedings of the IEEE*, *85*(1), 6–23. doi:10.1109/5.554205

Hall, D. L., McMullen, S. A., & Hall, C. M. (2015, September). New perspectives on level-5 information fusion: The impact of advances in information technology and user behavior. In *2015 IEEE International Conference on Multisensor Fusion and Integration for Intelligent Systems (MFI)* (pp. 214-219). IEEE. 10.1109/MFI.2015.7295811

Harris, C. J., Bailey, A., & Dodd, T. J. (1998). Multi-sensor data fusion in defense and aerospace. *Aeronautical Journal*, *102*(1015), 229–244.

Johnson, E. N., DeBitetto, P. A., Trott, C. A., & Bosse, M. C. (1996). *The 1996 Draper Laboratory small autonomous aerial vehicle*. Project Documentation.

Khaleghi, B., Khamis, A., Karray, F. O., & Razavi, S. N. (2013). Multisensor data fusion: A review of the state-of-the-art. *Information Fusion*, *14*(1), 28–44. doi:10.1016/j.inffus.2011.08.001

Li, Y., Dai, W., Ming, Z., & Qiu, M. (2015). Privacy protection for preventing data over-collection in smart city. *IEEE Transactions on Computers*, *65*(5), 1339–1350. doi:10.1109/TC.2015.2470247

Li, Y., & Wang, Y. (2018). Synergistic strategy for the geographical traceability of wild Boletus tomentipes by means of data fusion analysis. *Microchemical Journal*, *140*, 38–46. doi:10.1016/j.microc.2018.04.001

Li, Y. B., Li, L. L., Xu, B. B., Wu, W., Wu, R. Y., Wan, X., Cheng, Q., & Cui, T. J. (2016). Transmission-type 2-bit programmable metasurface for single-sensor and single-frequency microwave imaging. *Scientific Reports*, *6*(1), 23731. doi:10.1038rep23731 PMID:27025907

Li, Y. F., & Zhou, Z. H. (2014). Towards making unlabeled data never hurt. *IEEE Transactions on Pattern Analysis and Machine Intelligence*, *37*(1), 175–188. PMID:26353217

Liggins, M. II, Hall, D., & Llinas, J. (Eds.). (2017). *Handbook of multisensor data fusion: theory and practice*. CRC press. doi:10.1201/9781420053098

Liu, Y. C., Liu, S. Y., & Wang, N. (2016). Fully-tuned fuzzy neural network based robust adaptive tracking control of unmanned underwater vehicle with thruster dynamics. *Neurocomputing*, *196*, 1–13. doi:10.1016/j.neucom.2016.02.042

Llinas, J., Bowman, C., Rogova, G., Steinberg, A., Waltz, E., & White, F. (2004). *Revisiting the JDL data fusion model II*. Space and Naval Warfare Systems Command.

Lloret, J. (2013). *Underwater sensor nodes and networks*. Academic Press.

Luo, R. C., & Kay, M. G. (1988, August). Multisensor integration and fusion: issues and approaches. In *Sensor Fusion* (Vol. 931, pp. 42–49). International Society for Optics and Photonics. doi:10.1117/12.946646

Luo, R. C., Yih, C. C., & Su, K. L. (2002). Multisensor fusion and integration: Approaches, applications, and future research directions. *IEEE Sensors Journal*, *2*(2), 107–119. doi:10.1109/JSEN.2002.1000251

Majumder, S., Scheding, S., & Durrant-Whyte, H. F. (2001). Multisensor data fusion for underwater navigation. *Robotics and Autonomous Systems*, *35*(2), 97–108. doi:10.1016/S0921-8890(00)00126-3

Mitchell, H. B. (2007). *Multi-sensor data fusion: an introduction*. Springer Science & Business Media.

Murphy, J. B., Gutierrez-Alonso, G., Nance, R. D., Fernandez-Suarez, J., Keppie, J. D., Quesada, C., & Dostal, J. (2006). Origin of the Rheic Ocean: Rifting along a Neoproterozoic suture. *Geology, 34*(5), 325–328. doi:10.1130/G22068.1

Nicosevici, T., Garcia, R., Carreras, M., & Villanueva, M. (2004, November). A review of sensor fusion techniques for underwater vehicle navigation. In Oceans' 04 MTS/IEEE Techno-Ocean'04 (IEEE Cat. No. 04CH37600) (Vol. 3, pp. 1600-1605). IEEE. doi:10.1109/OCEANS.2004.1406361

Oxenham, M. G., Kewley, D. J., & Nelson, M. J. (1996, November). Performance assessment of data fusion systems. In *Proceeding of 1st Australian Data Fusion Symposium* (pp. 36-41). IEEE. 10.1109/ADFS.1996.581078

Paradis, S., Roy, J., & Treurniet, W. (1998). Integration of all data fusion levels using a blackboard architecture. *Proceedings of the 1998 International Conference on Data Fusion-EuroFusion'98, October 1998, Great Malvern, UK,* 195-202.

Pau, L. F. (1988). Sensor data fusion. *Journal of Intelligent & Robotic Systems, 1*(2), 103–116. doi:10.1007/BF00348718

Shchurov, V. A. (1991). Coherent and diffusive fields of underwater acoustic ambient noise. *The Journal of the Acoustical Society of America, 90*(2), 991–1001. doi:10.1121/1.401913

Song, Y., & Deng, Y. (2019). A new method to measure the divergence in evidential sensor data fusion. *International Journal of Distributed Sensor Networks, 15*(4), 1550147719841295. doi:10.1177/1550147719841295

Thomopoulos, S. C. A. (1991). Theories in distributed decision fusion. *IFAC Proceedings Volumes, 24*(5), 195-200.

Ubaldi, B. (2013). *Open government data*. Academic Press.

Waltz, E. L., & Buede, D. M. (1986). Data fusion and decision support for command and control. *IEEE Transactions on Systems, Man, and Cybernetics, 16*(6), 865–879. doi:10.1109/TSMC.1986.4309005

Zeng, C., King, D. J., Richardson, M., & Shan, B. (2017). Fusion of multispectral imagery and spectrometer data in UAV remote sensing. *Remote Sensing, 9*(7), 696. doi:10.3390/rs9070696

Zhang, Q., Yang, L. T., Chen, Z., & Li, P. (2018). A survey on deep learning for big data. *Information Fusion, 42*, 146–157. doi:10.1016/j.inffus.2017.10.006

Chapter 6
Congestion Control Protocols for UWSNs

Tanvika Garg
National Institute of Technology, Delhi, India

Manisha Bharti
National Institute of Technology, Delhi, India

ABSTRACT

UWSN is a grid of many purposes of self-operating nodes with various applications related to various disciplines such as hydrographic surveys, tactical surveillance, disaster prevention, and bathymetry. The process of transmission and reception of messages by propagating sound in an underwater environment is known as acoustic communication. Transmission of acoustic waves is the only method to communicate underwater, as radio waves get attenuated severely and there is severe scattering in optical transmission. Underwater wireless sensor networks (UWSN) have important applications in the exploration of underwater. UWSNs have various applications like in exploration of the sea, collection of data, monitoring of pollution, surveillance of tactics, prevention of disaster, in applications of ministry and surveying of mines.

INTRODUCTION

In UWSN, Underwater sensor nodes are designed in such a way so that it can tolerate the severe environment of the deep seas to work well for longer time without having to repair, service, or recharge. Since these nodes are incapable to show lateral movement independently, they should be arranged in space optimally.

Underwater sensor networks nodes are not fixed like ground-based sensor network nodes .Instead, they mobilize because of activities of underwater environment.

These waves are waves related to pressure that show transverse movement in medium. They suffer from attenuation losses and spherical spreading because of signal characteristics and medium .Underwater Sensor Networks has constraints like limitation in bandwidth, delay in propagation, 3Dtopology, and constraints in power and congestion .This chapter deals with congestion control in UWSN.(Jain 2015)

DOI: 10.4018/978-1-7998-3640-7.ch006

BACKGROUND

Issues, Controversies, Problems

A WSN contains many kinds of nodes. These nodes face various challenges. Some of them are: limitation in availability of energy, throughput and communication ability. UWSNs are complex. In this world, networks send data to the base-station. UWSNs have many characteristics which differ from TWSN (Terrestrial Wireless Sensor Network). This medium has limitation in bandwidth, consumes more energy for transmission and delay is longer. (Awan 2019)

Therefore, the characteristics of underwater acoustic sensor networks show that these networks need congestion control protocols. A transmission medium broadcasts to all neighboring nodes that can receive signal. In this a node may not get some packets at the time of transmission because transmitter and receiver perform in half duplex mode which communicates simultaneously on one channel. So the node may not detect a collision. Therefore protocols have to be designed otherwise collision will occur in the channel.

In UWSN, conservation of energy is an important concern because it causes difficulty in charging of the battery .Consumption of energy in UWSN is required at the time of transmission and reception of data which is done for the entire week. In UWSN loss of energy can occur by re-transmitting data due to conflict, implosion and overlap, control packet overhead, unnecessary usage of power in standby mode. Research is going on to lessen the energy wastage.

UWSN nodes are sparsely deployed and there is possibility of congestion. Network congestion takes place at the time of exceeding of capacity of load on the node, this can result into drop of packets, delay of queues or blockage of new connections and interferences of packets .This results in the reduction of throughput in the network. The result is the reduction in throughput. This might result in many retransmissions so that packet loss can be compensated which may lead to the congestion. For the accomplishment of the mission of the network, the QoS should be improved, and hence congestion control becomes important.

Congestion Control Schemes

The MAC Protocol for Single-Hop (Centralized)

This scheme has been suggested because of the various features of underwater channel which are

1. Battery supply is difficult
2. The propagation delay is long, and
3. The data transfer rate is low.

Media Access Control Protocol (MAC)

The single-hop (centralized) has been considered. This contains member nodes and single head node in the cluster. Here all nodes which are member of this network are connected to the head node (cluster) by a hop (single). In this scheme two phases are there which are shown in Figure. 1. (Byun 2013)

First phase is the "initialization phase "and the phase next is the "data transmission phase". During the "initialization phase", the member nodes in a cluster compete for the reservation of the channels to transmit the data. In this phase, the member nodes which are approved can send the data to the cluster head node.

The "Initialization Phase"

Firstly, the node (cluster head) does the broadcast of "ID request packet" to the other nodes (member nodes) of the network. The moment when every node (member node) receive the packet, the node which requires the channel for transmission responds with packet of ID response to the cluster head node. The "random waiting time", Ts results in reduction of the occurrence of member node at the head node. Ts occur within the interval 0 - T0 seconds. The interval of T0 affects the member nodes' collision ratio and the efficiency of channel. Thus, it should select T0 after proper optimization. After the "ID response packets" arrive at head of the cluster, the node (head node) makes a list of the member nodes (accepted) which has come to the node (head node) without being collided. N denotes accepted nodes' count. After this, a head node does the broadcast of the "Admission packet". In this time, the nodes which are not accepted should again do the transmission in the next "initialization phase".

Figure 1. Process of MAC protocol

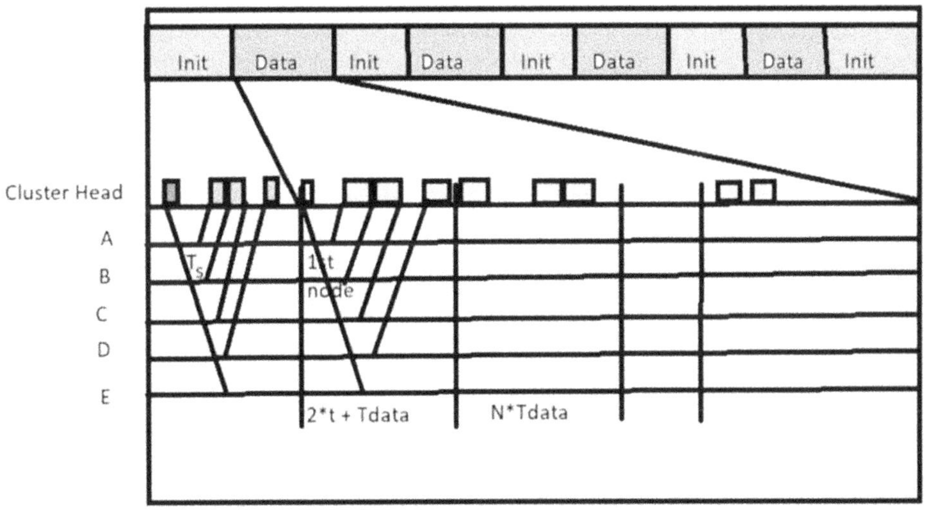

Phase of Data Transmission

The "admission packet" includes a list of nodes which are reserved in the channel in accordance with the distance and the time of sending of data.. The "data transmission phase" contains periods of transmission which is K in number. The time length of every period of transmission is the multiplication of Tdata (the data length) and N except the period of transmission which occurs for the first time. The nodes which are accepted could send the data for K number of times in the "data transmission phase".

When the data of transmission is over, each node should sleep till next transmission for the reduction of consumption of energy. When a node is finished with sending data, the sleep mode is achieved by it till data transmission phase ends . The data which is generated newly could make a node come out of sleep mode for sending of data in the coming period. Many nodes (accepted) are tough to fully accommodate the K reserved periods for transmission. The channel efficiency is effected by this condition. When K is small, data transmission phase is occupied by data, but data takes long because initialization process is frequent. Thus, the designing should be done as to do optimization of K.

Simulation and Results

It has been assumed that there are one head node and hundred other nodes (member nodes). They are arranged in regular intervals of about one kilometer. A head node can be reached by every member node in a hop (single). Ten is the maximum number of data packets which can be sent. Conditions of simulations are as mentioned below.

- **Cluster Size:** 10 kilometer x 10 kilometer
- **Rate of Transmission of Data:** 20 kilobits per second
- **Size of Data Packet:** 3000bits
- **Size of Control Packet:** 100bits

Simulation environment which has been used is MATLAB.

Collision Rate vs.T0

There may be collision of "response packets" among each other during the initial phase. The rate of collision is random. T0 (the random time) is shown in Fig.2 .The collision rate can be reduced by T0 which is long, but it might cause efficiency of channel to worsen. Thus, the right value of T0 should be selected for good performance of the system. T0 have to be above three millisecond so that collision rate is less than twenty percent .

Optimum K

In "data transmission phase", the number of repetitions K, of "data transmission period" is an essential quantity. Figure.3 depicts K (optimum).

There is Gaussian distribution of data load which has mean of five packets of data. In Fig. 3, the varying Gaussian distribution depicts the dispersion. When the dispersion is ten, each node's packet number is uniformly distributed between one and ten. The dispersion 1 has a meaning that the number of packet is near about the average. The "initial phase time" which is frequent causes the decrease in throughput if K is lesser than 5.The increase in "vacant time slots" in "data transmission phase" can cause decrease in throughput when K is more than 5 . The K (optimum) is close to the average packet number.

Figure 2. Rate of collision versus T0

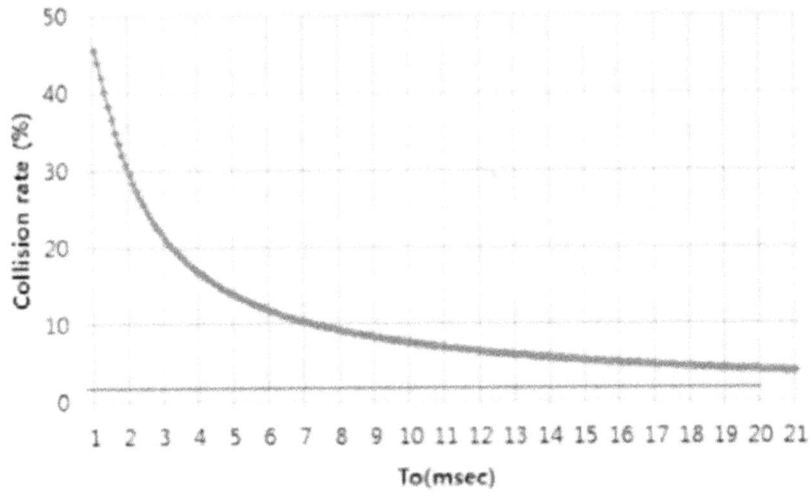

Figure 3. Throughput vs. K

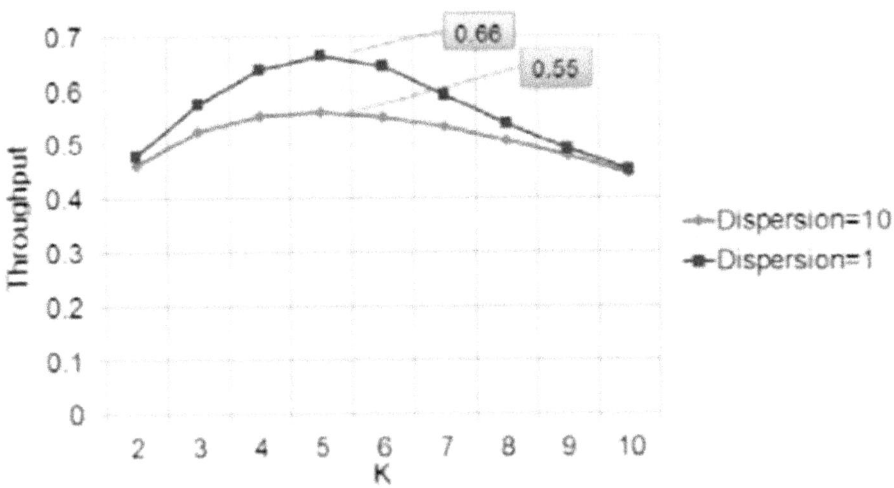

A MACA-Based Interference Avoidance MAC Protocol (Liu 2016)

The protocol called MACA-U was made for UWSN. In this section, the working of MACA in underwater conditions considering interference is briefly described. The definition of the "transmission range", "interference range" and "interference zone" is given as.

"Transmission Range": It is defined to be the range when there is correct decoding of received packet.

"Interference Range": It is defined to be the range where it can be made out by the nodes that the transmission is done by sender. And also interference can occur here. Interference range is larger than the transmission range, for example, double the transmission range. There is the dependence of "interference range" and "transmission range" on the level of power of transmission.

Figure 4. Shows the "transmission range", "interference range" and "interference zone" for C node

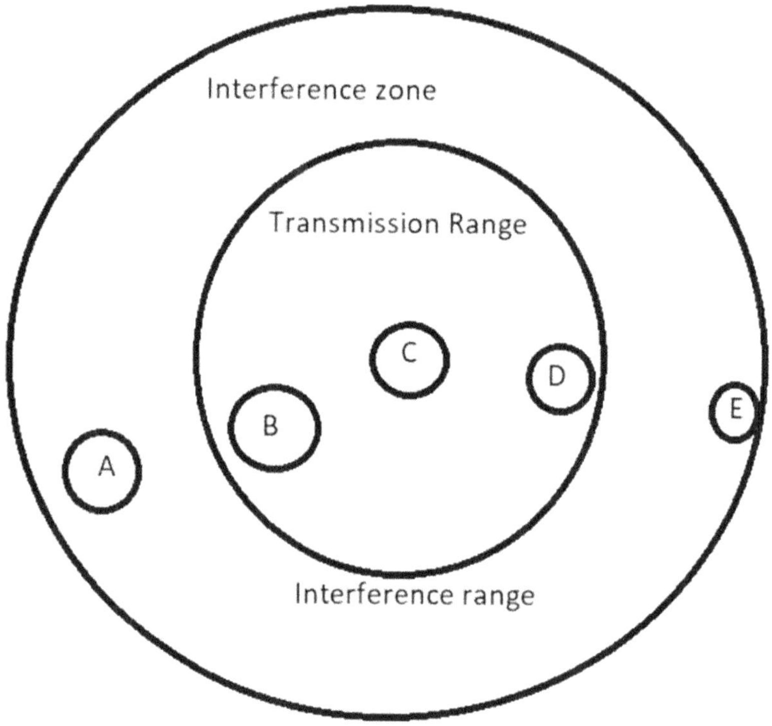

"Interference Zone": It is defined to be the range where the signal can be made out by the nodes but they cannot do its decoding .The transmission range is not included by the interference zone. Nodes can make out the transmission, and can also do its decoding correctly.

Let us now consider MACA including interference proposed for UWSNs. When a node tries to send data, an "RTS" packet is first sent and a timer is fixed for 2Tmax + TCTS, where Tmax is the propagation delay (maximum) and "TCTS" is the CTS packet's interval. When "RTS" is received by the receiver node, it immediately gives a response of a "CTS" packet. Just like this, the CTS is returned after that by the receiver node to the source node, the timer is set by the receiver node for 2Tmax + Tdata for the data (coming), where Tdata is the duration of the data packet. The data packet is sent by the sender as soon as the CTS is received.

The neighboring nodes' rules are presented here respectively in the "transmission range" and the "interference zone". In the "transmission range", the packets are received and correctly decoded by the neighboring nodes which mean the xRTS, xCTS and xDATA packets can be received by the node in the correct manner. As an xRTS packet is received, it is required to remain silent for the time "2Tmax + TCTS". And the time at which an "xCTS" packet is received, it is required to wait for "2Tmax + Tdata" till the arrival of xDATA. When node (neighbor) is absent in the "transmission range" but is present in the "interference zone", the received packets' types cannot be identified because they are not decoded correctly. This means that, this node does not know whether the packet is "xRTS packet", an "xCTS packet" or an "xDATA packet". These packets which are unknown are called as X. When an X packet is received by a node, for the avoidance of collision, there is requirement to wait for "2Tmax+TCTS" for its transmission. Every neighboring node remains silent when source is transmitting packets of data to the other end.

A MACA-based Collision Avoidance MAC Protocol

UWSN has been used in many situations, for example mineral exploitation, environmental monitoring and coastline protection. Because of, limitation of power and bandwidth, long propagation delay protocols made for TSN cannot be used directly for UWSN.

On designing MAC protocols for UWSN, we should not avoid the delay of propagation. Thankful to delay of propagation (long), two packets of data sent simultaneously might not collide. Utilization of this can be done for the improvement of the throughput of the network. Sensor nodes batteries most of the times are not charged and efficiency of energy must be taken into account. And also, sensor nodes might flow because of the currents; hence networks should be mobility tolerant.

Recently, proposals of many MAC protocols have been made, which can be categorized as: "collision free protocols" and "collision-based protocols".

FDMA, TDMA and CDMA are included in collision–free protocols. FDMA does not suit UWSN because bandwidth is limited. In CDMA, there is the problem namely near-far problem, and the receiver is of complex nature. TDMA protocols are well scheduled and mostly have high energy efficiency and throughput. In RETDMA, node (gate) has the responsibility to schedule the sub-nodes' transmission of data. More energy is consumed by gate- nodes than sub-nodes. RE-TDMA only suits a network which is centralized and sub-nodes are not able to communicate .

Mostly control packets for example "RTS/CTS" are sent for the avoidance of collision in collision –based protocols. Because propagation delay is long and bandwidth is limited, collision may occur (control packets with data packets) and hence may not able to be received on time by the neighboring nodes. This causes energy consumption and data re-transmission.

RM-MACA and FS-MACA, in both of them, the nodes (sensor) maintains a record in order to save the information of reservation which are inclusive in control packets, and record is helpful to the nodes (sensor) to take the decision of transmission of an "RTS/CTS". Hence collision avoidance is achieved. Search and maintenance of the record is not cheap. And also, packets of data and packets of control undergo modulation to various bands of frequency in RMMACA. Collision can be avoided by multi-channel protocols between data and control packets. Hardware is required to be modified which results into added cost.

T-LOHI

In case of T-Lohi nodes try to acquire the channel by sending a tone. The RTS in MACA has the same function as the function of tone, RTS is used instead of tone. (They differ in the sense that when a tone is sent it consumes less power).

If there is no data to be sent by sensor node, it continues to sleep. When it is made to come out of sleep by RTS (sent by other sensor nodes), it blocks till there is completion of transmission. If data is needed to be sent by sensor node, and any network transmission does not sense it, an RTS is sent and keeps on hearing the channel till the time (fixed). If other RTS is not received during this period, contention gets success and a packet of data is transmitted by it which does not vary till the termination of

Figure 5. MACA for UWSNs' timing diagram

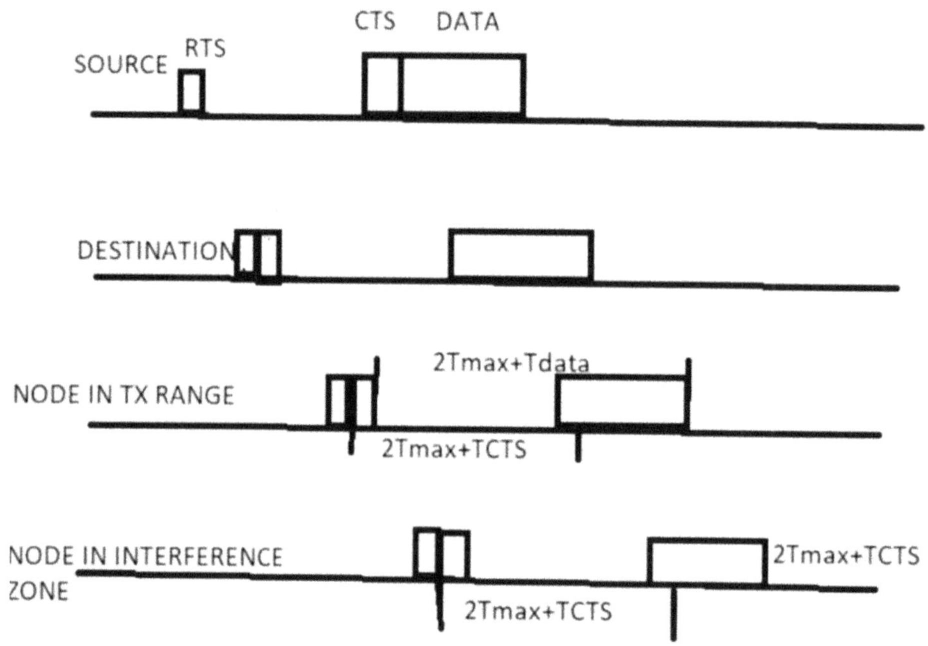

the channel listening period. Otherwise, the received RTS number is counted which is referred to as "contender counting" (CTC).

When handshake procedure is not present, the congestion state knowledge is very less. So the result is that, the collision, when hidden stations are present, is still there in T-Lohi.

Design of FI-MACA

RTS and CTS packets are exchanged in this protocol to get the channel. The interval between the data packet and RTS is not variable similar to T-Lohi. If there is no optimization of hardware then, wake-up and sleep does not occur.

Before the beginning of data transmission, an initialization phase is present in this scheme, in which information (local) is collected. The way sensor node's sender sends a data packet in which successful transmission of data is given by three stages: "Listening for RTS (LFR)", "Waiting for CTS (WFC)" and "Transmitting Data (TD)". In this scheme, at the end of LFR the RTS packet is sent; the CTS packet arrives at the node (sender); and, during TD stage it transmits the data packet. TFI is the time of "WFC stage". A sender firstly enters LFR stage. It keeps on hearing to the channel for Tps. During LFR stage if another RTS packet comes, then packet containing data reaches at the sender after "TFI "and lasts up to "Tps", which has the meaning that the packet containing data undergoes collision with the CTS packet which is sent by the sender. Otherwise, an RTS is sent by the sender and WFC stage is entered by it. The CTS packet is made in such a way so as to reach the end of WFC stage. At the termination of "WFC stage" if CTS packet is not present, contention gets failed and the sender surrenders owing to the "binary exponential back off algorithm". Handshake gets success otherwise and the "TD stage" is entered by the sender and then starts to retransmit.

Figure 6. Sender with three stages

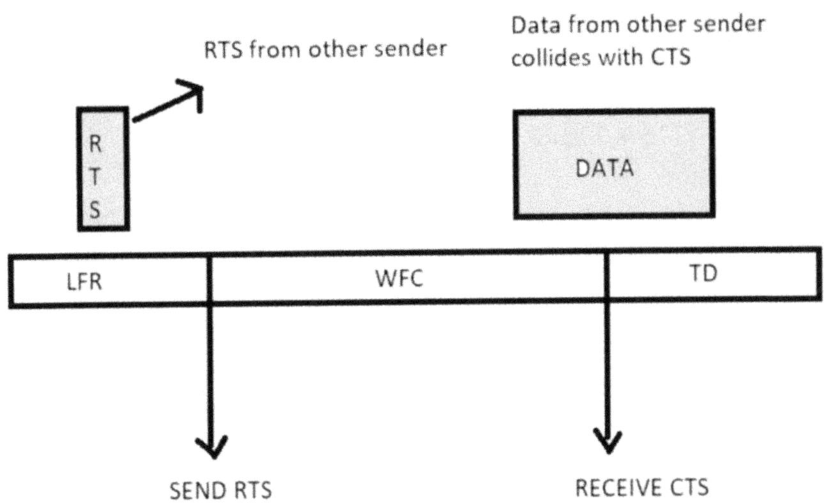

The Receiver's Way of Behaving

If an RTS packet is received by a sensor node and the below conditions are fulfilled:

- Another RTS packet has not been received for Tps;
- Another RTS packet has not been sent for Tps, the receiver keeps on hearing to the channel during this time. This duration is called as Continued Listening (CL) stage. Any other RTS packet does not arrive during this period, handshake gets success as no collision of data occurs with the data packet.

The Keeping Quiet Mechanism

Control packets are over heard by sensor nodes which should keep quiet for longer time. But exceptions are there.

- When a RTS packet is overheard a sender remains silent in LFR stage;
- Other control packets are ignored by a sender in WFC stage, yet it allows the response of the RTS if it satisfies above conditions;
- A RTS packet is overheard and so a receiver remains quiet.

Collection of Local Information

As the design of CTS packet is in a way that at the arrival at the last of "WFC stage", one node is required to have propagation delay knowledge of every neighbor.

Delay of propagation (coarse) is estimated in FI-MACA's "initialization phase", by using the proposed method. Messages are sent to their neighbor nodes by the sensor nodes. For example there is the presence of Node 0 and Node 1 which is its neighbor. The Node 0's clock IS T0x and Node 1's clock is T1x. Node0 sends a message at T01, Node 1 receives it at T12. At T13 a message is again sent by Node 1, that Node 0 takes at T04. The propagation delay between Node 1 and Node 0 is given by:

Tpd01 = (T04 – TO1 – T13 + T12)/**2**

Simulation Results

When the simulation is performed, ten homogeneous sensor nodes are present in a square of size of 2kilometre* 2kilometre. The speed with which it is propagating is 1500m/s.The value of communication range, data packet size and data rate are 1.5kilometre, 1024bytes and 16kbps, respectively. The time period of "WFC stage" must be two times as high as the delay of propagation, because the receiver must keep on hearing to the channel. The "FI-MACA" performance is given after comparison with "MACA".

Throughput

Figure. 7 depict the change in throughput changes according to network traffic. Traffic is meant by the average number of packets of data given by a node (sensor) in 1s. As the FI-MACA performance is not good when traffic is over 0.15 packet/s, this shows that a low load network suits FI-MACA.

Figure 7. Throughput vs. traffic

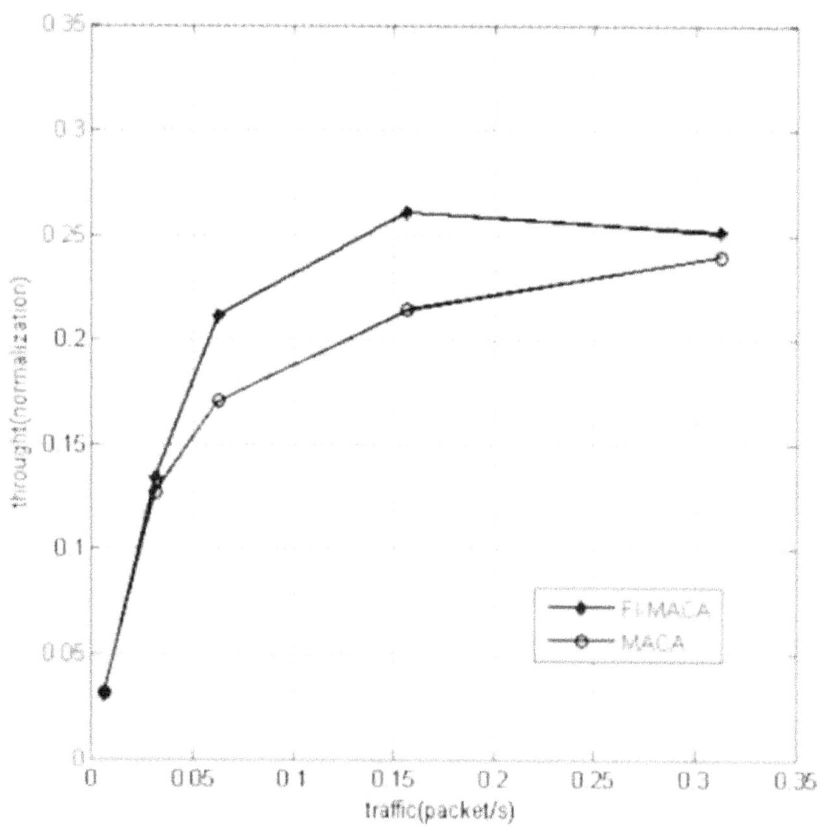

Efficiency of Energy

In Figure 8, as traffic goes over 0.05 packets per second, less power is consumed by "FI-MACA", this depicts that "FI-MACA" performs better than "MACA" in case of avoidance of collision.

Figure 8. Energy consumption vs. traffic

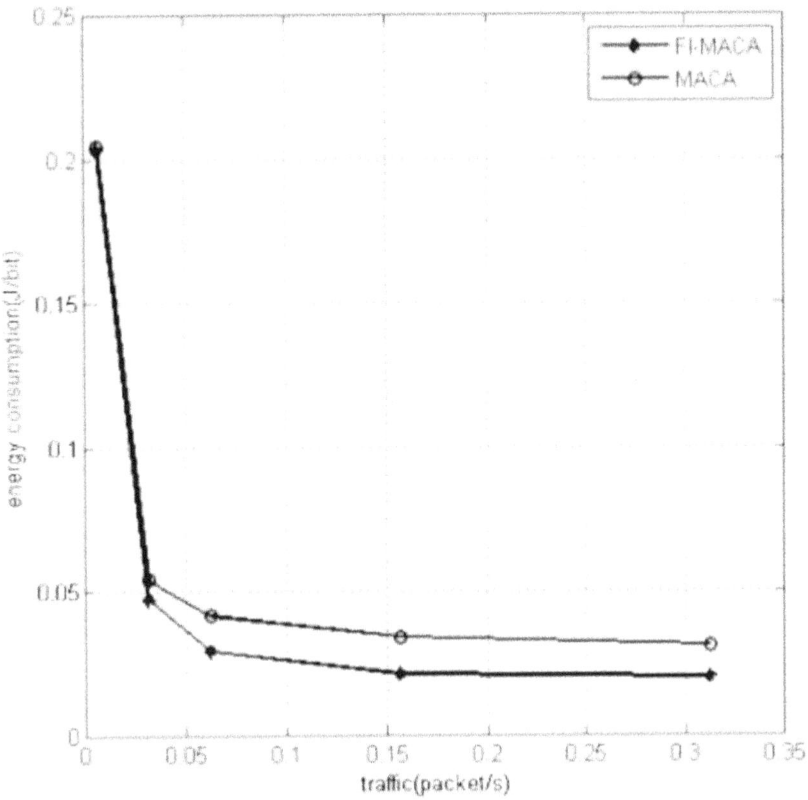

Performance Analysis

As FI-MACA concentrates on avoidance of collision. In this section, analysis of avoidance of collision is done. Packets can be categorized by three categories: packet of data; and, 2 packets of control which includes "RTS packet" and "CTS packet". As control packet is of very small size the interference caused by data packet is only considered here.

When WFC stage is entered by a sender, none of the "RTS packet" reaches in the time of "LFR stage".

When the sender is responded by a receiver, none of the "RTS packets" reaches in this period of time. Hence none of the data packets reach when the receiver is busy.

Therefore, the conclusion can be made that the time duration between the" RTS "and the data packet can be instrumental in avoiding collision.

Random Access where Collision is Low by Using Nodes in Group

Assumption is made that a network having N number of sensors are placed. There are random locations of all sensor nodes (SNs) .

Here is the description of the proposed random access protocol procedure. SNs are controlled by the FC regularly for the collection of sensors' data. A "beacon signal" is transmitted by the FC to nodes for waking up the nodes and for notifying the nodes which are transmitting values of sensor .As a beacon signal is received by the SNs, they wake up and the RSSI level is measured which relates the distance between SNs and FC.The division of the region is made by this scheme of the network into groups on the basis of the distance between SNs and FC. Hence, the SN belongs to a group that can be measured by determining RSSI level. The nodes which are farthest have large time interval because of propagation speed which is low. The same group collision is unavoidable. There is small number of SNs and the locations from the FC are different .They also have different RSSI. Therefore, the collision is reduced by the proposed random access .The proposed random access method also result into increase in efficiency of time.

The transmission time is aligned to the farthest SN by the sensor nodes of the same group. Their distance from the FC is estimated by them, and transmission times can be adjusted to the farthest Sensor node. The distances that are estimated are not accurate. OFDM signals are used and demodulated in this scheme. The subcarriers of OFDM distinguish each SN in the same group. Since there is no requirement of estimation of the distances among SNs, overhead and complexity of computation of this scheme are much lesser other conventional methods.

SIMULATION

Simulations in computer are done in order to depict the proposed method advantages. The "reception success rate (RSR)" and the throughput of the random access are compared with other conventional schemes.

An assumption is made that SNs' locations are random; the spectrum's bandwidth is taken to be 4 kHz. In this method, the group number is three and network area has been experimently used with two cases i.e. 2.5 kilometer and 3 kilometer. The subcarriers' number of OFDM is to be taken as 32, and there is a utilization of 21 subcarriers having null subcarriers of guard bands and DC .

The comparisons of RSR are shown in Figure 9 and 10 for 3 km and 2.5km, respectively. In Fig. 9, when there is small number of sensor nodes, there is low probability of collision and similar RSR is demonstrated by the conventional and the proposed scheme. As the SNs' number increases, the proposed method RSR is more than conventional method because the proposed method gives low collision rate than the conventional method. When the SNs' number is 21.The lower collision probability is shown by 3km radius than 2 km radius network because it has larger distance. Figure.11and 12 shows the throughput vs. number of sensor nodes for 3 km and 2.5 km by the proposed and the conventional methods. The larger throughput is shown by the conventional method when SNs' number is small. No collision occurs when the number of SNs is small. The sub carrier is only utilized by the proposed method, so the data rate is smaller as compared to the conventional method.

Figure 9. RSR vs. number of sensors for 3kilometer

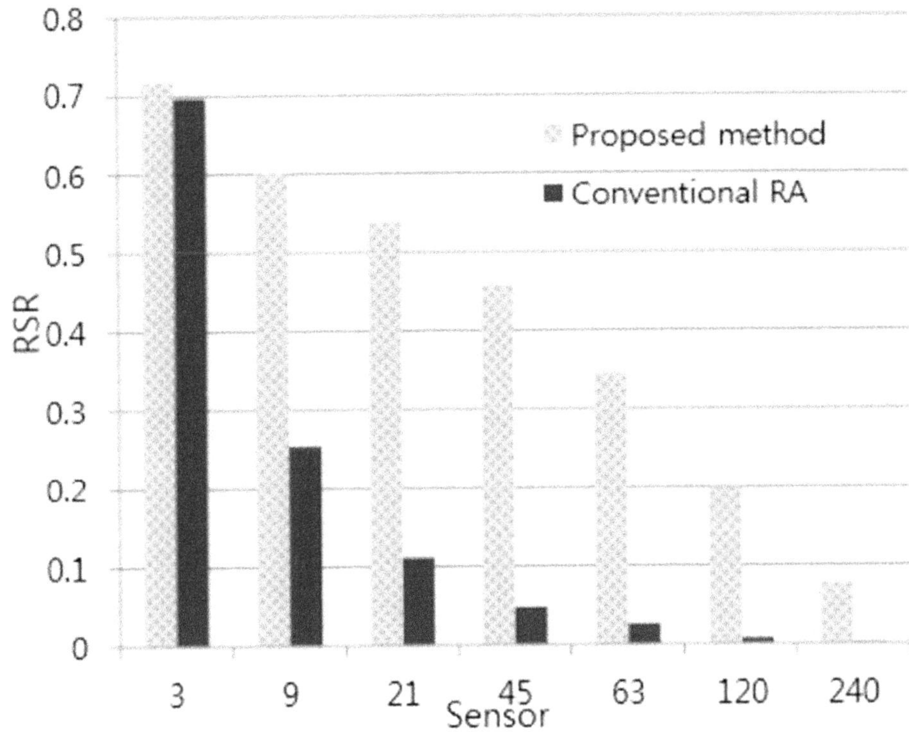

Figure 10. RSR vs. number of sensors for 2.5 kilometer

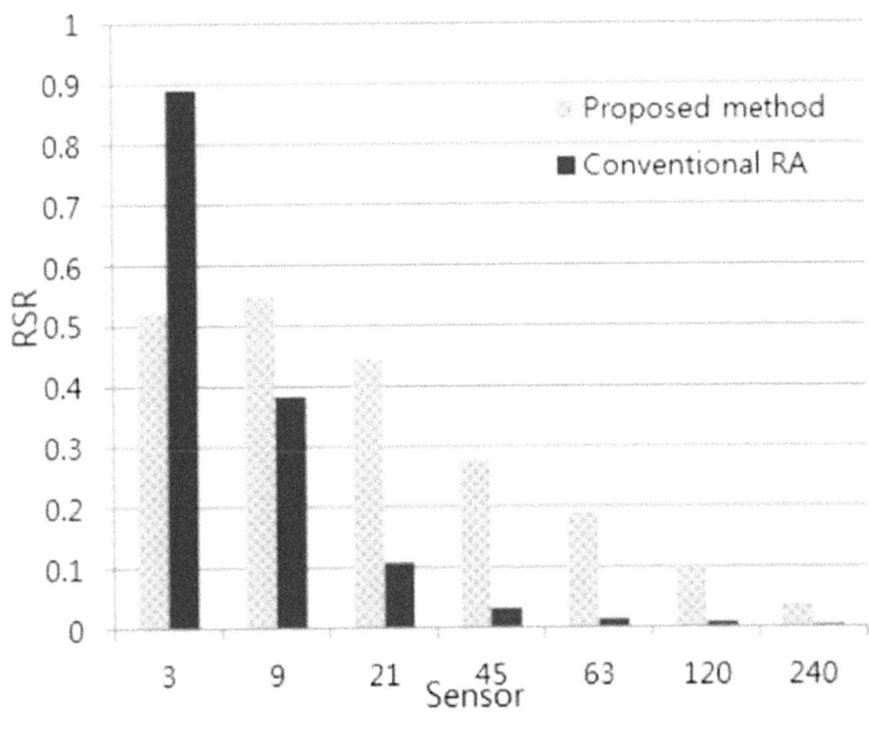

Figure 11. Throughput vs. number of sensor nodes in 3km

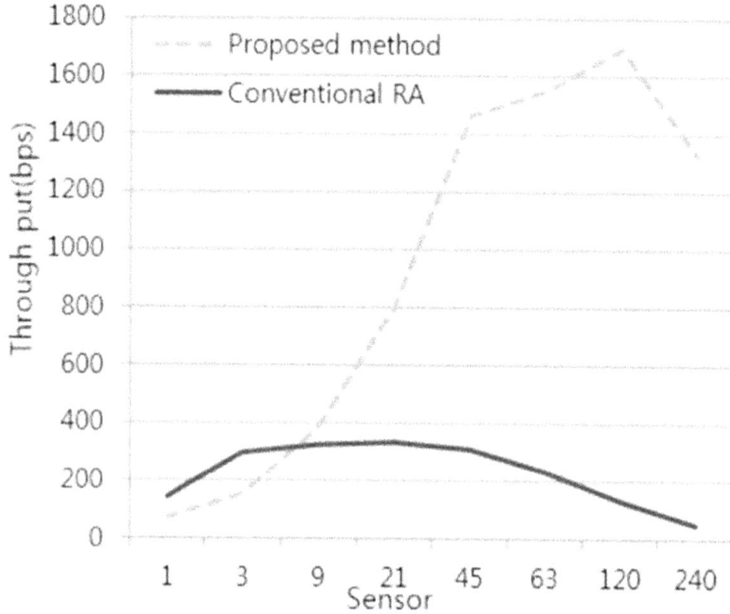

Figure 12. Throughput vs. number of sensor nodes in 2.5 km

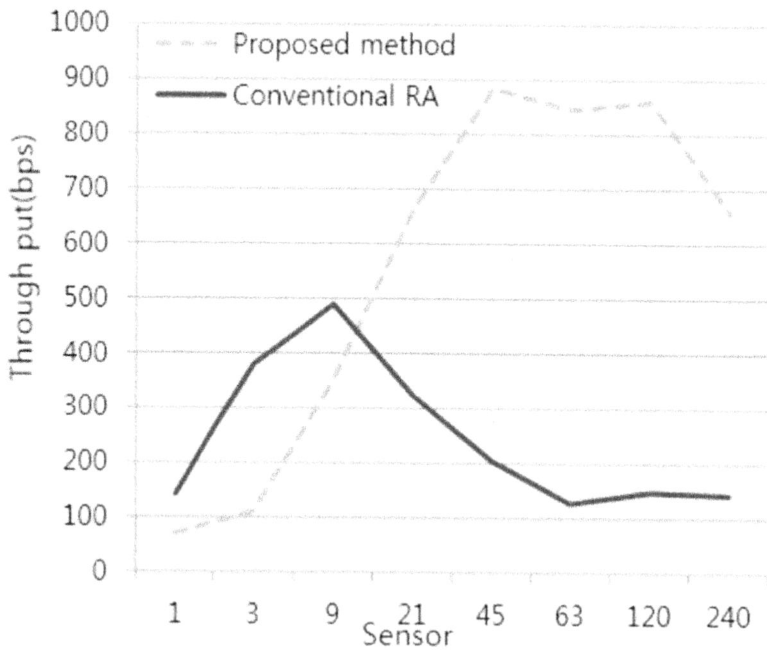

Why Not Traditional Congestion Control Techniques?

The terrestrial environment and the aquatic environment differ in many aspects. In TWSN the nodes are fixed while the nodes in UWSNs are not fixed. There are various other factors which make underwater wireless sensor networks different from terrestrial wireless sensor network such as limitation of bandwidth, delay in propagation, 3Dtopology and constraints in power.

Transmission of acoustic waves is the only available method for communication in underwater world, since radio waves get severely attenuated and optical transmission gets severely scattered .Since these waves are particularly waves of pressure that travel in the medium in transverse direction. The losses occur because of attenuation because of medium and properties of signal and spherical spreading. The water of sea causes severe conditions on efficiency of energy, on the ranges and elements' reliability.

Because of the above factors the congestion control schemes used in underwater wireless sensor networks differ from terrestrial sensor networks. The various schemes discussed above are not utilized in terrestrial sensor networks.

Recent Advancements in Congestion Control

There is research in progress on congestion control of underwater wireless sensor network. Many protocols are being developed to reduce congestion in underwater wireless sensor network. Although the underwater environment is not favourable, much advancement is done to reduce the drawbacks and help in promotion of the research in this field.

Conclusion and Challenges in Future

Various applications of this field decide the sensing and networking development in underwater world. Not so expensive computing, sensing and communications have provided the development of sensor networking in terrestrial environment in the past .There is expectation that inexpensive computing, in combination with reduced cost and advanced acoustic technology. Research and development on UWSN has shown advancement in these times, it has been shown that many challenges are yet not solved. With the advent of new technologies in communication, effectiveness in analysis, testing and integration of these ideas is significant—the field must cater to the development of insights, as well as understanding of practical issues. Because of these issues, we think that there is requirement of the development of new models related to this theory (both computational and analytical), and this much usage of experiments of field and test beds is necessary. On adding to it, testing and integration of recent ideas will lead to the issues which are not exposed to research going on in laboratories, such as requirement of energy, robustness in various situations and cost of entire system.

REFERENCES

Akyildiz, I. F., Pompili, D., & Melodia, T. (2005). Underwater acoustic sensor networks: Research challenges. *Ad Hoc Networks*, *3*(3), 257–279. doi:10.1016/j.adhoc.2005.01.004

Awan, K. M., Shah, P. A., Iqbal, K., Gillani, S., Ahmad, W., & Nam, Y. (2019). Underwater wireless sensor networks: A review of recent issues and challenges. *Wireless Communications and Mobile Computing*.

Byun, A. R., Son, K., Cho, H. S., & Jang, Y. S. (2013, July). Channel reserved MAC protocol for underwater sensor networks. In *2013 Fifth International Conference on Ubiquitous and Future Networks (ICUFN)* (pp. 110-111). IEEE.

Heidemann, J., Li, Y., Syed, A., Wills, J., & Ye, W. (2005). Underwater sensor networking: Research challenges and potential applications. *Proceedings of the Technical Report ISI-TR-2005-603, USC/ Information Sciences Institute*.

Inam, J. A., Ahmed, S., Ullah, N., Raza, A., Rehman, A. U., Khattak, A., & hahid Rahim, S. S. (2018). Congestion Control Reliability and Adaptive Cooperation for Efficient Underwater Sensor Networks. *EAI Endorsed Transactions on Scalable Information Systems, 5*(16).

Jain, S., Pilli, E. S., Govil, M. C., & Rao, D. V. (2015, February). Performance evaluation of congestion-aware routing protocols for underwater sensor networks with multimedia data. In 2015 IEEE Underwater Technology (UT) (pp. 1-6). IEEE. doi:10.1109/UT.2015.7108271

Liu, J., & Wang, J. (2016, January). A MACA-based collision avoidance MAC protocol for underwater acoustic sensor networks. In *2016 IEEE/OES China Ocean Acoustics (COA)* (pp. 1-4). IEEE.

Syed, A. A., Ye, W., & Heidemann, J. (2008, April). T-Lohi: A new class of MAC protocols for underwater acoustic sensor networks. In *IEEE INFOCOM 2008-The 27th Conference on Computer Communications* (pp. 231-235). IEEE. 10.1109/INFOCOM.2008.55

Zhang, J., Qiao, G., & Zhou, F. (2014, April). A relative clock based and energy sense MAC protocol for underwater acoustic communication networks. In *OCEANS 2014-TAIPEI* (pp. 1–7). IEEE. doi:10.1109/ OCEANS-TAIPEI.2014.6964341

Zhao, Q., Lambert, A., & Benson, C. R. (2014, November). FS-MACA: a future scheduling mac protocol in underwater acoustic networks. In *Proceedings of the International Conference on Underwater Networks & Systems* (pp. 1-5). 10.1145/2671490.2674470

Chapter 7
A Review of Recent Issues and Challenges of Fault Management Techniques in Underwater Wireless Sensor Network

Nonita Sharma
Dr. B. R.Ambedkar National Institute of Technology, Jalandhar, India

Krishna Pal Sharma
Dr. B. R.Ambedkar National Institute of Technology, Jalandhar, India

Rajneesh Rani
https://orcid.org/0000-0003-2104-227X
Dr. B. R.Ambedkar National Institute of Technology, Jalandhar, India

ABSTRACT

Underwater wireless sensor networks (UWSNs) are deployed to conduct cooperative surveillance and data gathering tasks in an acoustic region. Different nodes and ground-based stations use these networks interactively. UWSNs are currently facing problems and difficulties related to restricted bandwidth, substantial delay in propagation, 3D topology, control of media access, routing, use of resources, fault management, and energy limitations. The research community has given various methodologies over the past few decades to address these problems and challenges; however, owing to varying attributes of the underwater environment, some of them are still open to research. In this chapter, a survey of fault management techniques in UWSN regarding types of faults and their classification, environmental factors influencing the identification of faults, fault detection schemes, issues, and future directions is performed. This chapter outlines available fault management techniques and their pros and cons for further advancement in underwater sensor networks to highlight new research trends.

DOI: 10.4018/978-1-7998-3640-7.ch007

INTRODUCTION

The application of sound propagation techniques to send and receive messages in underwater environments is the basic premise of acoustic communication. The underwater telephone developed for the U.S. Navy during World War II was one of the first underwater communication devices (Headrick, Robert, & Freitag, 2009). Because of its numerous science, military, and commercial applications, acoustic communication has gained interest. These applications vary from strategic supervision to aquatic life analysis, including driverless vehicle activity, oil mining, mapping, oceanographic data processing, and aquatic environment commercial exploitation.

UWSNs consists of several mobile vehicles and sensors that work collaboratively to monitor and collect information in an aquatic area. As a matter of course, the surveillance of the ocean bottom is done through oceanographic sensors, which records data at a fixed place and perform analysis of the same after job completion. The main drawback of the traditional methodology is the absence of interactive communication between distinct ends, and therefore recorded data can never be obtained during any task, and information collected will be destroyed in the event of any failure (Goyal, Dave, & Verma, 2016)

For communication amongst underwater sensor nodes, electromagnetic, optical, and acoustic waves have been applied effectively (Stojanovic & Preisig, 2009). Nevertheless, radio waves are not considered viable for communication in the ocean due to high attenuation in water (Heidemann, John, Stojanovic, & Zo, 2012). Under the full constraints of UWSNs, the acoustic signals are considered viable as the sound speed is deemed continuous underwater. However, the underwater environment's factors like temperature, depth, and salinity influence the speed of sounds. Also, the variable characteristics of the acoustic channel pose several restrictions for using them for communication. For instance, multipath propagation causes fluctuations; the mobility of nodes causes Doppler effect. Also, nodes of underwater sensors are mobile, unlike nodes of ground-based sensor networks. Moreover, they move because of various underwater environment operations and conditions, generally 2-3m / sec with water currents (Mohamed, Jawhar, Al-Jaroodi, & Zhang, 2011)

Furthermore, underwater sensor nodes work under unfriendly and remote conditions. Faults in UWSNs, therefore, occur repeatedly and unpredictably as compared to conventional sensor networks (Benson, 2010). A fault is an unexpected change or malfunction, although it may not result in physical failure or breakdown. The provision of uninterrupted services to long-lasting aquatic applications is a significant challenge in UWSNs in the presence of these frequent failures. The erroneous information collected from faulty nodes may lead to wrong interpretations and false decisions. In some situations, the implications of a wrong decision can be served in terms of environmental effects, financial loss, and marine life. For example, in underwater warfare, if faulty nodes embedded around the suspicious area fail to give early warning correctly, the purpose of deploying the UWSNs vanishes (Goyal, Dave, & Verma, 2017).

The faults span from a single node behaving maliciously to the node crash, where it becomes completely inactive. As faults in nodes significantly reduce the data quality and decision capability, the faulty nodes need to be diagnosed correctly for filtering erroneous data and improving data quality. This chapter's significant contributions are not only to illustrate the classification of different faults but also to describe the impact of faults in the operating aquatic environment. Besides, the transmission of erroneous data not only reduces the decisive capability, but it harms in terms of wastage of constrained network resources like energy, bandwidth, etc. This chapter explores in-depth to facilitate the best use of available resources in the underwater environment under various faulty conditions. In Table 1, the most important terms related to fault diagnosis is elaborated so as to quickly explicate the current state of the art.

Table 1. Terminology related to fault handling in UWSNs

Term	Meaning
Terms related to Faults Classification in UWSNs	
Fault	An unsolicited change in the network or malfunction, although it may not result in physical failure or collapse.
Communication Failures	Failure causes the interruption in services generally due to environment imbalances like communication lost, hardware problems etc.
Transient Faults	Faults that cause the UWSNs to breakdown temporarily
Permanent Faults	Faults that are continuous in nature and malfunction the network when fully executed
Intermittent Faults	Faults that are transient and keep on occurring in the network on and off
Tampered Nodes	Nodes that are attacked by a malicious attacker and tampered nodes behave malignantly after that
Terms related to Faults Diagnosis in UWSNs	
Fault Diagnosis	Consists of four steps mainly: 1) Fault Finding 2) Segregation 3) Detection 4) Recovery
Fault Finding	Detecting whether the fault is present in the network
Fault Segregation	Segregating the faulty nodes in the network
Fault Detection	To find the type of fault in the UWSNs
Fault Recovery	Recovery from the faulty node
Active Diagnosis	Diagnosis obtained through active and consistent tracking of the current state using ordinary reference status
Passive Diagnosis	Passive diagnosis is done through occasional tracking of the current state using ordinary reference status
Failure Detection Rate	The rate of finding existing failures in the network
Isolation	Finding faulty nodes in the network
Faulty Values	The faulty values typically occur in networks due to transmission strategies. A fault in the hardware leads to measuring and receiving faulty values.
Fault Prevention	Minimizing the effect of fault in the network by ensuring network connectivity though redundant nodes/ edges in the network
Metrics related to Fault Management in UWSNs	
Detection Accuracy	Total number of detected faulty nodes/Total number of Faulty nodes
Classification Accuracy	Total number of faulty nodes detected in a particular class/Total number of faulty nodes in that class. The class of the faults: 1) Permanent Faults 2) Intermittent Faults 3) Fault Free Class
Time Complexity	The maximum time required to find a faulty node in the network
Space Complexity	Memory Space required to execute the algorithms to find the faults in the network
Message Complexity	Maximum number of messages communicated in order to uncover the faulty nodes in the network
False Alarm Rate	The number of times a non-faulty node is classified as a faulty node
Fault Hypothesis	Introducing a dormant fault in the network; analyzing its effect and the ways to deal with it
Model-based	Finding faults in the network using a model
Model less	Finding the faults in the network without using a model
Online Detection	Finding faults in the network real-time
Offline Detection	Finding faults in the network when the network is non-operational

This chapter is organized as follows. Section 2 elaborates UWSNs architecture. Section 3 outlines the fault management in UWSNs covering types of faults and their classifications, factors influencing the fault identification process, and various fault detection schemes. Section 5 generates insights by discussing issues and Future Directions. In the end, section 6 presents the conclusion and future scope. Figure 1 elaborates on the overall organization of the work presented in this chapter.

Figure 1. Organization of the chapter

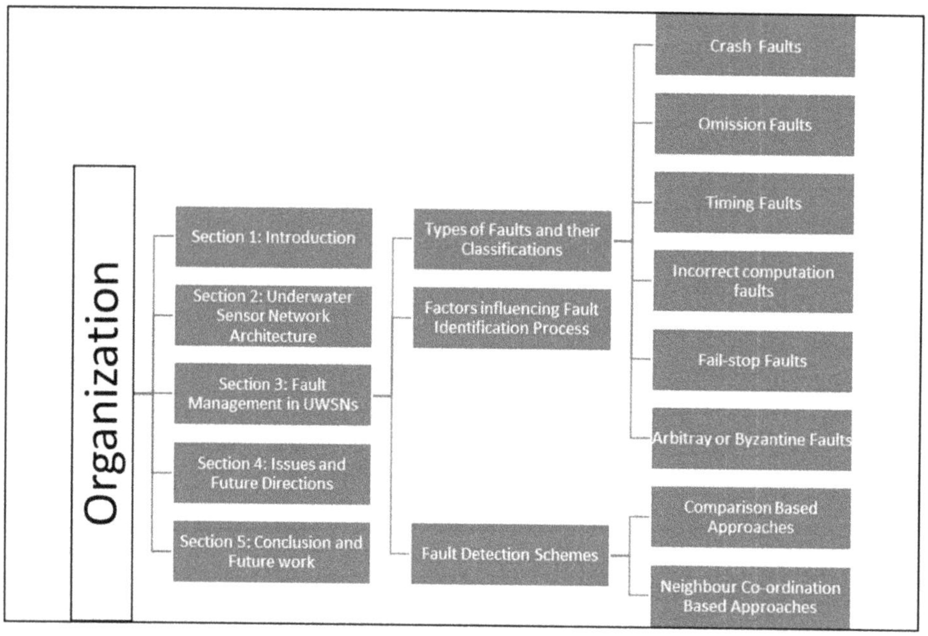

UNDERWATER SENSOR NETWORK ARCHITECTURE

Each system must be based on a properly designed architecture. Underwater Wireless Sensor Networks (UWSNs) differ substantially in many ways from Terrestrial Wireless Sensor Networks (TWSNs), for example, increased delay, greater odds of node malfunction, and higher transmission costs (Akyildiz, Pompili, & Melodia, 2005). The essential components that are the part of UWSNs architecture are elaborated below:

Underwater Channels

UWSNs employ basic methods such as sound waves to broadcast signals. TWSN network stacks could not be applied for UWSNs in the same way as minimal bandwidth and higher latency results in elongated end-to-end interruptions. There are distinctive characteristics of acoustic signals. Primarily, due to attenuation and spatial spreading, Acoustic signals inevitably result in high losses in transmission. Because the acoustic energy is transformed to heat, the attenuation is caused by absorption. Spatial spreading means the distribution of sound energy due to wavefront expansion. Secondly, noise can easily affect acoustic signals. Noise involves human-made noise from ships and background noise from ecological events like waves, currents, and hurricanes. Speed is a function of temperature, salinity, and depth in marine water, as expressed in equation (1) (Min, Hong, Cho, & Heo, 2012):

$$Speed(\alpha,\beta,\gamma) = 1448.96 - 4.59\gamma - 5.3*10^{-2}*\gamma^{2} + 2.374*10^{-4}*\gamma^{3} + 1.34(\beta-35) + 1.63*10^{-2}*\alpha + 1.675*10^{-7}*\alpha^{2} - 1.025*10^{-2}*\gamma(\beta-35) - 7.139*10^{-13}\gamma\alpha^{3} \quad (1)$$

In the above equation, α, β, γ refers to depth in meters, the salinity in parts per thousand, and the temperature in degrees Celsius respectively.

Network Protocol Stack in UWSNs

The design of a protocol stack plays a vital role in improving the efficiency of acoustic signal based data transmission. The core requirement for UWSNs is to control power and boost collaboration among the nodes in the underwater environment. The protocol stack consists of various layers, as demonstrated in Figure 2.

Figure 2. Network Protocol Stack in UWSNs

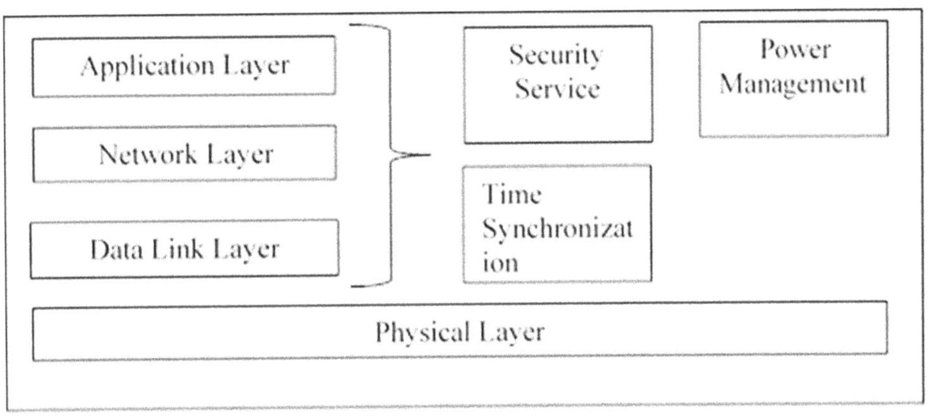

The network stack for UWSNs consists of four layers: The physical layer, Data link Layer, Network Layer, and the Application layer. The physical layer is responsible for the physical connections in the network. The stack consists of the four service planes provided by the other three layers in the network stack. The security service is to safeguard the underwater sensor nodes from any external sources. Network functionalities and other optimization services such as reducing energy consumption are taken care of by management plane. The time synchronization sets the time of the sensor nodes in the network. The localization service is used to provide the location information of the nodes in the network.

Network Architecture

Figure 3 presents the UWSNs network architecture. The sensor nodes are secured with the ocean base and are further connected to the cluster nodes via wireless acoustic communication. Cluster nodes are the nodes that relay data from the nodes at the ocean base to the base station.

Figure 3. Network architecture in UWSNs

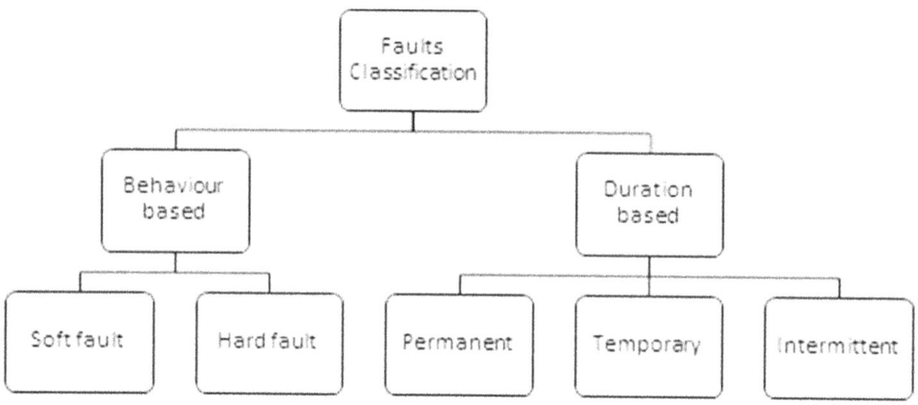

FAULT MANAGEMENT IN UNDERWATER WIRELESS SENSOR NETWORK

Primarily, UWSNs operate in unattended and unfriendly terrains such as deep water for specific applications that monitor the scenarios for months to years. Due to the random deployment of a large number of sensor nodes in such harsh and remote regions, faults in UWSNs occur more frequently than in other traditional networks. The fault in a system or sensor node can be thought of as an unexpected change or malfunctioning in its intended operation, although always it does not need to lead to physical breakdown. Thus providing uninterrupted services in the presence of frequent faults to such long-lasting applications in underwater scenarios is a big challenge in UWSNs. The erroneous information collected from faulty nodes may lead to wrong interpretations and false decisions.

In some cases, the consequences of a false decision can schemes in terms of environmental impact, economic loss, and human life. For example, in a railway bridge monitoring system, if faulty nodes embedded around beams of the bridge fail to give early warning of bridge deterioration correctly, the bridge may collapse suddenly at the time when a train is passing over. The transmission of erroneous data not only reduces the decisive capability, but it harms in terms of wastage of constrained network resources like energy, bandwidth, etc. The faults span from where a node behaves maliciously to the crash of the node, where it becomes completely inactive. As faults in nodes significantly reduce the data quality and decision capability, the faulty nodes need to be diagnosed correctly for filtering erroneous data and improving data quality.

Propagation Phenomenon in Underwater Sensor Networks

The propagation phenomenon in UWSN is a multifaceted communication due to the involvement of several environmental factors. The various factors are noise, Doppler effect, attention,and multipath effect, and are described as following:

- **Noise:** Noise is the degradation of signal strength and hence affecting the quality of the underwater acoustic channel. Underwater noises can be classified as background noise and noises by human beings as described below:
  - **Noise by Human Beings:** Human beings do several activities underwater, such as shipping, fishing, sonar, aircraft, and several other underwater activities, and these activities involve sending and receiving signals and hence causing heavy data traffic leading to disturbance and interference during acoustic communication (Akyildiz, Pompili, & Melodia, 2005).
  - **Background Noise:** Background noise, also known as ambient noise, is a blend of multiple sources that cannot be exclusively established (Ocean ambient noise: its measurement and its significance to marine animals) (Cato, 2008). The primary sources of the ambient noise are underlying, wind, shipping, thermal, and the turbulence (Ismail & Hussein, 2010). Underlying noise is the noise present in the absence of other sources such as self-noise. Wind noise is due to the propagation of air in the water. Ships while traveling in the water produce high traffic and hence disruption in the water communication (Cato, 2008). Turbulence is the disturbance due to surface roughness (Harland & Richards, 2006).
- **Doppler Spread:** Doppler shift occurs due to the fluctuation in the frequency due to the transmitter and receiver motion. The fluctuations in the frequency occur due to various flaws in the channel, and hence triggers a range of degradations. The flaws could be interference, reflections, and attenuation (Perrine, Nieman, Henderson, Lent, & Evans, 2010). Doppler spread affects the underwater communication in two probable ways: first is pulse width that will be compressed or stretched, and second is frequency offset as a result of frequency offset compressing or expending of signal time-domain occurring (Zhang, Han, Yin, & Sheng, 2013).
- **Attenuation:** Attenuation also known as path loss, occurs when sound while traveling underwater gets converted into heat and hence loses its strength. There are three main reasons for the path loss, as mentioned below:
  - **Geometric Spreading:** Geometric spreading occurs due to the propagation of the acoustic signal in the form of wavefronts. It depends on the distance covered by the wavefront and is not dependent on frequency. Depending upon the communication channel characteristics, Geometric spreading can be classified as spherical spreading and cylindrical spreading depicting deep ocean communication and shallow water communication, respectively (Lanbo, Shengli, & Jun-Hong, 2008)
  - **Path Loss due to Conversion to Heat Energy:** The wave energy converts into heat energy due to the propagation of waves in underwater. The path loss occurs due to the heat energy is absorbed by the underwater environment. This path loss is directly affected by frequency and distance (Lanbo, Shengli, & Jun-Hong, 2008).
  - **Path Loss due to Scattering:** The path loss occurs when the wave changes its path and deviates from the line of sight during communication. This type of path loss depends on the roughness of the surface (Lanbo, Shengli, & Jun-Hong, 2008).

- **Multipath:** When a sound wave gets reflected due to surface or the bottom obstacles, it causes significant and variable communication delays in acoustic communication. This primary cause that makes acoustic signal weak is called multipath effect that becomes the cause of intersymbol interference, which also makes acoustic data transmission difficult and erroneous. The vertical acoustic channel is less affected by the multipath effect as compared to the horizontal acoustic channel (Akyildiz, Pompili, & Melodia, 2005).

Factors Influencing Fault Identification Process

Due to the distributed design of the network and resource-constrained nodes, efficient fault detection is challenging in such networks. The traditional approaches cannot be applied directly to UWSNs due to the following reasons (Blough, Brown, & Hongying W, 1999):

- **Limited Hardware Resources:** Sensor nodes have limited processing, limited memory, low power communication unit, and restricted power supply, hence the fault diagnosis approaches should consider these issues during the fault detection process. Most of the traditional fault detection approaches are developed for wired networks and pay less attention to these unique issues associated with the sensor network. In order to identify faulty units, available fault detection schemes generally need to exchange too many messages. In WSNs, most of the energy of nodes is consumed in communication. Thus, the fault detection schemes should involve minimum inter-node communication. Another issue is limited memory storage on nodes. The fault diagnosis schemes cannot save too much historical data at the nodes to find suspicious behavior.
- **Dynamic Environments:** Sensor nodes are typically deployed in volatile and rural areas where faults will frequently occur. Nodes are deployed either by throwing them randomly in massive numbers or placing one by one. Thus, the density of nodes may be sparse in some regions and dense in other regions. Also, different phenomena have different behaviors. Hence, in such varying environments, developing a robust scheme that can work in all scenarios is challenging.
- **Dynamic Network Topology:** There are few mobile nodes in the network, few mobile nodes are also considered for some applications. These mobile nodes are free to travel at different speeds to any spot arbitrarily. Thus, in such dynamic networks where connectivity and density control are a big issue, diagnosing a faulty unit is challenging. Due to significant variations in Spatio-temporal correlation, detection accuracy in such scenarios gets reduced.
- **Signal Loss and Attenuation:** Some phenomena deteriorate monotonically with the distance. Thus, the nodes sensing an event and spatially apart from each other may be treated uncorrelated because of the significant variation in readings (Guo, Zhong, & He, 2009)Additionally, the different phenomena may deteriorate differently, and hence developing a single model to negotiate the impact of attenuation is challenging.

Types of Faults and Their Classifications

Faults in UWSNs can be classified based on their frequency of occurrence, the underlying cause of fault, and the behavior of the failed component (Elhadef, Boukerche, & Elkadi, 2008)The main classification of faults can be based on behavior and duration based as shown in Figure 4.

Figure 4. Broad Fault classification

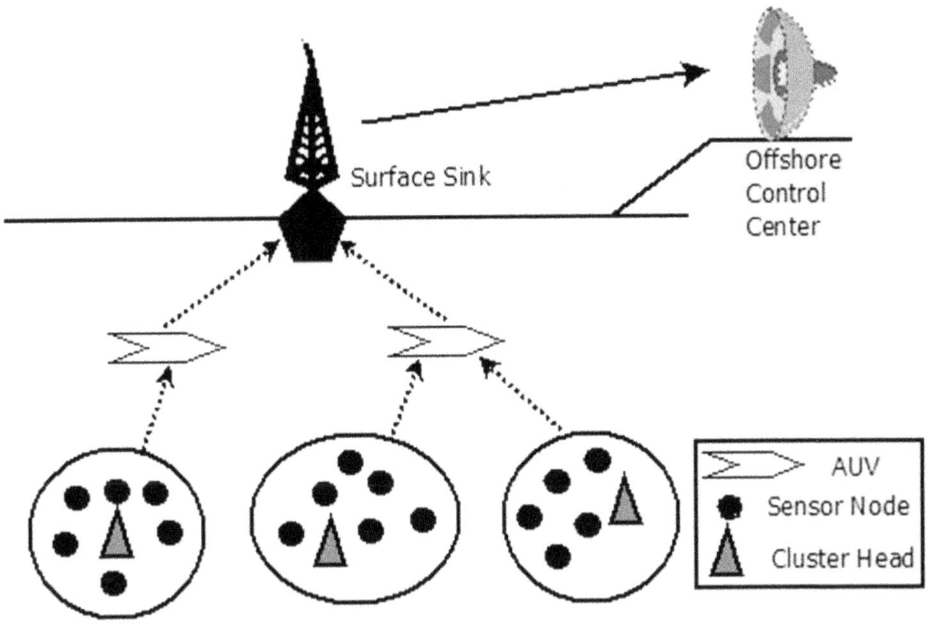

Behaviourly, a Fault in a sensor node can be either soft or hard based. A sensor node is said to be in the category of hard fault if it can not connect with the other parts of the network; otherwise, it belongs to the category of soft faults. The faults in a sensor node are categorized as permanent, intermittent, and transient on the basis of the duration of the altered behavior. A node said to be permanent faulty if it always produces errors in measurement after the error appears once. Permanent faults may be either in software or hardware part of the node (Barborak, Dahbura, Malek, & Miroslaw, 1993). The transient faults are unpredictable and sometimes distinguished as external faults, whereas the intermittent faults are the problematic type of transient faults. They reoccur regularly if they once appear and eventually tend to become permanent (Bondavalli, Chiaradonna, Di Giandomenico, Grandoni, & Fabrizio, IEEE). Based on the classification presented in (Avizienis, Laprie, Randell, Landwehr, & Carl, 2004), the faults can also be classified into several categories as outlined below. Also, table 2 summarizes the various fault classifications.

- **Crash Faults:** Such faults generally occur due to damage in the communication unit, complete battery depletion, and if the node is fully damaged. A crashed node is not able to participate in any activity of the network. Such faults occur due to natural incidents without any human intervention.
- **Omission Faults:** An omission faulty node extends the response time with the sink or any node to which it must respond. In WSNs, the nodes may be compromised to make it omission faulty for the malicious objective. Sometimes, such faults occur naturally.
- **Timing Faults:** Timing fault is behavior by which a node responds either too late or too soon to its natural response. This kind of behavior is considered faulty in synchronous systems. Such faults can occur due to human intervention or due to natural phenomena.

- **Incorrect Computation Faults:** Sometimes sensor nodes do not transmit the correct measurement even they receive correct readings. In that case, the node is considered as incorrect computation faulty. Like timing and omission faults, such faults may be either human-made or natural.
- **Fail-Stop Faults:** Such faults are similar to crash faults and occur due to complete battery depletion, node crash, and failure of the communication unit. Such type of faulty nodes sends alerts to their one-hop neighbors about their faulty behavior.
- **Arbitrary or Byzantine Faults:** Some faulty node behaves arbitrarily where the behavior of node is not covered in any above categories. The nodes with those irresponsible or undefined behaviors come under arbitrary faults or Byzantine faults. Many security attacks fall in this category, such as data corruption, censorship, misrouting, freeloading, etc. (Haeberlen, Kouznetsov, Druschel, & Peter, 2006).

Table 2. Classification and grouping of different fault types

Faults	Permanent	Transient	Intermittent	Soft	Hard
Crash	ü	X	X	X	ü
Omission	ü	ü	ü	ü	X
Timing	ü	ü	ü	ü	X
Incorrect Computation	ü	ü	ü	ü	X
Fail-Stop	ü	ü	ü	ü	X
Byzantine	ü	ü	ü	ü	X

FAULT DETECTION SCHEMES

Comparison Based Approaches: In these approaches, a test task is prepared and executed on a pair of nodes. It is a test task outcome comparison-based approach where task sender and receiver both execute the same task and test results are compared. The faulty and non-faulty behavior is found based on the comparison of outcomes. This section presents some comparison-based fault detection schemes. While sensor nodes are more prone to failure than wired networks and failures do not only impact the accuracy, it adversely affects the network lifetime also. Thus, the development of solutions for the fault-tolerant sensor networks has been a focused research area for the last few decades and was primarily explored by Preparata et al. (Preparata, Metze, Chien, & Robert T, 1967) particularly for wired inter-connected networks. In the proposed model, a scheme is presented for the automatic detection of multiple faults in a system. The complete system is viewed as it is containing n units or portion where fault can occur, i.e. the system is decomposed into n such units where a unit is a well-identified portion of the system which can not be further decomposed for fault identification. In this model, some units collaboratively test other units and outcomes are compared to find the correct status of a unit. After this model, many other strategies have been proposed based on collaborative comparison of test outcomes. For identifying faulty units, the comparison-based fault identification strategies are considered most effective and realistic (Duarte Jr, Ziwich, Albini, & Albini, 2011). Such approaches utilize the correlation exist in the data to find the fault. Thus, these more suitable for WSNs as the data sensed by sensor nodes is highly correlated temporally and spatially. In (Malek & Miroslaw, 1980), M. Malek et al. suggest a similar approach for fault identification in system having n connected processing elements. In order to find a faulty unit, the system is modeled as a connected graph. In the graph, the problem of the minimum number of edges required for fault detection is solved using a covering algorithm. In this scheme, a mismatch beyond a threshold limit during comparison of data which is identified by any two adjacent elements suggests that either or both units are defective.. The variations of this approach are proposed for wired as well as wireless networks. For WSNs, the approach is utilized by assigning a test to a pair of sensor nodes and results are compared to find the faulty node. In the following section, some other related schemes are listed.

In (Chessa, Santi, & Paolo, 2001), S. Chessa et al. proposed two implementations of the Ad-hoc Network fault diagnosis scheme. In this scheme, a fault free node s_i tests its adjacent node by sending them a message $M(s_i, q, TT_q)$ at time t, where TT_q is the test plan. At the same time, s_i executes the message and finds results as $Result(s_i, q)$. On receiving the message M from s_i, each adjacent node s_j executes TT_q and makes test result as $Result(s_j, q)$. After that, each node s_j respondes to s_i by transmitting another message $M_1(s_i, q, Result(s_j))$ at time t_1. Here, the t_1 should be less than the t_{out}, where t_{out} is maximum allowed response time for a node. After obtaining results from adjacent nodes, s_i compares results. When a node s_l in the neighbor of s_j receives M_1, then it first checks if $s_i = s_l$. If condition comes out to be true, it further evaluates $Result(s_j, q)$ with the expected result $Result(s_i, q)$ and create comparison results. If the comparison result is same, node s_j is detected fault free. If $s_l^1 s_j$ and s_l receives message M from s_i, it compares $Result(s_j, q)$ with $Result(s_l, q)$ (Figure 5). If both results are same, s_j is declared as fault free. For $s_l^1 s_j$ and s_l is not a neighbor of s_j, the result is compared with other results received for the same task if any (Figure 6).

The comparison-based techniques proposed in (Elhadef, Boukerche, Elkadiki, & Hisham, 2006) are based on the model presented in (Chessa, Santi, & Paolo, 2001). In (Elhadef, Boukerche, & Elkadi, 2008) M. Elhadef et al. propose three strategies for faulty mobile host identification in MANET with variable characteristics. The first is the Adaptive distributed self-diagnosis protocol (Adaptive-DSDP) with aiming to diagnose faulty mobile nodes in a diagnosable fixed-topology mobile ad hoc networks (MANETs). This method does not include the evaluation function when a test query is answered. In order to test a neighboring node, nodes in its vicinity broadcast evaluation function either intermittently or on the detection of suspicious behavior. The second self-diagnosis approach is Mobile-DSDP which is based on a comparison based diagnostic model and developed mainly for such networks the mobility is comparatively higher. Adaptive-DSDP utilizes a spanning tree approach while Mobile-DSDP adheres to the model of (Chessa, Santi, & Paolo, 2001)by incorporating the evaluation fucntion with the response message. In Mobile-DSDP, if the node s_i is not in the reach of its old neighboring node s_j, any other node that is now in its neighborhood is able to diagnose the correct status of s_j. The scheme uses a flooding mechanism to broadcast results evaluated locally. In mobile and adaptive DSDP, each node needs to reply to the request message which it received. While in Dynamic-DSDP, each node responds to exactly m request messages in an m-connected network.

An adaptive failure detector for ad hoc and sensor network is presented by M. Elhadef et al. (Elhadef, Boukerche, & Azzedine, 2007) over the same model presented in (Chessa, Santi, & Paolo, 2001), (Elhadef, Boukerche, & Elkadi, 2008). The scheme is based on an adaption of a gossip-style failure detecting protocol and the heartbeat message-based failure detection. In order to detect faults, the scheme predics the onset of heartbeat messages. For example, when node s_i is estimating the heartbeat messages' arrival times from another node s_j, and if s_i does not receive these messages on time, then s_i suspects node s_j as faultsy. The scheme satisfies both properties: strong completeness and eventual strong accuracy.

Figure 5. Node sl received the test request message M from si

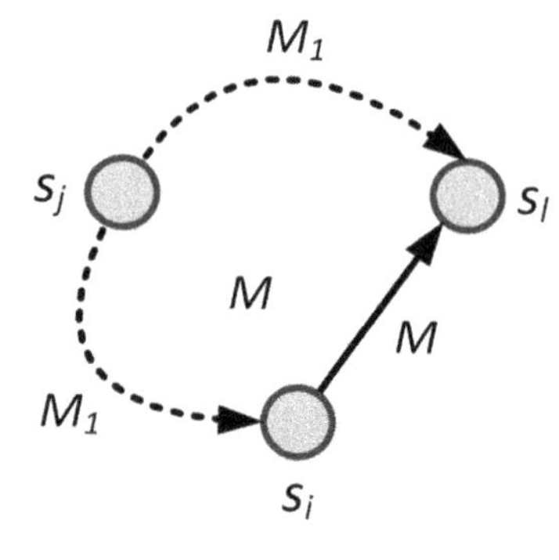

Figure 6. Node sl received response M1 and M2 concerning the test request M.

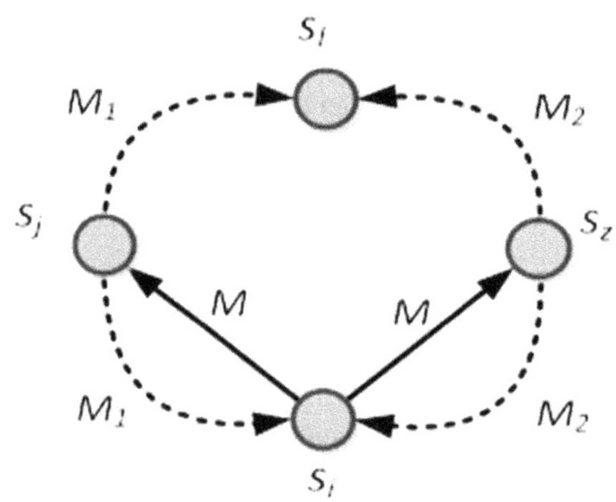

Neighbor coordination-based approaches: In such approaches, a node compares its reading with the neighboring nodes and outcome of the comparison decides whether its behavior is faulty or not. The node can simply exchange readings with its one hop neighbors, or the region can be decided based on the sensing phenomena's characteristics. In such schemes, nodes find their state locally, thus these approaches reduce message overhead significantly and conserve nodes energy. They are considered more suitable for wireless sensor networks.

The neighbor coordination-based approaches exploit the fact that non faulty nodes in proximity measures correlated readings while faulty nodes measure uncorrelated readings. For example, suppose s_i and s_j are two neighboring nodes and measure readings X_i^t and X_j^t respectively at time t. Both sensor readings are considered correlated if $| X_i^t - X_j^t | < \theta$, where θ is application dependent threshold. For example, in temperature recording application the variation in temperature measured by two neighboring nodes may be too less, then θ is expected lesser. In event detection applications, the measurement is discrete i.e. sensor measures 1 if an event happens otherwise 0. In such applications the θ is selected as 0. For event detection applications, each node s_i finds its correlation C_{ij} with each of its neighboring nodes s_j as: C_{ij}=0, if $| X_i^t - X_j^t | < \theta$. Otherwise, C_{ij} is set as 1. Thereafter, each node transmits this locally found correlation to each of its neighbors. If a node finds that most of its neighbors have C_{ij}=1, then it considers its status as faulty otherwise good. The Figure 6 depicts the concept more clearly.

Figure 7. Fault Detection using neighbour coordination approach

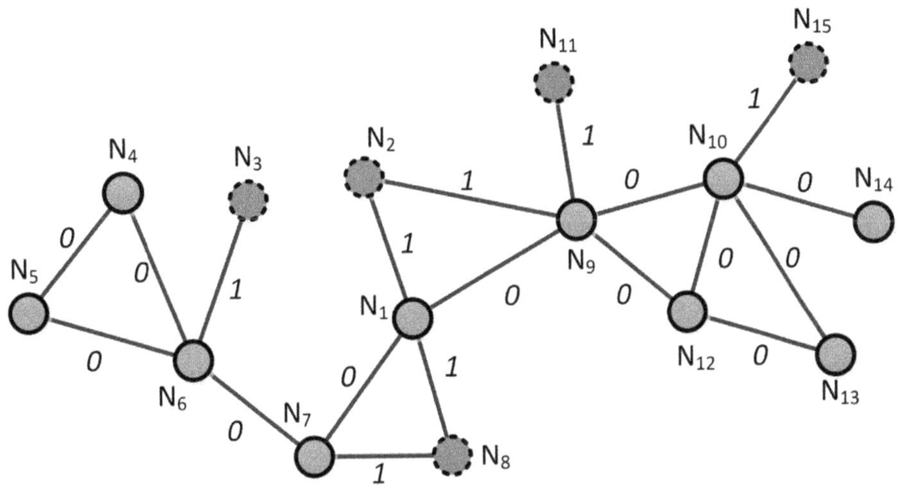

Jinran Chen *et al.* (Chen, Kher, Somani, & Arun, 2006) propose a distributed fault detection (DFD) scheme. The main aim of DFD is to find faulty sets of sensor nodes in a WSN locally with comparatively low processing complexity and hight detection accuracy. In this scheme, the sensed data is communicated amongst adjacent nodes intermittently for some rounds. The status of nodes decided based on the correlation between the data of the node and its neighboring nodes. The scheme has four phases. In first phase, comparison C_{ij} is obtained based on the readings of node s_i and readings of each of its neighbors s_j. The approach finds difference in readings at time t and t+1 as ΔX_{ij}^{t} and ΔX_{i}^{t+1} respectively, where $\Delta X_{ij}^{t} =| X_i^t - X_j^t |$ and $\Delta X_{ij}^{t+1} =| X_i^{t+1} - X_j^{t+1} |$. Subsequently, a combined outcome of temporal and spatial correlation is found as: $\Delta X_{ij}^{\Delta t} = | \Delta X_{ij}^{t+1} - \Delta X_{ij}^{t} |$. The $\Delta X_{ij}^{\Delta t}$ is calculated, if $\Delta X_{ij}^{t} > \theta_1$. The C_{ij} is set as 1, if $\Delta X_{ij}^{\Delta t} > \theta_{2,}$ otherwise it is set as 0. If a node s_i finds most of its neighboring nodes correlated (i.e. C_{ij}=0), s_i marks the status as locally good (LG) if not then it marks it as possibly faulty (PF). In the next round, all comparisons are done with the nodes in LG state and faulty nodes are diagnosed based on the comparison outcomes. Similarly an approach proposed by Peng Jiang in (Jiang & Peng, 2009) also focusses on distributive fault detection scheme. It evaluates that if the determinant condition of DFD is too harsh, it can alert a false alarm rate. Thus, in order to reduce false alarm rate, the author presented this approach by improving the determinant condition. In this approach, the node s_i initially sets C_{ij}=0 for each node s_j. If $\Delta X_{ij}^{t} > \theta_{1,}$ s_i calculates ΔX_{ij}^{t} and C_{ij} is set as 1. Otherwise, if ΔX_{ij}^{t} £ θ_1, $\Delta X_{ij}^{\Delta t}$ is calculated. The C_{ij} is set to 1 if $\Delta X_{ij}^{\Delta t} > \theta_2$. After that, the majority voting approach is applied to find the faulty nodes in the network.

J.-Y. Choi *et al.* (Choi, Yim, Huh, Choi, & Yoon-Hwa, 2009) suggest a fault detection approach very similar to (Chen, Kher, Somani, & Arun, 2006). The scheme utilizes the temporal data correlation to find the transient faults. In this approach the degree of the node and the threshold θ are updated dynamically. The degree is calculated by reducing the count of faulty nodes from the initial degree of network. Whereas, the threshold θ is updated as θ=*max* (x, *degree*/2), where x is randomly selected. The scheme executes many rounds in order to find the faulty nodes, hence, has comparatively more message overhead.

Yang *et al.* (Yang, Gao, Zhou, & Qiu, 2014) analyze the approaches proposed in (Chen, Kher, Somani, & Arun, 2006) and find that the false alarm rate and detection accuracy of the sensor node is poor when a node at doubt has same count of nodes having Locally Good or Possibly Faulty status in its neighbor. The high false alarm rate and low detection accuracy with some missing measurement values seriously influenced the performance of the fault detection approaches. Thus, the author presented the Uncertainty-based Distributed Fault Detection algorithm (uDFD). The scheme applies Markov decision processes in order to recover the missing measurement. In addition to the Markov decision process, it enhances the accuracy of the process by applying entropy based and disagrrement function based evidence fusion rules to detect missing data

There are also numerous correlation-based schemes reported in the literature that consider other information such as remaining node power, node degree, and node distance in addition to sensed data to improve the accuracy of error detection. M. N. Sahoo et al. (Sahoo & Khilar, 2014)suggest a Fault Diagnosis Algorithm (FDA) which is based on remaining node power and measured data to handle both permanent and intermittent faults. Likewise, fault detection algorithm centered on the median of recorded data is proposed by M. Ding *et al.* (Ding, Chen, Xing, & Cheng, 2005). The suggested approach is localized and incurs less computational overhead. In (Guo, Zhong, He, & Tian, 2009), authors suggest a similar approach called FIND. In this approach, sensor readings and the distance of the node from the event are utilized to calculate the ranks and further assign it to the remaining nodes in the network. According to the scheme, if there is a significant discrepancy between the sensor data rank and the distance rank, a node is considered defective.

There are few centralized schemes that have also been proposed in the literature. The approach proposed by C. Jaikaeo *et al.* addresses the implosion condition during fault diagnosis in sensor networks (Jaikaeo, Srisathapornphat, & Shen, 2001, June). Similarly, in (Lee, Datta, & Cardell-Oliver, 2006), W. Lee et al. suggest a supplementary approach but the suggested approach faces the problem of saclability, failing to be deemed appropriate for larger networks. In Ruiz *et al.*, management architecture based approach is proposed (Ruiz, Siqueira,, Wong, Nogueira, & Loureiro, 2004). Similar to Lee *et al.*, this scheme follows the idea of fault manager (Lee, Datta, & Cardell-Oliver, 2006). In this system, the manager is outside the network and has regional knowledge.

Most of the schemes available in the literature communicate the messages in order to look at the measured data leading to higher energy consumption and bandwidth. To detect faulty nodes, after each defined interval, there is an exchange information amongst all nodes in the network that suddenly increases the network traffic and can affect the network's underlying performance. Often, if the network has a high ratio of faulty nodes, maximum schemes lack accuracy in the detection of faults. Table 3 provides a comprehensive overview of different approaches to the identification of faults.

ISSUES AND FUTURE DIRECTIONS

1 The available approaches to detect faults are generally conducive to high accuracy of detection of faults with low false alarm rate. Many schemes use the spatial and temporal correlation in sensed information. Nodes with correlated and uncorrelated data readings are found by communicating messages with neighboring nodes. Thus, these schemes generally have high message overhead. The number of messages exchanged is also independent of the probability of node fault. Therefore, too many messages are always exchanged in order to find the right node status even if no node in

Table 3. Comparison of various fault detection approaches

Reference and year	Major contributions	Diagnosed faults	Centralized/ Distributed
J. Chen *et al.* (Chen, Kher, Somani, & Arun, 2006)(2004)	· The message complexity is $n\ (r + d + 4) - 2$, where r depicts the number of rounds, d is the network's degree. · The scheme uses both temporal and spatial correlations.	Crash, omission	Distributed (Neighbor Co-ordination)
P. Jiang et al., 2009	· Extension of [6]. · Ability to detect faulty nodes in networks having less node degree. · Exhibits 0.992 detection accuracy.	Inaccurate calculation	Distributed (Neighbor Co-ordination)
Guo *et al.* (Guo, Zhong, & He, 2009)	· Considers the monotonically attenuating phenomena over distance. · Needs no prior knowledge about the sensed phenomena.	Omission, Incorrect computation	Centralized (Comparison based)
(L.B. Ruiz *et al.* 2004)	· The failure identification system uses an event-driven WSN management model. · Supports self-alignment, self-identification and self-recovery.	Crash	Centralized
(W.L. Lee *et al.*, 2006)	· Ability to handle network dynamics. · Fault management data is disseminated and collected through a schedule driven MAC protocol. · Systematic resource transfer for allowing resources sharing in various parts of networks.	Crash	Centralized
(S. Chessa *et al.*, 2001)	· Ability to diagnose system level faults in ad-hoc networks. · Can deal with time varying and fixed topology networks. · Scheme is k-1 diagnosable in a k-connected network.	Crash, omission, incorrect computation	Distributed (Comparison based)
(M. Elhadef *et al.*, 2006)	· Two separate approaches adaptive-DSDP and Mobile-DSDP for fixed and time varying topology respectively. · Adaptive-DSDP message complexity is $5n - 6 + nd$, where d is the nodes degree. · The message complexity of Mobile-DSDP is $n\ (n + d)$	Crash, omission, incorrect computation	Distributed (Comparison based)
(C. Jaikaeo *et al.*,(2001)	· Three approaches presented to handle the response implosion. · Adaptive probabilistic response.	Crash	Clustering approach
(M. Ding et al.,2005)	· Uses real numbers to model the abnormal event behaviors. · Exhibits 0.91 detection accuracy when fault rate is 0.2 and d =30.	Incorrect computation	Distributed (Neighbor Co-ordination)
(X. Xu *et al.*,2014)	· A low energy-based scheme. · Ability to detect faulty nodes in sparse networks.	Crash Omission Incorrect computation	Distributed (Neighbor Co-ordination)
(M.N. Sahoo *et al.*, 2014)	· Utilizes nodes energy · Ability to detect compromised nodes.	Crash Fail-stop Omission	Distributed (Neighbor Co-ordination)
(Y. Yang *et al.*,2014)	· An uncertainty-based scheme is proposed · The scheme corrects the missing data in channel by using historical data · Low false alarm rate and high accuracy · Uses Markov decision process to fill the missing data.	Crash Omission Incorrect computation Fail-stop	Distributed (Neighbor Co-ordination)
(M. Elhadef *et al.*, 2007)	· Uses gossip-style failure detection. · Heartbeat message-based failure detection. · Dynamically updates the arrival time of heartbeat messages.	Crash Fail-stop	Distributed (Comparison based)
(J.-Y. Choi *et al.*, 2009)	· Closely follows the conditions of [6]. · The scheme dynamically changes the network degree and threshold value to adopt present network's scenarios	Incorrect computation	Distributed (Neighbor Co-ordination)

the network is faulty. Therefore, an effective scheme that involves less exchange of messages for a few faulty nodes in the network needs to be developed.

2 Fault detection accuracy sharply dips in existing fault detection schemes as the likelihood of faulty nodes increases. Also, some sensing phenomena like heat, RF emission, audio waves, etc. attenuate with increasing distance and hence in such scenario's nodes at different locations from source may read different values for the same phenomenon at a given time. Hence, readings from such nodes sensing the same phenomena may falsely be treated as uncorrelated by existing fault detection approaches. Thus, a scheme that can negotiate the impact of signal attenuation due to distance to reduce the false alarm is needed.

3 The quality of failure recovery schemes depends entirely on error handler selection criteria. Most of the available schemes like DARA, PADRA, RIM and LeDiR decide failure handler based on either 1-hop or 2-hop information. Based on this partial information, these schemes only can find the optimal solution if the failure handler is available within 2-hop range.

4 Most approaches suggested in literature only address the issue of connectivity while addressing node failures and ignore the sensing coverage. Due to dense deployment of nodes, WSNs have overlapping sensing and communication regions. Consequently, there are risks that sensing coverage of a failed node overlaps with some nearby neighboring node's coverage. In such cases, if neighbor node is selected as a back-up node, it can cover failed node's sensing region and thereby avoiding the movement of a mobile node from some remote location. In the absence of such nodes, the coverage can be recovered by exploiting movement.

SUMMARY

In this chapter, the researchers discussed related approaches for fault identification for underwater wireless sensor networks. Various faults are discussed and grouped together based on various parameters. In the last section, the node failure and topology reconstruction techniques are discussed and classified. Based on the literature review some issues and research direction are listed at the end of the discussion. In Conclusion, it can be stated that UWSNs face several challenges due to the design constraints and hence, effective schemes that involve less exchange of messages for a few faulty nodes in the network needs to be developed. Also, underwater channel suffers from attenuation severely, hence, a scheme that can negotiate the impact of signal attenuation due to distance to reduce the false alarm is needed. Most approaches suggested in literature only address the issue of connectivity while addressing node faults and ignore the sensing coverage.Hence, a new scheme of fault detection considering sensing coverage can be devised.

REFERENCES

Akyildiz, I. F., Pompili, D., & Melodia, T. (2005). Underwater acoustic sensor networks: Research challenges. *Ad Hoc Networks*, *3*(3), 257–279. doi:10.1016/j.adhoc.2005.01.004

Avizienis, L., Laprie, J.-C., Randell, B., & Landwehr, C. (2004). Basic concepts and taxonomy of dependable and secure computing. *IEEE Transactions on Dependable and Secure Computing*, *1*(1), 11–33. doi:10.1109/TDSC.2004.2

Barborak, D., Dahbura, A., & Malek, M. (1993). The consensus problem in fault-tolerant computing. *ACM Computing Surveys*, *25*(2), 171–220. doi:10.1145/152610.152612

Benson, B. (2010). *Design of a Low-cost Underwater Acoustic Modem for Short-Range SensorNetworking Applications* (Ph.D. Thesis). University of California, San Diego, CA, USA.

Blough, B., & Brown, H. W. (1999). The broadcast comparison model for on-line fault diagnosis in multicomputer systems: Theory and implementation. *IEEE Transactions on Computers*, *48*(5), 470–493. doi:10.1109/12.769431

Bondavalli, C., Chiaradonna, S., Di Giandomenico, F., & Grandoni, F. (2000, March). A., Di Giandomenico, S., Grandoni, F., & Fabrizio. (IEEE). Threshold-based mechanisms to discriminate transient from intermittent faults. *IEEE Transactions on Computers*, *49*(3), 230–245. doi:10.1109/12.841127

Cato, D. (2008). Ocean ambient noise: Its measurement and its significance to marine animals. *Proceedings of the Institute of Acoustics. Institute of Acoustics (Great Britain)*, 30.

Chen, K. J., Somani, S., & Arun. (2006). Distributed fault detection of wireless sensor networks. In *Proceedings of the 2006 workshop on Dependability issues in wireless ad hoc networks and sensor networks* (pp. 65-72). ACM. 10.1145/1160972.1160985

Chessa, S. S., & Paolo. (2001). Comparison-based system-level fault diagnosis in ad hoc networks. In *Proceedings 20th IEEE Symposium on Reliable Distributed Systems* (pp. 257-266). IEEE. 10.1109/RELDIS.2001.970776

Choi, Y., J.-Y., Huh, S.-J., Choi, Y., & Yoon-Hwa. (2009). A distributed adaptive scheme for detecting faults in wireless sensor networks. *WSEAS Transactions on Communications*, *8*(2), 269-278.

Ding, M., Chen, D., Xing, K., & Cheng, X. (2005). Localized fault-tolerant event boundary detection in sensor networks. In *IEEE 24th Annual Joint Conference of the IEEE Computer and Communications Societies.* (pp. 902-913). IEEE.

Elhadef, B. M., Elkadiki, A., & Hisham. (2006). Diagnosing mobile ad-hoc networks: two distributed comparison-based self-diagnosis protocols. In Diagnosing mobile ad-hoc networks: two distributed comparison-based self-diagnosis protocols (pp. 18-27). ACM.

Elhadef, B. M., & Azzedine. (2007). A failure detection service for large-scale dependable wireless ad-hoc and sensor networks. In *The Second International Conference on Availability, Reliability and Security (ARES'07)* (pp. 182-189). IEEE. 10.1109/ARES.2007.3

Elhadef, Boukerche, M., & Elkadi, A. (2008). A distributed fault identification protocol for wireless and mobile ad hoc networks. *Journal of Parallel and Distributed Computing,* (3), 321-335.

Goyal, N., Dave, M., & Verma, A. K. (2016). Congestion control and load balancing for cluster based underwater wireless sensor networks. In *2016 Fourth International Conference on Parallel, Distributed and Grid Computing (PDGC)* (pp. 462-467). IEEE. 10.1109/PDGC.2016.7913240

Goyal, N., Dave, M., & Verma, A. K. (2017). Trust model for cluster head validation in underwater wireless sensor networks. *Underwater Technology, 34*(3), 106–113. doi:10.3723/ut.34.107

Guo, S., Zhong, Z., & He, T. (2009). FIND: faulty node detection for wireless sensor networks. In *Proceedings of the 7th ACM conference on embedded networked sensor systems* (pp. 253-266). ACM. 10.1145/1644038.1644064

Haeberlen, Kouznetsov, A., Druschel, P., & Peter. (2006). The Case for Byzantine Fault Detection. *HotDep.*

Harland, E., & Richards, S. (2006). *SEA 7 Technical report: Underwater ambient noise (Non-technical summary).* Report by QINETIQ.

Headrick, R., & Freitag, L. (2009). Growth of underwater communication technology in the US Navy. *IEEE Communications Magazine, 1*(47), 80–82. doi:10.1109/MCOM.2009.4752681

Heidemann, J., Stojanovic, M., & Zorzi, M. (1958). Stojanovic, M., & Zo, M. (2012). Underwater sensor networks: Applications, advances and challenges. *Philosophical Transactions of the Royal Society of London. Series A, Mathematical and Physical Sciences, 370*(1958), 158–175. doi:10.1098/rsta.2011.0214

Ismail, N.-S., & Hussein, L. (2010). Analyzing the performance of acoustic channel in underwater wireless sensor network (UWSN). In *2010 Fourth Asia International Conference on Mathematical/Analytical Modelling and Computer Simulation* (pp. 550-555). IEEE. 10.1109/AMS.2010.111

Jaikaeo, C., Srisathapornphat, C., & Shen, C. (2001, June). Diagnosis of sensor networks. In *IEEE International Conference on Communications. Conference Record (Cat. No. 01CH37240)* (pp. 1627-1632). IEEE.

Jiang & Peng. (2009). A new method for node fault detection in wireless sensor networks. *Sensors, 9*(2), 1282-1294.

Jr, Duarte, Ziwich, E., Albini, R., & Albini. (2011). A survey of comparison-based system-level diagnosis. *ACM Computing Surveys, 43*(3), 22.

Lanbo, L., Shengli, Z., & Jun-Hong, C. (2008). Prospects and problems of wireless communication for underwater sensor networks. *Wireless Communications and Mobile Computing, 8*(8), 977–994. doi:10.1002/wcm.654

Lee, W. L., Datta, A., & Cardell-Oliver, R. (2006). *Winms: Wireless sensor network-management system, an adaptive policy-based management for wireless sensor networks.* Academic Press.

Malek & Miroslaw. (1980). A comparison connection assignment for diagnosis of multiprocessor systems. In *Proceedings of the 7th annual symposium on Computer Architecture* (pp. 31-36). ACM.

Min, H., Cho, Y., & Heo, J. (2012). Enhancing the reliability of head nodes in underwater sensor networks. *Sensors (Basel), 12*(2), 1194–1210. doi:10.3390120201194 PMID:22438707

Mohamed, N. J., Jawhar, I., Al-Jaroodi, J., & Zhang, L. (2011). Sensor network architectures for monitoring underwater pipelines. *Sensors (Basel), 11*(11), 10738–10764. doi:10.3390111110738 PMID:22346669

Perrine, K., Nieman, K., Henderson, T., Lent, K., & Evans, B. (2010). Doppler estimation and correction for shallow underwater acoustic communications. In *2010 Conference Record of the Forty Fourth Asilomar Conference on Signals, Systems and Computers* (pp. 746-750). IEEE. 10.1109/ACSSC.2010.5757663

Preparata, M., Metze, G., & Chien, R. T. (1967). On the connection assignment problem of diagnosable systems. *IEEE Transactions on Electronic Computers*, *6*(6), 848–854. doi:10.1109/PGEC.1967.264748

Ruiz, L. B. S. I., Wong, H. C., Nogueira, J. M., & Loureiro, A. A. (2004). Fault management in event-driven wireless sensor networks. In ACM international symposium on Modeling, analysis and simulation of wireless and mobile systems (pp. 149-1567). ACM.

Sahoo, M. N., & Khilar, P. M. (2014). Diagnosis of wireless sensor networks in presence of permanent and intermittent faults. *Wireless Personal Communications*, *78*(2), 1571–1591. doi:10.100711277-014-1836-6

Stojanovic, M., & Preisig, J. (2009). Underwater acoustic communication channels: Propagation models and statistical characterization. *IEEE Communications Magazine*, *47*(1), 84–89. doi:10.1109/MCOM.2009.4752682

Yang, Y., Gao, Z., Zhou, H., & Qiu, X. (2014). An Uncertainty-Based Distributed Fault Detection Mechanism for Wireless Sensor Networks. *Sensors (Basel)*, *14*(5), 7655–7683. doi:10.3390140507655 PMID:24776937

Zhang, X., Han, X., Yin, J., & Sheng, X.-l. (2013). Study on Doppler effects estimate in underwater acoustic communication. In *Proceedings of Meetings on Acoustics ICA2013* (p. 070062). ASA. 10.1121/1.4799512

Chapter 8
Firewall in Underwater Wireless Sensor Networks

Manni Kumar
Chitkara University Institute of Engineering and Technology, Chitkara University, Punjab, India

Osho Gupta
Chitkara University Institute of Engineering and Technology, Chitkara University, Punjab, India

Shikha Rani
National Institute of Technology, Kurukshetra, India

ABSTRACT

In this technological arena, a firewall is a major tool for network security system to predetermine its basic rules so that it can control and monitor incoming and outgoing network traffic. The firewall generally restricts the faulty and unusual data entering the system. It can be categorized as host-based firewalls and network-based firewalls. There are various applications of firewall and how these firewalls can help the underwater wireless sensor network (UWSN) in protecting the sensor nodes where a huge amount of sensitive data is communicated among different countries. Firewall mainly aims to secure the network from different attacking devices and the attackers in a very efficient way. Firewall creates a barrier in which only the authorized data can be passed through after continuously checking by the firewall policies. To protect a local system and network of system from threats which are generally network based firewall is the most effective way. The design of the firewall is required to be accurate as it acts as a filter at higher protocol layer and some levels of IP-packets.

INTRODUCTION

It has been observed a steady evolution in the information system of corporate sector, government agencies or any private organizations. Some of the developments that came to see are to support directed connected terminals in which there is a main data processing system which have its own central mainframe. Use of LAN's helps to interconnect PC's in an organization and also all the terminal units

DOI: 10.4018/978-1-7998-3640-7.ch008

(Alves et al., 2012). Network premises which consists of interconnected LAN's, PC's, servers, and the connected mainframe unit. Wide Area Network (WAN) consists of wide networks that are interconnected to each other. Every employee working in an organization need internet services to fulfil their task and if it is not provided by the company they can get a dial-up connection from the Internet Service Provider (ISP). Internet gives every facility that is required by the users but the main problem is data security on the internet (Barbeau et al., 2014; Goyal et al., 2016; 2017). There is huge amount of data present on the internet the internet service providers must ensure for the proper use of that data. Similarly in case of Underwater Wireless Sensor Network (UWSN) as network providers are increasing day by day, they don't have enough space to install their devices and communication lines so for that a unique concept of communication have been introduced that is UWSN. In UWSN a network is installed deep under the sea water to provide communication facility to different people among different countries. In this chapter we will discuss about firewalls and how firewalls can help to protect UWSN from various attacks (Chouiten et al., 2012; Goyal et al., 2018).

Generally firewalls works on the rule defined in them when a data enter in the system it checks the data with its defined table, if the data is secure which meets the requirements policy of the firewall it allows the data to enter. Every firewall has its default policy and every default policy has three actions (reject, accept and drop).Suppose when there is no rule defined to the server about the SSH connection the firewall will follow the default policy. Only the default policy allows the system to establish a connection to the server, if it is set to allow the other users can easily establish a connection to the server line therefore always set default policy as drop which is a great exercise (Dee et al., 1995; Goyal et al., 2019a).

The mentioned authors also discuss about various filtering techniques or types that can be implemented to protect data entering the sensor nodes and also how each firewall types have their own features of protecting the data. The main area to discuss is to protect UWSN from various kinds of external unusual activities or noisy data present in the data stream, this can only be secure with the help of the various firewalls which help to protect the noisy data to enter or pass through the nodes making a good way to transfer data in a decent way (Goyal et al., 2019b; Feldstein et al., 2008). This firewall also helps to protect against various attacks that can be enforced or performed in UWSN. Some of the attacks like attaching viruses or malware with the data packet, hijacking the sessions and get fully access to the nodes in which data is transmitted can be discontinued with the help of the firewall. So firewall plays a vital role to protect the network. Sometimes the attacker tries to overload the traffic by adding more number of data packets to a particular path that leads to slow down of the network. In this case firewall helps to discard the unwanted data packets from the path making it easy for the actual data packets to pass easily. The necessity of firewall is increasing as it provides the single point protection where inspecting and security of data can be imposed. It also helps to provide the log details about each and every event occurring within the nodes and that log contains information about what type and how much traffic is there and can only be accessible to administrator. Thus this is an important point as the firewall serves the same purpose as armed guard.

THE REQUIREMENTS OF THE FIREWALL

The giant cyber security professionalism of NCSA studied that only 4 percent of Americans have a proper knowledge of firewalls and rest around 90 percent they are completely unaware of the firewall. Sometimes without having the enough knowledge about firewall they set there firewall to off or sometimes they even not know where to make changes of the firewall which makes the system more vulnerable to these attacks like viruses, malicious and phishing. When your PC connects with the computer there are potentially more chances that your PC was harmed by some cyber threats like hackers, key loggers and Trojans which put a deep impact on the system. Firewall act as a barrier between the user system and the internet which allows you for the transmission and receiving of data in the form of small block called packets (Forrest et al., 2010; Goyal et al., 2019c). Each packet while entering at a receiving end was filtered by the firewall and only the actual data will be allowed to enter the system, the filtration process of the data has been taken place by various set of rules defined by the firewall and if the data meets the requirements of that rules or policy it will be marked as the authentic data. By this firewall helps people of not stealing their sensitive data like bank account numbers and passwords by the hackers. Firewalls we use in windows like windows XP, windows VISTA only control the data entering into the system but the best practices for the firewall is that it can control data coming and outgoing in both the direction. The basic firewall only filters data that is entering the system but there is not security of outgoing data like when you are doing some type of online transaction that transaction will not be considered to be safe. As shown in the diagram given blow the use of firewall in UWSN in this various sensor nodes that are deployed deep under the sea water sent data in form of packets to the base station or receiving node which is called as Autonomous underwater vehicle (AUV) the firewall which act as barrier between them only sent actual data or the data which meets the requirements of firewall policy. By this way it becomes very difficult for the attackers to attacks and they are having very low chances of sending any type of malicious or Trojans to the receiving system (Jacka et al., 2012; Khurana et al., 2012).

As in UWSN there are many sensor that are deploy deep under the sea water in which huge amount of sensitive data is transferred or received at a particular intervals of time. For the proper security of data, firewall plays a vital role in it. Firewall helps to secure data from the various attackers that are trying to fetch data travelling in the sensor nodes.to make sure each sensor gets the actual data that is received from the other sensors firewall filters the each data packet and according to the policy setup by the firewall administrator only that data which fulfill the security policy can allow to pass through the firewall else it will be stated as the garbage or effected data (Kamoun et al., 2019; Khurana et al., 2016; Goyal et al., 2020).

In UWSN various autonomous vehicles can also be used which helps for the proper communication of the data from the source to the user end. Submarines are one of the examples of autonomous vehicles that help to receive data from the various devices which are present deep under the sea water and send response accordingly. But it is quite a major task to make sure that the data received in the system network that is present in the submarines is freed of any type of malicious and viruses. So for that software based firewalls have been used these firewalls have been installed on a separate system instead of using it in the hardware. The best part of the software firewall is, it isolates each and every system endpoints present in that network creating a wonderful and efficient defense system (Liu et al., 2018; Kumar et al., 2014).

Packet Filtering is the type of the firewall that can used to filter the data similarly when this firewall is placed between various sensor nodes it will helps in data filtering during the time of filtering various process can be taken place firstly when a data packet enters the filtering firewall the header attached with data packet which consists of sensitive information of the data packet is processed and get matched with policy implemented by the firewall admin. If the data present in the header fulfil the required admin policy it allows the data packet to pass the next neighboring sensor nodes if still due to some reasons the data packet will not able to pass it will be stated as effected data (Nightingale et al., 2002; Lisa 2013). But there exist some limitations of using software firewall in AUV like

1. Not every firewall is compatible with the system software so their might be some compatibility problems for some of the systems.
2. It is not easy to maintain, it might be difficult to install and maintain firewall on the systems.
3. It is time consuming process.

Figure 1. Requirements of firewall in UWSN

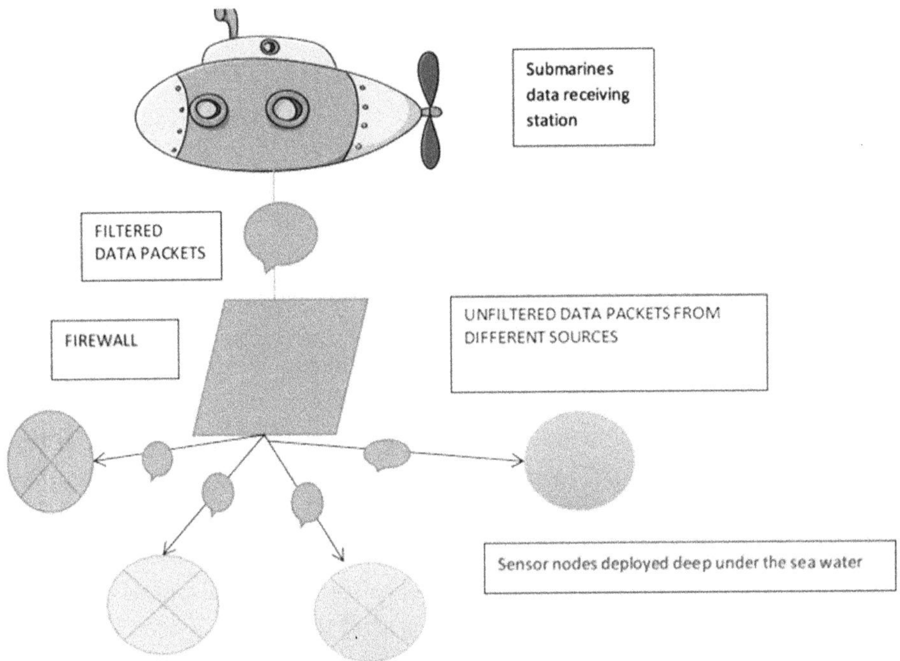

FIREWALL CHARACTERISTICS

For controlling the access and enforcement of various sites present on the internet firewall uses four techniques which are listed below (Peng et al., 2011):

- **Service Control:** Service control helps to determine which type of internet services can be easily accessed or which are blocked according to the firewall policy. Two parameters are used by firewall to filter traffic named as TCP port number and IP address. Sometimes proxy software has been used to illustrate and receives request before passing it on or may use particular hosting server software like web services and mail services.
- **Direction Control:** Every service request that is generated must flow in some particular direction to the firewall which is always determined or decided by the direction control.
- **Behavior Control:** It generally control the behavior of every process that has been taken place in firewall during the filtering of data packets.

CAPABILITIES OF FIREWALL

A firewall helps to secure various services from the unauthorized users and prevent to enter various potentially vulnerable services that can harm systems up to a larger extent. Routing attacks and IP spoofing can also be cured by the firewall. Every security related event can be easily monitored with the help of the firewall we can also implement alarms and audits on the firewall system. Basically the main capability of the firewall is to act as barrier between two systems and only allow the authorized data to enter the system without making any harm to it (Pelizzari et al., 2004). There exist mainly 3 types of firewalls

1. Packet filtering router
2. Application level gateways
3. Circuit level gateways

Packet Filters: It generally works on the network layer of the OSI model with the application of certain rules on every packet based on the results it allows or discard the packets entering in the system. The three types of packet filtering firewalls are (i) Static filtering (ii) Dynamic filtering and (iii) Stateful inspection

Static filtering is considered to be the one of the oldest filtering method that depends or operates on the network layer. It generally scans the underwater data packets IP header and TCP header part. It is advantageous because of low impact on network performance and less cost of maintenance. Although it is disadvantageous because it examines less information of the data packet (i.e. only IP and TCP information) and packet payload cannot be easily examined. It is very less protective technique.

Dynamic filtering firewalls can be considered to be the more advance and efficient than the static firewalls. It generally makes the decision to allow or deny the packet based on the information like IP and protocol header. In UWSN when the data packet is send through one node to another node before entering the another node the dynamic firewall makes must check the IP and protocol information of the data packet header attached to it. If the firewall decision requirements are fulfill then it must pass the packet to the other node without denying it. It is advantages because of low cost, increase in performance, security levels and maintenance. It is disadvantageous because of less information gathering and less security in certain cases.

Stateful inspection filtering firewall is similar to static filtering with the only difference that this firewall checks or inspect the data packet information more accurately and in a proper way, making it possible to enter only required and noisy free data to pass through the nodes.

Security threats to packet filtering:

IP address spoofing: IP address spoofing is the kind of attack in which the attacker tries to send packet to a company network with the same IP address of the source system and internal computers working in an organization.

Source routing attacks: These types of attacks are often performed in which there is a predefined route is given to the packet to fool the firewall.

Tiny fragmented packets: When the data is transfer in the form of packets sometimes it happens that the size of the IP data packet is to large that it is not able to enter to a particular network to resolve that the IP packet is divided into smaller packets similarly in this way the attacker breaks the data packet into smaller fragments such that they are able to fool the firewall.

Figure 2. Packet filtering router

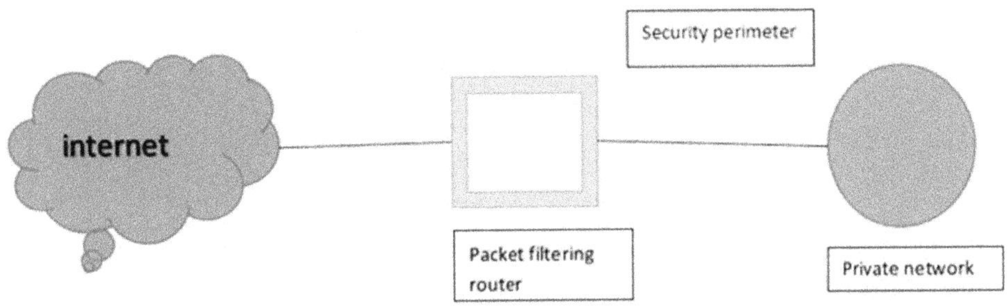

Application level gateway: Another name of the application level gateway can also be termed as Proxy server. All the application level traffic is controlled by the application level gateways. The TCP/IP application helps the user to interfere with the gateways, at that time for the purpose of security gateway ask the user to enter valid details of the user they want to access. While entering the contact details by the user the user then able to access the remote user he wishes to access and there is a proper arrangement of TCP segments which contains data between the two end points. Application level gateway is considered as more secure as compared to packet filters because it is easy to audit and log all application level incoming traffic.

Figure 3. Application level gateway

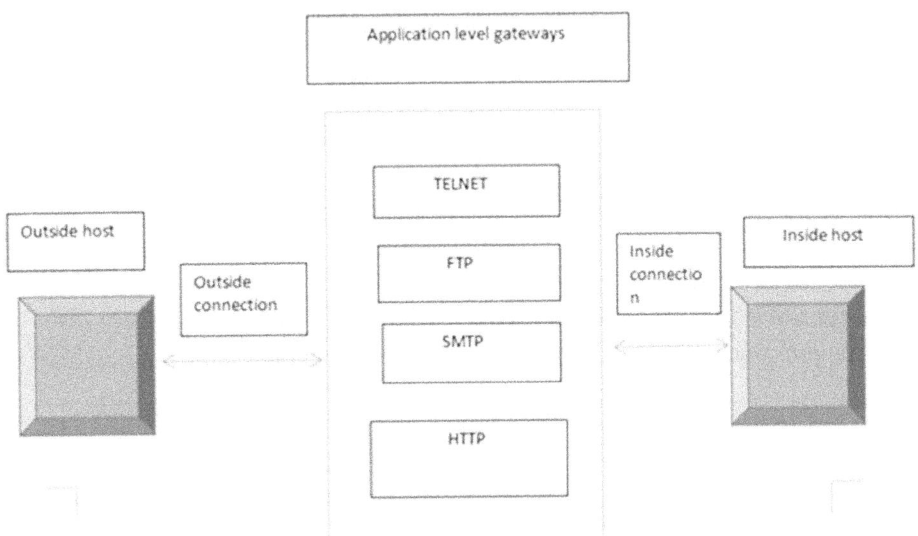

Circuit level gateway: In circuit level gateway two TCP connections has been established and there is no application of end to end TCP connection. The two TCP connections have been established between itself and TCP outer user and another with same but the outer user. A circuit level gateway can also be considered a situation where the admin trust the internal users.

Figure 4.Circuit level gateway

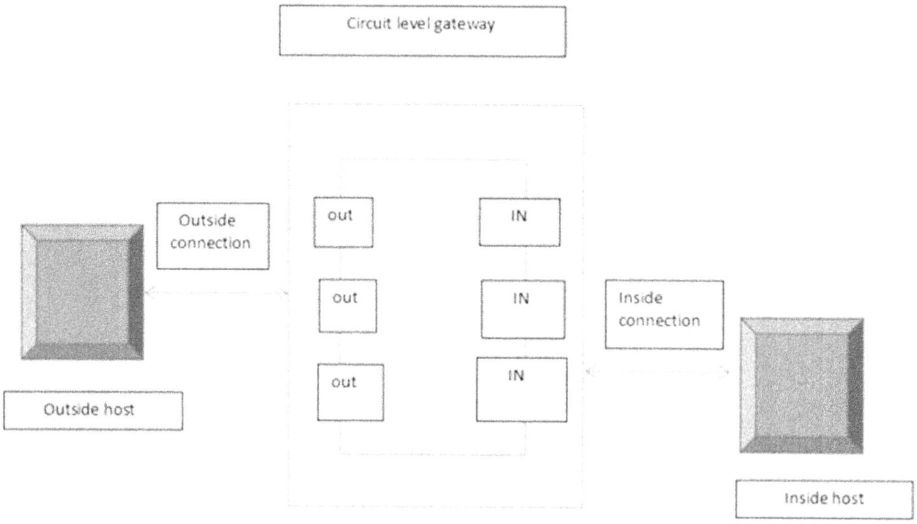

FIREWALL BASING

It is very basic to implement a firewall on a machine or a particular type of operating system such as Linux or Unix but we can also implement the firewall on any type of software module like switches and routers. Some of the basic firewall considerations are (Tuna et al., 2017):

Bastion host: It acts as a platform for the circuit level and the application level gateway. To identify a critical point in a network system a bastion host is used by the firewall administrator. Some of the characteristics of this are:

- The hardware platform of the bastion host uses such a secure operating system which helps to increase its security level and making it more hardened.
- The network administrator could install only those services on the bastion host which are needed frequently and of more uses on daily bases or which are considered as essential.it includes proxy application for SMTP, HTTP, FTP and DNS.
- To access the proxy services the bastion host must have authentication data in order to access the services.
- The configuration of the proxy is done in such a way that only the allowed users can access to the network.

ADVANCEMENT IN FIREWALL SYSTEM

In UWSN, secure transmission of data is foremost required. This network consists of a secure path, a parent node that controls all other sensor nodes. For the security of parent node the concept of DMZ (Demilitarized zones) is introduced. DMZ consists of zones in the firewall system. These zones are secure enough that even an attacker cannot bypass unauthorized packets through these zones. DMZ established secure policies for the data making it encrypted and secure. Basically DMZ consists of a layer structure system and each layer has its own confidential levels like high, medium or low. Some of the systems are also using numeric values ranging 0 to 100. DMZ helps to protect the parent nodes from various attacks making it secure to tackle various problems occurring at the time of transmission. Due to its layered structure, it becomes very difficult for the hacker to add noisy data in the nodes or the network. The attacker have to pass all these 3 layers to reach the parent node that makes more challenging or time consuming process to access the parent node. When the data packet enters the node, packet has to undergo various types of processes defined in the firewall system. DMZ process defined the 3 layer structure for every packet as detailed below stepwise:

1. While passing the first layer of the DMZ system by the data packet, it is required to put into general scanning process. The system check for the type of data in the packet.
2. After passing from the first layer it goes to the second layer where it undergoes the processing of the header part of the data packet that includes the sensitive information about the data packet.
3. Finally the packet reaches the third layer, this is the most secure layer and also considered as decision layer that decides whether the packet have to pass or deny and also check the unwanted data present in the data packet.

So this method concludes that if a packet is not able to fulfil the requirement of all the 3 layers, it will be discarded from the packet queue and is considered to be as the unauthorized data. The below figure represents the above process of DMZ.

THREATS AND ATTACKS IN FIREWALL BASED SYSTEM

If firewall is not installed or doesn't exist in the system then there may occur various attacks in UWSN as listed below:

1. DoS Attack
2. Port Forwarding
3. Security of IoT Devices

So to avoid these attacks, firewall is a necessity to install.

CHALLANGES IN FIREWALL SYSTEM

There exist various challenges for the firewall system as below:

1. The time consuming process
2. Compatibility issues of firewall w.r.t to operating systems
3. Power variations in the system can cause damage to firewall
4. Timely updating the firewall system is much more required

Figure 5. DMZ layer structure process

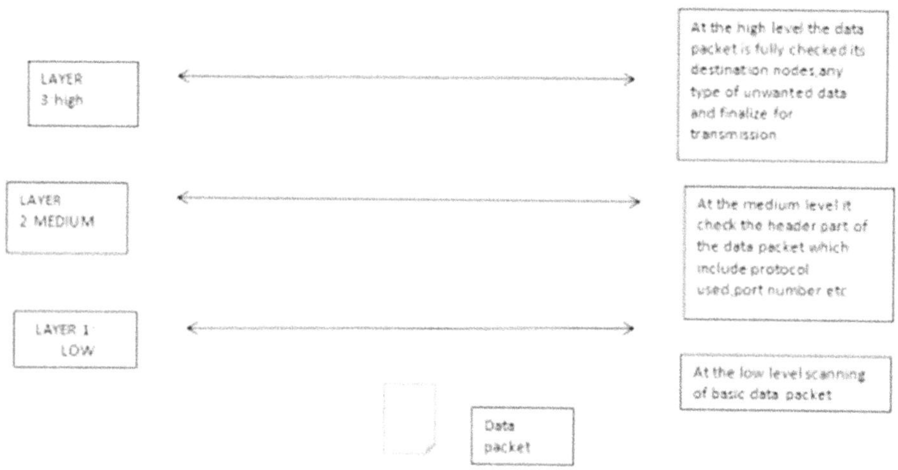

CONCLUSION

In the given paper we are trying to understand the importance of firewall in underwater sensor networks and also examined that how the firewall helps to secure our network services from various types of unusual attacks. In the above paper we also discuss about various types of firewall and how each firewall can be used in different purposes to secure the network. Since it considered to be the most important part of the network in the terms of network security but still with the help of keen research it seems that every firewall that we are using in our network have some type of limitation generally based on the compatibility level or any other issues like in UWSN it is very difficult to install hardware or any other type of firewall so to resolve that we are using software based firewalls which can be easily install to the system and we can easily protect our data that is continuously flowing deep under the sea water via nodes.so at last it is very important to understand the basic network configuration and the basic steps that have to be taken first to secure our network from various kinds of the attacks.

Table 1. Basic set of rules

Type	Source IP	Source Port	Destination IP	Destination Port	Action
TCP	178.168.19.2.1	8080	178.168.19.2.4	1024	accept
TCP	178.168.19.2.2	8082	178.168.19.2.5	2020	reject
UDP	178.168.19.2.3	8084	178.168.19.2.6	80	reject

REFERENCES

Alves, J., Potter, J., Zappa, G., Guerrini, P., & Been, R. (2012, October). A testbed for collaborative development of underwater communications and networking. In *MILCOM 2012-2012 IEEE Military Communications Conference* (pp. 1–8). IEEE. doi:10.1109/MILCOM.2012.6415691

Barbeau, M., Blouin, S., Cervera, G., Garcia-Alfaro, J., & Kranakis, E. (2014). *Appendix to simulation of mobile underwater communications*. Academic Press.

Chouiten, M., Domingues, C., Didier, J. Y., Otmane, S., & Mallem, M. (2012, March). Distributed mixed reality for remote underwater telerobotics exploration. In *Proceedings of the 2012 Virtual Reality International Conference* (p. 1). ACM. 10.1145/2331714.2331716

Dee, J. S. (1995). *U.S. Patent No. 5,408,740*. Washington, DC: U.S. Patent and Trademark Office.

Feldstein, M. (2008). How to Help People Whose Home Values Are Underwater—The Economic Spiral Will Get Worse Unless We Do Something about Negative Equity. *Wall Street Journal*, 18.

Forrest, A. L., Laval, B. E., Lim, D. S. S., Williams, D. R., Trembanis, A. C., Marinova, M. M., Shepard, R., Brady, A. L., Slater, G. F., Gernhardt, M. L., & McKay, C. P. (2010). Performance evaluation of underwater platforms in the context of space exploration. *Planetary and Space Science*, *58*(4), 706–716. doi:10.1016/j.pss.2009.08.007

Goyal, K. N. (2016). An Optimal Scheme for Minimizing Energy Consumption in WSN. *Global Research and Development Journal for Engineering*, *1*(8), 1–7.

Goyal, N., Dave, M., & Verma, A. K. (2017). Improved data aggregation for cluster based underwater wireless sensor networks. *Proceedings of the National Academy of Sciences, India Section A: Physical Sciences, 87*(2), 235-245. 10.100740010-017-0344-y

Goyal, N., Dave, M., & Verma, A. K. (2018, August). Adaptive Error Control Technique for Cluster-Based Underwater Wireless Sensor Networks. In *International Conference on Wireless Intelligent and Distributed Environment for Communication* (pp. 269-280). Springer. 10.1007/978-3-319-75626-4_20

Goyal, N., Dave, M., & Verma, A. K. (2019a). Data aggregation in underwater wireless sensor network: Recent approaches and issues. *Journal of King Saud University-Computer and Information Sciences.*, *31*(3), 275–286. doi:10.1016/j.jksuci.2017.04.007

Goyal, N., Dave, M., & Verma, A. K. (2019c). Protocol Stack of Underwater Wireless Sensor Network: Classical Approaches and New Trends. *Wireless Personal Communications*, *104*(3), 995–1022. doi:10.100711277-018-6064-z

Goyal, N., Dave, M., & Verma, A. K. (2020). SAPDA: Secure Authentication with Protected Data Aggregation Scheme for Improving QoS in Scalable and Survivable UWSNs. *Wireless Personal Communications*, *113*(1), 1–15. Advance online publication. doi:10.100711277-020-07175-8

Goyal, N., Sandhu, J. K., & Verma, L. (2019b). Machine learning based data agglomeration in underwater wireless sensor networks. *International Journal of Management. Technology And Engineering*, *9*(6), 240–245.

Jacka, L., & Booth, K. (2012, December). What about the firewall? Creating virtual worlds in a public primary school using Sim-on-a-Stick. *Australian Computers in Education Conference*.

Kamoun-Abid, F., Meddeb-Makhlouf, A., Zarai, F., & Guizani, M. (2019). DVF-fog: Distributed virtual firewall in fog computing based on risk analysis. *International Journal of Sensor Networks*, *30*(4), 242–253. doi:10.1504/IJSNET.2019.101242

Khurana, K., & Goyal, N. (2016). A Survey on Deployment Strategies and Energy Efficiency of Wireless Sensor Networks. *International Journal of Electrical Electronics & Computer Science Engineering*, *3*(3), 16–20.

Khurana, T., Singh, S., & Goyal, N. (2012). An Evaluation of Ad-hoc Routing Protocols for Wireless Sensor Networks. *International Journal of Advanced Research in Computer Science and Electronics Engineering*, *1*(5).

Kumar, M., & Goyal, N. (2014). Reviewing underwater acoustic wireless sensing networks. *International Journal of Computer Science and Technology*, *5*(2), 95–98.

Lisa, G. (2013). The great underwater firewall. *The World of Chinese*, *3*(6), 71–74.

Liu, W., Weaver, J., Weaver, L., Whelan, T., Bagrodia, R., Forero, P. A., & Chavez, J. (2018, May). APOLL: Adaptive Polling for Reconfigurable Underwater Data Collection Systems. In *2018 OCEANS-MTS/IEEE Kobe Techno-Oceans (OTO)* (pp. 1-9). IEEE.

Nightingale, D., Troxel, S., & Libby, B. (2002). Use of the internet in maintaining and operating the Intermediate Scale Measurement System (ISMS) underwater test facility. *The Journal of the Acoustical Society of America*, *112*(5), 2328–2328. doi:10.1121/1.4779397

Pelizzari, U., & Tovaglieri, S. (2004). *Manual of Freediving: Underwater on a single breath*. Idelson-Gnochi Limited Publishers.

Peng, Z., Le, S., Zuba, M., Mo, H., Zhu, Y., Pu, L., . . . Cui, J. H. (2011, June). Aqua-TUNE: A testbed for underwater networks. In OCEANS 2011 IEEE-Spain (pp. 1-9). IEEE. doi:10.1109/Oceans-Spain.2011.6003661

Tuna, G., & Gungor, V. C. (2017). A survey on deployment techniques, localization algorithms, and research challenges for underwater acoustic sensor networks. *International Journal of Communication Systems*, *30*(17), e3350. doi:10.1002/dac.3350

Chapter 9
Error Control and Reliable Data Delivery in UWSNs

Sneha Singh
National Institute of Technology, Delhi, India

Manisha Bharti
National Institute of Technology, Delhi, India

ABSTRACT

Underwater wireless sensor networks (UWSNs) contains many components such as vehicles and sensors that are deployed for cooperative monitoring and data collection tasks in a particular acoustic environment various nodes and ground-based stations use these networks interactively. Presently, UWSNs face the problems and obstacles regarding limited bandwidth, high propagation delay, 3D topology, media access control, routing, resource utilization, and power constraints. The research community has developed different methodologies over the past few decades to address these issues and challenges; but, due to complex characteristics of the underwater environment, some of them are still open to research. The main drawback of the traditional approach is the lack of direct interaction between different ends, recorded information can never be accessed during any mission, and data recorded will be lost in the event of any failure.

INTRODUCTION

A utilization of the 'Wireless sensor network' for the process control has been pulled in a great deal of enthusiasm for years ongoing (Gungor & Hancke, 2009), (Song, Han, & Mok, 2008), (De Biasi, Snickars, Landernäs, & Isaksson, 2008), (Saifullah, Xu, Lu, & Chen, 2010). The essential reasons for the WSNs can be effectively and adequately conveyed, sensors are used where the wires can't reach and also cost of the wires and facilities can be reduced altogether. However, upgrading Wireless control with sensor driven with small batteries hubs requires huge energy for skill. In the industry of process, much of the times the plants have a huge number of control loops. For instance, consider an industrial plant with 4,000 battery-driven sensor hubs. In the event that the hub lifetime is consistently appropriated somewhere in

DOI: 10.4018/978-1-7998-3640-7.ch009

the range of one and two years, at that point in enduring state one will, all things considered, need to interchange the battery on around ten hubs every day. This clearly isn't an alternative in maximum scenarios. Consequently, it's not clear presently as if the use of battery- sensor controlled hubs can be achieved for process control. It is important to put the sensor hub radio to sleep for long life, however much it might be expected. Any path this will be in struggle with the necessity of little delay in the system and the utilization of unnecessarily fast sampling, much of the time happening in the process industry. Thusly, it turns out to be progressively critical to utilize the radios' real-time effectively, also to pick intervals of sample and procedures for transmission sensibly. The structure of latency, energy effective control of error plans assumes a significant job. Error control can, for the most part, be acknowledged via (ARQ) Repeat Automatic Request and Error Forward Correction (FEC), or a mix between two Hybrid-ARQ (HARQ) (Akyildiz & Vuran, 2010). Nowadays, ARQ is absolutely founded on retransmissions and is the actualized system in control of process systems (HART Communication Foundation, 2007).

Problem Description

FEC (Forward Error Correction) presents excess bits, which are because of expanded packets length increments the both energy use and latency. Then again it diminishes retransmissions, which works the other path. Thus, there is an exchange between off the rate of the code and re-transmissions. And how this exchange off ought to be made in industrial situations will be examined in this postulation. Besides, the researchers will examine the upsides and downsides of ARQ and HARQ as far as energy productivity and latency. The HARQ plot depends on FEC with BCH codes. Such plots have been utilized beforehand in the writing. Here the researchers will utilize a ARQ-Hybrid-FEC-Adaptive scheme.Execution of the 'Hybrid ARQ-Adaptive' FEC as far as energy utilization, latency, and packet loss will be explored and contrasted with ARQ (Vuran & Akyildiz, 2009), (Howard, Schlegel, & Iniewski, 2006), (Kleinschmidt, Borelli, & Pellenz, 2007), (Balakrishnan, Yang, Jiang, & Kim, 2007).

The comparison done here depends on experiences assembled from a radio channel estimation crusade linearized at a plant paper in Sweden. From the estimations acquired the researchers saw that the channel qualities were very extraordinary. While a few channels were for all intents and purposes static others were dependent upon serious shadow fading, with the signal quality inconstancy of as much as of 25 dB (Björnemo, Ahlén, & Gidlund, 2010). Not with standing, on the span of a packet the channels were consistent. In Section 5 the researchers will utilize both AWGN channels just as insights of fading channels, acquired from this estimation crusade, in the comparison of energy consumption, latency, and packet loss for the 'ARQ-Hybrid-FEC-Adaptive' and ARQ plans. The comparison will be done by means of numerical assessments in MATLAB. The Purpose of this proposition is to give a better comprehension of the described exchange-off, for example between code rate and retransmissions, an understanding that will be moved to industry and continuous institutionalization work on WSNs for process control.

Work Related

(Howard, Schlegel, & Iniewski, 2006) has considered the proficiency in energy of explicit (ECC), for a several decoders usages, in Wireless Sensor networks WSNs. The researchers concluded that energy is saved by coding even for a short distance, but the researchers did not represent the packet size extension, which prompts longer on-schedule radio. Just the energy devoured by the coding and translating forms is considered. Only the energy used by the modes of coding and translation is considered. A variety of error control techniques are considered in (Sankarasubramaniam, Akyildiz, & McLaughlin, 2003), but the comparison is centred around the topic of packet ideally of size for WSNs. (Balakrishnan, Yang, Jiang, & Kim, 2007) assesses the power utilization of distinct three error control codes (ECC), yet for explicit stages that is the FPGA and ASIC.

In (Shih, Cho, Lee, Calhoun, & Chandrakasan, 2004) and (Min, o.a., 2002), just the expense in the decoder is incorporated at the collector side and not the all-out preparing cost, which increments with bigger packets. Besides, their comparison depends on convolutional codes, whereas the researchers centre around block codes (Vuran & Akyildiz, 2009). In a comparison of error control schemes in Wireless Sensor networks (WSNs) is introduced, which analyses ARQ, FEC, and HARQ as far as energy utilization, dormancy, and PER. See again (Akyildiz & Vuran, 2010). Here the researchers use an alternative organizing model, and the researchers also consider the use of an adaptive the authors use of adaptive FEC has already been considered (Cho, 2000) where the display of an adaptive error control conspires in integrated Terrestrial Satellite wireless systems for multimedia applications. They infer that under continuous application, the adaptive convention beats the static FEC conventions concerning packet miss probability (Shiozaki, Okuno, Suzuki, & Segawa, 1991).

WirelessHART

Wireless High path Addressable Wireless Transducer Protocol (wirelessHART) dependent wireless sensor networking innovation (De Biasi, Snickars, Landernäs, & Isaksson, 2008). It is basically the main standard released in the process industry for wireless communication and is the only open standard explicitly intended for process control (Song, Han, & Mok, 2008). The sketch of Wireless HART is shown in Figure 1

Figure 1. Structure of WirelessHART

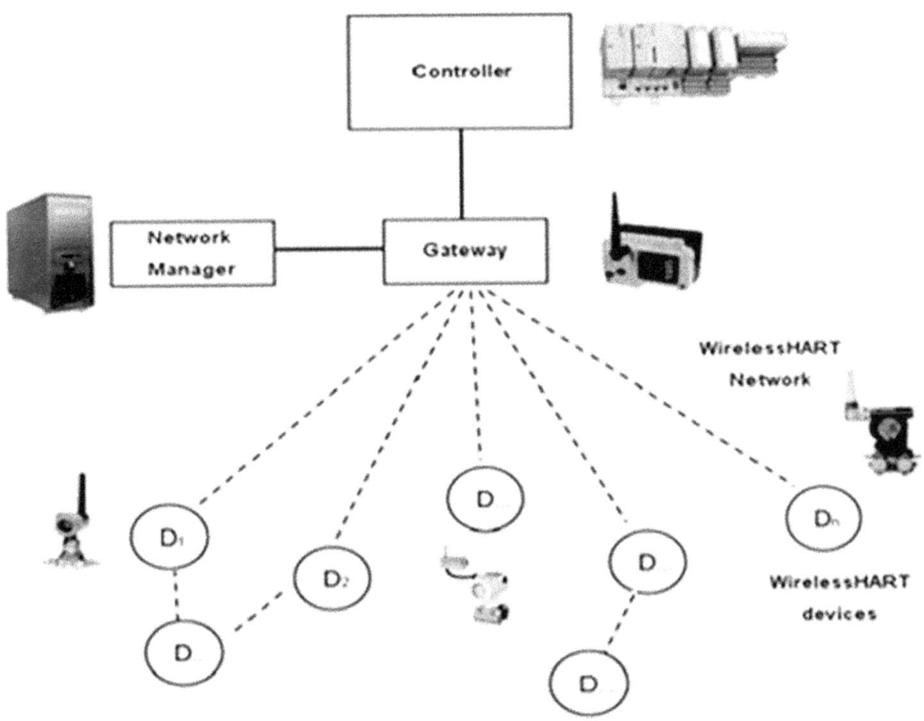

WirelessHART is a technology focused on TDMA that facilitates networking of mesh. This uses the radio band of 2.4GHz ISM, has a bit rate as stated of 250 kbps per channel over 15 channels. ARQ is used for error control. The physical layer is obtained from the standard IEEE 802.15.4 and is therefore used to change the Offset Quadrature Phase-Shift Keying (OQPSK). The (PPDU) is represented, (IEEE, 2006).

Figure 2. (PPDU) Physical Layer Protocol Data Unit

		Octets		
		1		variable
Preamble	SFD	Frame length (7 bits)	Reserved (1 bit)	PSDU
SHR		PHR		PHY payload

The synchronization header i.e. SHR and the Physical Layer header i.e. PHR are expected to be overload for the payload PHY. The SHR empowers the accepting gadget to discover the start of the packet through a known sequence, the preamble. The Start-of-Frame Delimiter (SFD) is a 1-byte value of the preamble's end. The PHR contains data on the packet's full size. The overhead length for the 2.4 GHz network, e.g. the SHR and the PHR, is of the 6 bytes and the PHY container, or the PSDU that is stand for Physical Layer Service Data Unit, comprises inward layers of packets with additional overhead. The SFD that is Start Frame Delimiter is a 1-byte denoting the finish of the preamble. The PHR contains data related to the complete size of the packet. The complete length of the overhead, for example the SHR and the PHR, for the 2,4 GHz channel, is 6 bytes (IEEE, 2006). The PHY payload, include the packets of inward layers, extra overhead just like the actual bits of data. The size of the payload therefore depends on the size of the transmitted data. In Wireless HART, the entire packet size is controlled by the time-slot structure that endures the evaluation of the packet to 133 bytes.

ERROR CONTROL

Automatic Repeat Request

Automatic Repeat Request-based error management, if errors are detected, the packet will be retransmitted. Such packets will be retransmitted until error is released. The position of the error is usually performed by (CRC) 'cyclic redundancy Code' test. Before transmission, a simple technique for error detection is applied to the packet and a checksum is calculated on the receiver side to conform that there is no error in the packets occurs. If unable to contain the correct value in checksum, then the packet will be transmitted again. The ARQ conspire uses (ACK) positive affirmation i.e. positive Acknowledgement or negative affirmation (NACK) i.e. negative Acknowledgement to blame the transmitter for the success of the transmission. If it is unlikely that ACKs will be used, the transmitter will then again transmit the ACK packet that has been not received inside a pre-indicated time allotment. The extra cost brought about from retransmissions is the significant downside with ARQ.

Forward Error Correction (FEC)

Channel coding or FEC is a method which is used to extend the error management screen. This is accomplished through the use of 'error correction codes' (ECCs) that add redundancy into the packet, allowing the receiver side to recognize and correct a specific amount of bit-errors. The main demerit is the price of the surplus, the parity bits, which reduces the size of the packet. FEC also has the cost of encoding and disconnecting. FEC is therefore commonly used in situations where again transmissions are costly relatively. There are FEC codes of many types in which the commonly used are isolated into two classifications, block codes and convolution code. FEC is therefore commonly used in situations where retransmissions are relatively costly There are FEC codes of many types in which the commonly used are isolated into two classifications, block codes and convolution code. Firstly, block codes partition the message of a predefined length to be sent to littler frames. Such blocks are then stored in code words separately. Instead, encoded each bit as a component of a pre-defined going number before bits due to convolutional codes. Bose Chadhuri Hocquenghem (BCH), Reed-Solomon and also Hamming codes are some of the block codes that are most commonly used. In 1950, the Hamming code is recognized as one of the spearheading codes developed by Richard Hamming, which is still 'Error-redressing code' memory (ECC-memory) as used in applications today. For example, Reed-Solomon and BCH have been through all the more dominant codes over the years.

Block codes are ordinarily triple as represented, where code word length, is the number of data bits in a code word, and is the changing error ability regarding the bits' number that can be redressed. Figure below shows how increase gain parity bits a FEC size of blocks. Together with the parity bits, the data bits of length then make up a word of length (n) from which the code frequency is defined as Rc= k / n, code-explicit. In this direction, the encoded packet will consist of (m/Rc) bits for a given code level Rc and a message with m data bits. The below diagram depicts the FEC block as: *k* information bits *n-k* parity bit

There are even more efficient codes to benefit from conventional block-and convolutionary codes, like 'Low-Density-Parity-Check' (LDPC) codes and turbo codes. They are limited to applications that can handle the computational complexity required due to their high level encoding and decoding processes or can say advance level encoding and also due to high complexities.

There are still increasingly amazing codes (Goldsmith, 2005), such as Low-Density-Parity-Check (LDPC) codes and turbo codes, despite the usual frame, and convolution codes. Due its high complexities and propelled forms of deciphering and encoding, they are restricted to applications that can be tackle the computational unpredictability required (Vuran & Akyildiz, 2009). BCH codes include a cyclic, linear block coding group. These are effective codes that consistently outperform all remaining block codes at high speed or high rates for the moderate to high Signal to noise ratio. BCH codes do, now and again, beat convolutional codes as far as energy productivity (Sankarasubramaniam, Akyildiz, & McLaughlin, 2003). The consequences show that outperform of BCH codes is the energy effective convolutional codes by about 15 (Balakrishnan, Yang, Jiang, & Kim, 2007). In Table 1 shows the (n, k, t) the qualities of 70 BCH codes are exhibited.

Table 1. BCH codes (IEEE,1964)

BCH Codes (IEEE, 1964)

n	*k*	*t*	*n*	*k*	*t*	*n*	*k*	*t*
7	4	1	127	92	5	255	163	12
15	11	1	127	85	6	255	155	13
15	7	2	127	78	7	255	147	14
15	5	3	127	71	9	255	139	15
31	26	1	127	64	10	255	131	18
31	21	2	127	57	11	255	123	19
31	16	3	127	50	13	255	115	21
31	11	5	127	43	14	255	107	22
31	6	7	127	36	15	255	99	23
63	57	1	127	29	21	255	91	25
63	51	2	127	22	23	255	87	26
63	45	3	127	15	27	255	79	27
63	39	4	127	8	31	255	71	29
63	36	5	255	247	1	255	63	30
63	30	6	255	239	2	255	55	31
63	24	7	255	231	3	255	47	42
63	18	10	255	223	4	255	45	43
63	16	11	255	215	5	255	37	45
63	10	13	255	207	6	255	29	47
63	7	15	255	199	7	255	21	55
127	120	1	255	191	8	255	13	59
127	113	2	255	187	9	255	9	63
127	106	3	255	179	10			
127	99	4	255	171	11			

Adaptive Coding

The authors (Gungor & Hancke, 2009) expect to give an occurring took at the state present of the edge in modern WSNs. Author present a rundown with "Structure Goals" of eight to fill in as a rule for future work in research. One target within the list is "Adaptive Network Operation," where adaptive conventions criteria are worried in order to adapt to evolving channel conditions. Adaptive coding takes into account the quality and unpredictability of code differences (Goldsmith, 2005). E.g. a basic code could be made similarity with other block codes with the equivalent and utilized for a decent channel while even extra dominant codes could be implemented on it when the poor channel is there. Puncturing and Multiplexing are two different paths for adaptive code execution. Multiplexing adds codes with different capacities for errors adjustment, whereas puncturing fluctuates the value of the code using recalling almost the coded bits for the transmission, in the event that the coded packet was effectively gotten or not.

Hybrid Automatic Repeat Request (HARQ)

ARQ offers efficient interaction by retransmissions, which is going to be exorbitant in weak networks where retransmissions occur a lot of time. Through weak channels, FEC performs better, while repetitive bits then come out as an unnecessary expense when conditions of channel are high. HARQ plans misuse both merits through strategies to join ARQ and FEC. There are forms of HARQ-I, HARQ and HARQ-II for the most part. The HARQ-I plot essentially sends a coded packet that is again transmitted if the recipient is not ready to address all errors and also HARQ-II is increasingly unpredictable, and during the main transmission endeavour just a constrained measure of parity bits FEC are incorporated. The added parity bits are attached to the recipient side with the past transmission and which allows to attempt the decoder to address the actuated errors again, but now with more repeat data. Whereas HARQ-II reduces the utilization of data transfer capacity, HARQ-I does not require path to be placed on the recently sent packets (Vuran & Akyildiz, 2009).

WSN Error Control

When planning a WSN error control plot, energy performance is fundamental. To this end, it is necessary to contrast the gain energy costs induced by FEC coding and what can be spared to reduce in retransmission. The primary expense to assume is the all-encompassing lengths of packets, resulting in larger on-time radio. Such expenses are cited in (Vuran & Akyildiz, 2009) about figures from the SA-1100 processor, and specifically the disengagement costs. Considering, for example, a low-control processor, the MSP430, (Björnemo E., 2009) makes the presumption that "the use of energy for coding and interpretation is insignificant in connection with other cost of preparation. "A FEC or HARQ-dependent error control configuration will have the upside of having the option to address a specific error measure within a packet. Fundamentally the advantage is a reduced

(packet error rate) PER in correlation with ARQ at the expense of bigger packets. Taking into consideration a multi-hop organizes, the lower PER can be abused in 3 different paths, depicted straight path (Howard, Schlegel, & Iniewski, 2006).

Avoid Retransmissions

Essentially, the lower FEC or lower HARQ PER can be used to stay away from again transmissions. The PER would increase with a basic ARQ plan at the point when the packet is transmitted or sending over the channel which is poor, and retransmissions would occur. The retransmission costs as far as the use of energy and latency are concerned, if a FEC or HARQ plan were to be updated, could be stayed away from something.

Hop Length Extension

Hop length extension is defined by making the long length hops in a multi-hop arrangement that could abuse the lower PER (Akyildiz & Vuran, 2010). This would change in a lower amount of hops required for a transmitted packet reach its goal. Therefore, lesser hops would cause increased productivity of lower latency and energy.

Length Extension of Hop

Figure 3. example of length extension of hop strategy. By utilizing FEC& HARQ the green represent the one addition hop can be avoided.

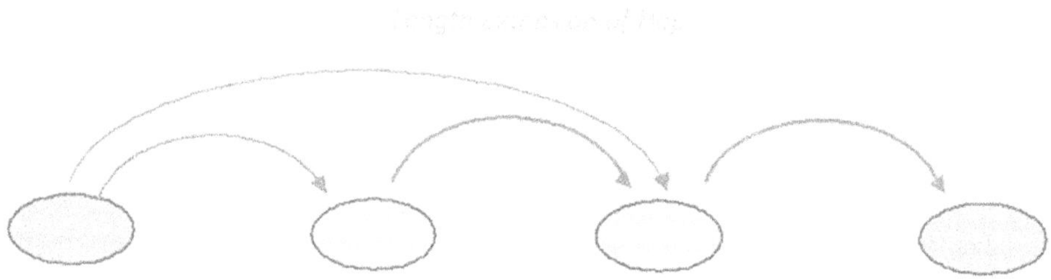

Power Transmit Control

let a framework structured with an objective BER (bit error rate). With HARQ or FEC the power transmit can be decreased contrasted with ARQ, yet at the same time accomplish the objective BER. In that manner energy utilization could be compelled. In theory of coding, it is depicted as *coding gain*, which actually is the contrast in the signal to-noise ratio (SNR) levels required to arrive at a similar objective PER, with and without FEC.

Loss in Path

Loss in path is the decrease in an electromagnetic wave's power density as it proliferates into the wireless radio channel. Path loss could be attributed to various impacts, such as landscape shapes, not similar conditions (outdoors or indoors, urban or rustic), and length between receiver transmitter. There are a few models utilized to produce rough signals, such as loss of space free path, propagation of tracing ray, and loss of distance log path models.

The distance log path loss model is a disengaged model, typically used to analyse various system designs for general trade-off (Goldsmith, 2005). A model that exactly depicts the multifaceted complexity of a particular signal proliferation is difficult to acquire, but without falling back on confused loss path models, the distance log path loss model captures the pith of signal generation.

$$P_r = P_t K \left(d_o / d\right)^{Y}$$

Here K is a steady one that relies on channel attenuation and antenna characteristics, and d0 is a distance of reference. The exponent of the loss path is resolved depending on the condition of generation. Values typically vary from 1.6 to 3.3 in a processing plant environment.

ERROR PROBABILITY

Bit Error Probability

For example, when the signal is produced through the wireless channel, it depends on unsettling pressure, interference and noise, which can cause the bits to be modified. Errors and constraints could have a similar impact in the transmitter, or the receiver, for example bad synchronization. The bit error rate (BER) shows the amount of bits that were adjusted at the time of the transmission.

The probability of error bit can be calculated for non-similar mixes of channels and advanced balance plans. Because of an AWGN stream where the BPSK modulation schemes are modified and perfectly synchronized is expected, a precise expression of the error bit probability is given by

$$p_b = Q\left(\sqrt{\frac{2E_b}{N_0}}\right)$$

Where Q is a scaled type of the *complementary Gaussian error function (Goldsmith, 2005).*

The probability of a bit error for Quadrature Phase-Shift Keying (QPSK) is equal to that of BPSK, as QPSK modified signal can be perceived on both the in-stage and quadrature sections of the signal as a signal with BPSK balance. Balance QPSK (OQPSK) has the equivalent of BPSK and QPSK (Sklar, 2003) theoretical bit error execution.

Packet Error Probability

The probability of packet error can be determined as for an un-coded transmission of a length bits' information packet

$$p_p = 1 - (1 - p_b)^m$$

Where p_b is the bit error probability, for QPSK and BPSK given as above.

Error Probability of the Code Word

For the situation of a transmission with an actualized error control conspire that utilizations FEC codes, condition can be communicated as far as *code word error probability.*

$$p_p = 1 - (1 - p_w)^w$$

Where w in a packet is the number of code words. The probability of code word error can therefore be communicated as a component of the probability of bit error. (Proakis, 2001), for a particular instance of a linear binary block code with features (n, k, t), a hard-decision decoder will have the option to fix any number of errors up to

$$t = \left| \frac{1}{2}(d_{min} - 1) \right|$$

This is a lower limit on the correction capacity t as it is a portion of the minimum Hamming distance (dmin) between any two different code words. From now on, even with more than bugs, a code word can be corrected. The probability of errors in a code expression of n bits, as indicated by the binomial hypothesis, is gainful.

$$p(\mu, n) = \binom{n}{\varepsilon} p_b^{\mu}(1 - p_b)^{n-\mu}$$

By this, block codes are valid for the particular instance. For example, for different types of FEC codes, convolutionary codes, the definition of a code word does not occur in a similar sense, because blocks are not used.

Finally, there is an upper limit on the probability of code word error, by which all the probabilities for all un-correct number of errors are given

$$p_w \leq \sum_{\epsilon = t+1}^{n} \binom{n}{\epsilon} p_b^{\varepsilon} (1 - p_b)^{n-\epsilon}$$

System Model

The researchers present the model of process and path loss that will be used in the spin-off in this region. In addition, we present the ARQ-Hybrid-FEC-Adaptive conspire, which in Section next will be contrasted with ARQ.

Network Model

Latency specifications in process control applications constantly limit the amount of hops that can be performed between sensor hubs and a shared Gate path. This encourages us to present a basic model of the system consisting of three equidistant hubs associated with figure. Every core, which can hand-off packets and then empower multi-hoping, is supposed to have the ability to re-code. Packets are sent by single-hop or multi-bounce to the ultimate a path from the centre. There are three separate courses for the; through and, or through three types of hops.

Notwithstanding the elective courses, there will be an alternative to using different codes because of the ARQ-Hybrid-FEC-Adaptive conspire. To improve, the researchers make the assumption that steering data exists in all hubs just as data on what codes to use. Practically speaking this could be accomplished by the utilization of a brought together plan where the is liable for error control, including steering. The would have the vital CSI as far as signal to-noise ratio (SNR), here characterized as ¤ 6, from past system action, settle on decisions what the course and the codes which are to be utilized, just as keeping the system refreshed with the data through occasional telecom of synchronize packets. At the

Figure 4. Model of network

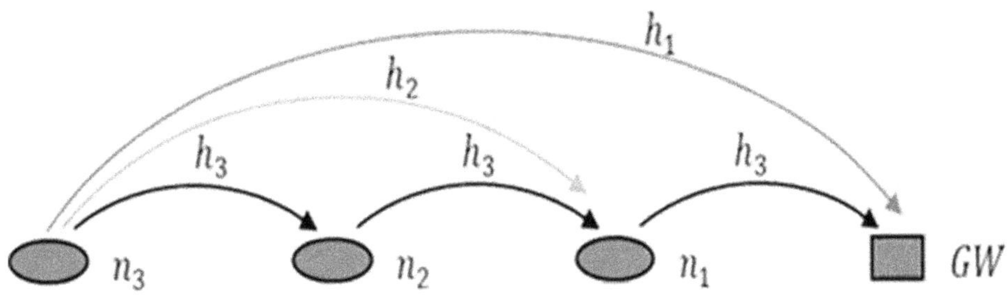

end of the day; in view of CSI, the transmission course and codes will be chosen, and this information will also be transmitted to the relevant hubs. Note, this is a dis- embroidered model that would require further comparison, yet it is profoundly relevant despite everything and gives significant bits of knowledge about how to use distinctive FEC plans. Unlike all the error control decisions are in the option is to use a decentralized plan where the code rate decision is handled locally in the hubs by encouraging back-channel state data.

Path Loss Model

Considering a modern situation where the quality of the channel can change due to the location of hubs, as the researchers will be as the good paths from different hubs and the GW, the numerical calculation is carried out as an element of signal to noise ratio. In order to assess how the SNR changes between the hops, a path loss model is used, which approximates the impacts of the wireless channel spreading signal. Here the researchers look at the template of path loss. Because the signal power (Pr3) has been obtained in the wake of transmitting over the hop(h3) that is short hop, signal command over the more drawn hops can be expressed or h1 or h2.

$$P_{rj} = P_{r3}\left(\delta_j\right)^{\gamma}$$

Here δ_j is the relative bounce distance over h_j as for h_3, that is, $\delta_j = \frac{d_3}{d_i}; j = 1,2$, and d_j is the distance over hop . For straightforwardness, the constants K, d_0 and the path loss model, ϒ, are thought to be the equivalent for all hops. Condition can besides be communicated as far as signal to-noise ratio as () .

$$SNR_j = SNR_3\left(\delta_j\right)^{\gamma}$$

where is the $SNR_j, j = 1,2$ at hub n_j.

ARQ-Hybrid-FEC-Adaptive Scheme

Energy consumption and latency, we will analyse the advantages and disadvantages of ARQ and ARQ-Hybrid-FEC-Adaptive. The ARQ-Hybrid-FEC-Adaptive plot used here is a HARQ-I conspire. The researchers mean transmission by Adaptive-FEC where the code rate depends on (CSI). As it happens, a more powerful code will be used in case of a bad channel based on CSI, while codes with less multifaceted character will be used for a better channel. Therefore, needless repetition in the packet will remain path from under great conditions of the network, while the ability to change further formed codes can be misused under terrible condition.

In addition, the configuration of 70 BCH codes shown in Table above is used for forward error correction. BCH codes are spoken by (n, k, t) straight block codes. The BCH codes to be used fluctuate in length, and their ability to change varies from t=1 to t=63.The researchers have selected BCH codes for FEC evaluation where the BCH are implemented as ground-breaking block codes that outperform convolutionary codes as far as energy efficiency is concerned. At the same time, they are not as intricate as turbo codes or LDPC codes and fit a lot of the time used in the sensor hubs for the processors in this way conventions utilized for without wire i.e. wireless procedure control.

Numerical Analysis

In this section, the researchers describe the comparison model that will be used to look at the energy utilization and latency of ARQ and ARQ-Hybrid-ARQ-FEC-Adaptive. Both the use of energy and the latency is relying on the probability of packet error, which can be approximated for a specific channel and balance. Try transmitting data using OQPSK tweak over an AWGN network., OQPSK is used in the IEEE standards of 802.15.4 and WirelessHART (HART Communication Foundation, 2007) (IEEE, 2006). Genuine industrial plant estimates, show that the researchers can foresee block fading in the flow, see e.g., (Björnemo, Ahlén, & Gidlund, 2010). The stream is roughly consistent over the length of the packet. As a result, our AWGN model is right and the researchers can use AWGN in our study with flawless CSI.

The researchers will use the deduction of the probability of packet error. Let all channels connected to each bounce be autonomous and indistinguishably Gaussian disseminated. Further, let the comparing bit error probability be indicated p_{bj}. The researchers would then be able to express the packet error probability over hop h_j, for the ARQ plot as

$$p_{pj}^{ARQ} = 1 - \left(1 - p_{bj}\right)^{m}$$

Therefore, for the ARQ-Hybrid-FEC-Adaptive conspire, accepting hard-translating, the likelihood of a packet error is given by, as

$$p_{pj}^{HAF} = 1 - \left(1 - \sum_{\epsilon = t+1}^{n} \binom{n}{\epsilon} p_{bj}^{\epsilon} \left(1 - p_{bj}\right)^{n-\epsilon}\right)^{W}$$

Where is the length of the coded word, where is the quantity of code words in a packet, where is the redressing ability, and where, for an AWGN channel using OQPSK adjustment

$$p_{bj} = Q\left(\sqrt{2SNR_j}\right)$$

the quantity of code words, w, in a packet is controlled by the quantity of data bits, and the code word length, k, i.e., the ceiling function

$$w = \left|\frac{m}{k}\right|$$

The *ceiling function* é ù is utilized because of that N_w must be a number in a viable usage.

Latency Analysis

A broadly inclusive latency model would have numerous parts, such as the capacity preparation hubs, the amount of hop in the connection, the topology of the system and size, and the conventions for correspondence. In systems with multi-hop, for example, WSNs, the quantity of retransmissions is a signalling latency segment. The impact to the extent that packet delay starts to end depends in particular on how the system is intended to the handling of retransmissions. Here the researchers will model latency(L) as the number of transmissions to get a packet that has been effectively decoded. Clearly, this is a rearrangement, but above all it provides an indication of how the distinctive error management plans work as far as dormancy is concerned. In addition to increasing dormancy, retransmissions can else have negative impacts, such as blocking dedication, which is a specific issue in networks, such as WSNs.

Since each time a packet is unrecoverable by the FEC, a retransmission occurs, the researchers can calculate the amount of transmissions from the likelihood of packet error. The probability of effectively transmitting a packet, on the endeavour is represented by the geometric probability distribution

$$p_s = \left(1 - p_{pj}\right) p_{pj}^{(i-1)}$$

from which the normal latency, i.e. number of transmissions, h_j is given over hop

$$E\left[L_j\right] = \frac{1}{1 - p_{pj}}$$

In order to figure the normal number of transmissions starting to finish, say [], the researchers add the normal qualities as indicated by for the various hop designs. For instance, for the situation when the researchers have three hops start to finish the researchers get a $E[L_{tot}]=3*E[L_3]$. For two hops $E[L_{tot}] = E[L_2] + E[L_3]$ etc.

Energy Consumption Analysis

The researchers determine the energy utilization model to be used in this segment. The costs of the modes of coding and deciphering are deemed to be immaterial in conjunction with other costs of planning. There is a possibility that an unsuccessful transmission will result in a retransmission. Therefore, the use of energy is based on the total transmissions number, here referred to as the dormancy, L, required to transmit the effectively of packet. In addition, the use of power is based on the time (t) that the transmitter or send and the receiver should be turned on for solitary transmission. The researchers can adequately express the normal use of energy over hop h_j, for a fixed transmit control, as

$$E\left[\in_j\right]=E\left[L_j\right]\left(E\left[P_{TX}\right]t_{TX}+E\left[P_{RX}\right]t_{RX}\right)$$

P_{TX} and P_{RX} are the all out power devoured by the sending or transmitting and getting hub separately. Here and are the transmitting and getting times individually. To rearrange the condition the researchers expect $t_{TX}=t_{RX}=t$. Subsequently the researchers can rework as

$$E\left[\in_j\right]=E\left[L_j\right](E\left[P_{TX}+P_{RX}\right]t$$

The approximation $t_{TX}=t_{RX}=t$ is, obviously, fairly coarse since $t_{TX}<t_{RX}$ in many situations.

The theory is that before packet is arrived at the receiver end the recipient will start listening a while. Nevertheless, the measure of time burned by tuning in before packet are appears is not dependent on the type of forward error code used. Along these lines, the estimate will only present a scaling factor, i.e. a steady increase in energy consumption, per transmission effort in addition.

Figure 5. Radios on time parts

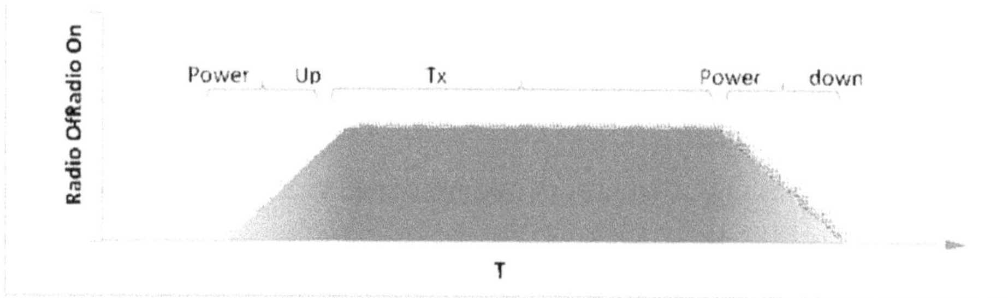

On the unlikely chance that the researchers ignore the time required to control and shut down the radios, the researchers can express as much as the quantity of bits in a transmitted packet duplicated when a solitary bit is expected to be transmitted. The packet must contain some synchronization overhead. The sum of bits in a packet is the whole of the coded message and the overhead in this way. Let be the length of the message and the length of the overhead. At that point, for a FEC code with altered code rate, the researchers can express as

$$t = \left(R_c^* m + m_{oh}\right) t_b$$

The changed code rate R_c^* is given by

$$R_c^* = \frac{m}{n * w}$$

Here 'n' is the length of a code word for a specific FEC code and w is the quantity of code words. Note that, for encoded transmission $R_c^* = 1$.

The researchers do not mean a comparison of energy use for a particular radio, but rather a general analysis of the display of the two error management plans, ARQ and ARQ-Hybrid-FEC-Adaptive FEC. The researchers can therefore standardize the condition with the use of energy for a solitary bounce, a single transmission of an uuencoded packet $\left[\left(E\left[P_{TX} + P_{RX}\right]\left(m + m_{oh}\right)t_b\right)\right]$, which in mix with Equation

$$E\left[E_j\right] = E\left[L_j\right] \frac{\frac{m}{R_c^*} + m_{oh}}{m + m_{oh}}$$

That is $E\left[E_j\right]$ is the normal estimation of the standardized energy utilization, E_j, over h_j hop. Eventually, the usual use of energy for each course is obtained as the summation of the normal quality for each hop in the process, as the researchers will as the normal full latency.

Numerical Evaluation

In this region the researchers can demonstrate the results of the quantitative analysis carried out in MATLAB. It depends on the model of the framework. The findings will delineate differences between ARQ and ARQ-Hybrid-FEC-Adaptive with respect to energy usage, latency and loss of packets. The software and courses on energy skills were optimized for all outcomes. The un-coded packet length, including the overhead size, is selected by WirelessHART's packet design. The overhead. is a 5-byte synchronization header (SHR). Since impeccable matching is accepted, this sequence cot changes or effect the determined PER, but it will add to the radio receiver and the radio transmitter energy required.

The set-up of 70 BCH codes shown in Table is used. The show assessment measures the uniform expected use of power, [$E\left[E_j\right]$], as an element of the signal to-noise proportion (SNR_3), over the short hop h_3. In the spin-off SNR_3 is reffered to simplicity. What's more, in story myths, ARQ-Hybrid-FEC-Adaptiveis truncated as HAF. The different modules used in the display

Figure 6. Block graph showing the numerical assessment, where [$E\left[E^{HAF}\right]$ speak to the energy consumption with upgraded course and BCH code.

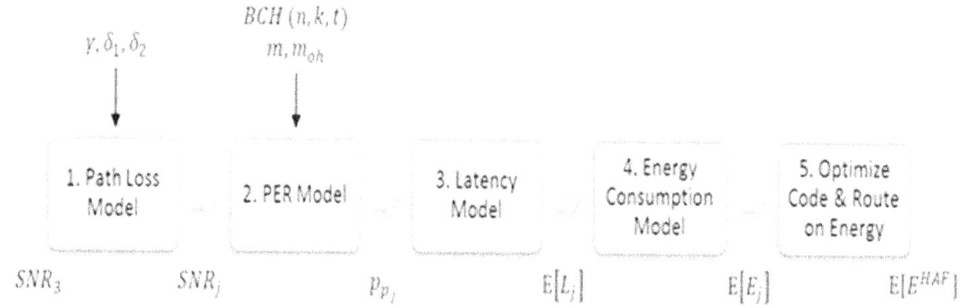

Figure 7. Graph showing (a) error probability packet, and the (b) energy expected consumption.

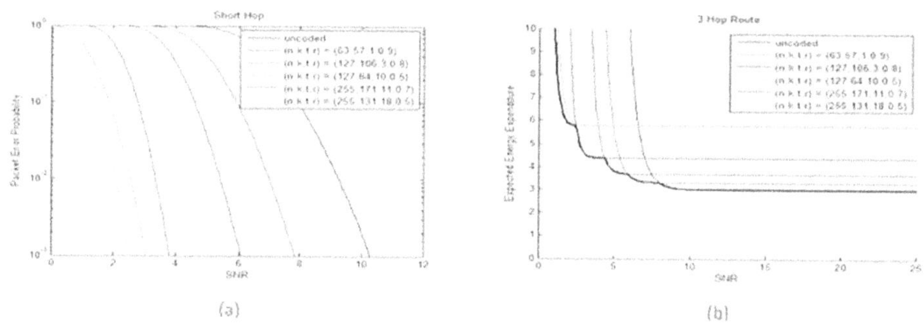

Energy Consumption

Normal energy uses of the ARQ and the ARQ-Hybrid-FEC-Adaptive plans as a component of SNR. The bars above the figure refer to the courses chosen for each project by the enhancement, and demonstrate that hoping is invaluable at what SNR. When the researchers think about the bars for ARQ-Hybrid-FEC-Adaptive and ARQ, the researchers see that ARQ-Hybrid-FEC-Adaptive requires only one hop down to a $SNR \approx 16dB$. For this SNR two hops are required for this ARQ. Nonetheless, with deference, ARQ-Hybrid-FEC-Adaptive is stronger than energy consumption as the researchers will as latency. While ARQ needs three hops from $SNR \approx 16dB$ and down, ARQ-Hybrid-FEC-Adaptive will bounce at $SNR \approx 11dB$. In other words, if the researchers ~$5dB$ use ARQ-Hybrid-FEC- Adaptive-FEC instead of ARQ, the researchers benefit. For two hops, similar holds.

In the SNR intervals 21dB-16dB and 16dB–11dB, the efficiency gains for ARQ-Hybrid-FEC-Adaptive varies from 0 percent to 100 percent individually. The benefit is a transmission at equal intervals of latency, see. In addition, the bend of the code frequency reflects the codes used for the ARQ-Hybrid-

FEC-Adaptive story. Notice that the code rate bend in the middle area, where the two-hop course is used, refers to the code level applied to the more comprehensive of the two hops for example $^{h}2$.

It is ideal not to use coding for the shorter hops in this locale. Figure 5.3 also shows that ARQ-Hybrid-FEC-Adaptive energy use is never more noteworthy than ARQ's. Despite this, for terrible channels, with *SNR* < 7.5*dB*, the researchers note that ARQ-Hybrid-FEC-Adaptive beats ARQ drastically, and for *SNR* < 6 *dB* the energy use of ARQ, due to such a large number of retransmissions, it reaches extraordinary qualities. In Figure 5.4, the 95th percentile delineates the changeability in energy usage. The fluctuation in the expected use of energy is obviously larger for ARQ than for ARQ-Hybrid-FEC-Adaptive.

Latency

Normal latency as a portion of the received SNR, where latency is defined as the total number of transmissions beginning to end, including retransmissions. Note that the energy efficiency codes and courses are streamlined. Latency, shown here, is the result of this progress. The normal dormancy shows a similar behaviour IN as the normal energy. The researchers note that for a wide range of SNRs, ARQ-Hybrid-FEC-Adaptive beats ARQ to latency. As gained in Figure, the ARQ conspire is unable to deal *SNR* < 6 *dB*, while ARQ-Hybrid-FEC-Adaptive can adapt to a latency of <4 transmissions starting to finish for *SNR* > *0.4 dB*. The two uncommon changes in code rate, that show up at around 11dB and 16dB, relates to a switch in course. the Hybrid-ARQ-Moreover, the tops in the ARQ-Hybrid-FEC-Adaptive bend are due to a higher retransmission. The results state what is normal, that the more prevalent codes are used for lower SNR channels.

Therefore, if the researchers observe the ARQ-Hybrid-FEC-Adaptive bend from right to left, which implies a decrease in channel performance, and note that the effects are optimized in terms of energy, the researchers see how the ARQ-Hybrid-FEC-Adaptive conspires to stifle the quantity of retransmissions by using increasingly incredible codes until it reaches a breakpoint where it is more energetic. See that the tops in the bend of latency relate to a code rate adjustment. It suggests the cost, as far as fuel, which is included by the use of forward error correction codes, is not necessarily the cost of retransmissions.

Constraint on Latency

The two plans are set to streamline energy efficiency in the results, implemented up to this point. This means choosing the ideal course for ARQ, while for ARQ-Hybrid-FEC-Adaptive it means choosing the ideal combination of both the code and course. Once latency criteria are laid, the researchers demonstrate the layout of the two plans. Here, despite everything the researchers improve the productivity of energy, but now with the condition that the 99th percentile for latency should never exceed one retransmission. Going from right to left in Figure 10 below, the researchers see that ARQ is forced to switch to courses with a higher number of hops earlier than in Figure 8, eventually collapsing at around 8dB. Strikingly, the ARQ-Hybrid-FEC-Adaptive applies even more powerful codes to conform to the requirement, and the hop shift focuses are not really affected at the cost of a slight increase in energy consumption. This is a desirable aspect of ARQ-hybird-FEC-Adaptive.

Figure 8. (a) Expected consumption of energy versus SNR consumption of energy is normalized to that of a single hop single transmission for a packet un-coded. (b) Shows the spread of the consumption of energy.

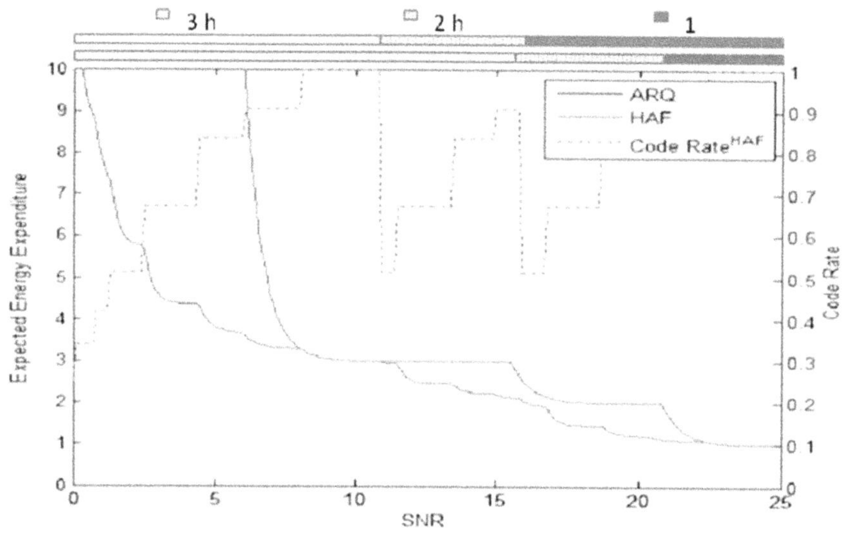

Expected consumption of energy versus SNR Consumption of energy is normalized to the of a single hop

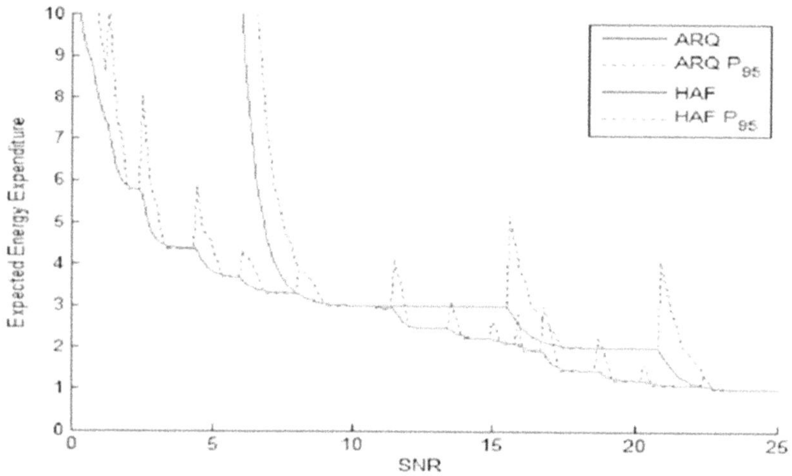

spread of the Consumption of energy

Fading Channel

The fading stream used is from a state of a paper factory. where an estimation crusade was performed. In Figure 10, the linear is delineated, standardized at. It takes after a fading stream from Rayleigh.

Packets are distributed through the paper plant to various sensor hubs. That hub logged the performance of the signal that are transmitted and received over a duration of 14 hours. The increase here is defined as the value of the signal over the transmitted power. A handful of hubs have shown more horrible channels than others, and the divert used in this analysis points to a run of the channel of the mille state. The gain dispersion has been reported from above figures.

Figure 9. (a) latency versus SNR. The number of transmission end to end, including retransmissions is latency. (b) shows the spread of the latency via the 95th percentile.

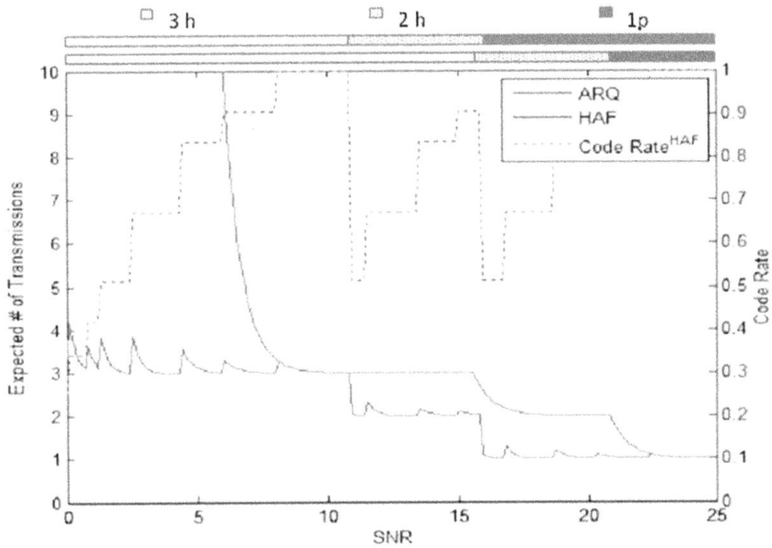

latency verses *SNR*.

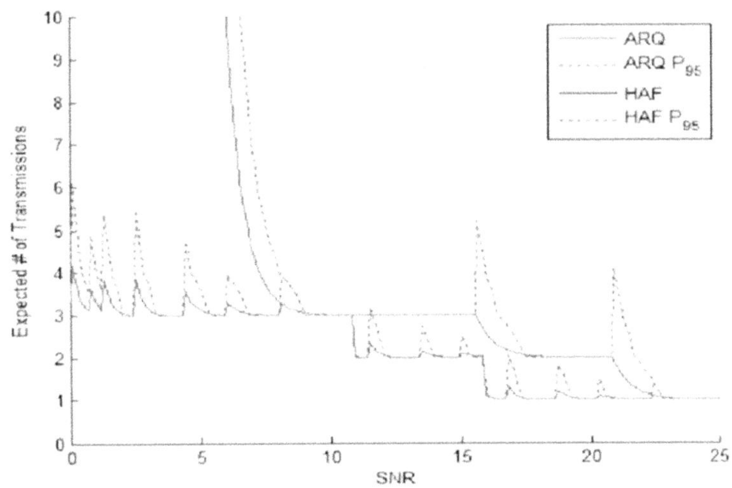

latency via the 95th percentile.

In this contrast, the researchers have placed the latency imperative on the absolute number of retransmissions. Figure below indicates the energy bends over the fading appropriation for the fading case. the normal packet loss for an AWGN, probability of energy and a fading situation in Figures below. On the off chance that the researchers think about ARQ for AWGN and fading channels, the researchers note that the probability of packet loss begins to expand from almost zero at *~8 dB* in the case of AWGN while this occurs at the *~17 dB* in the case of fading. The related SNRs are and *~1 dB* and *~10 dB* are individually for the Hybrid ARQ-Adaptive-FEC case. In other words, the researchers lose some 9dB

Figure 10. Channel Fading conducted in a factory from a radio channel campaign estimation.

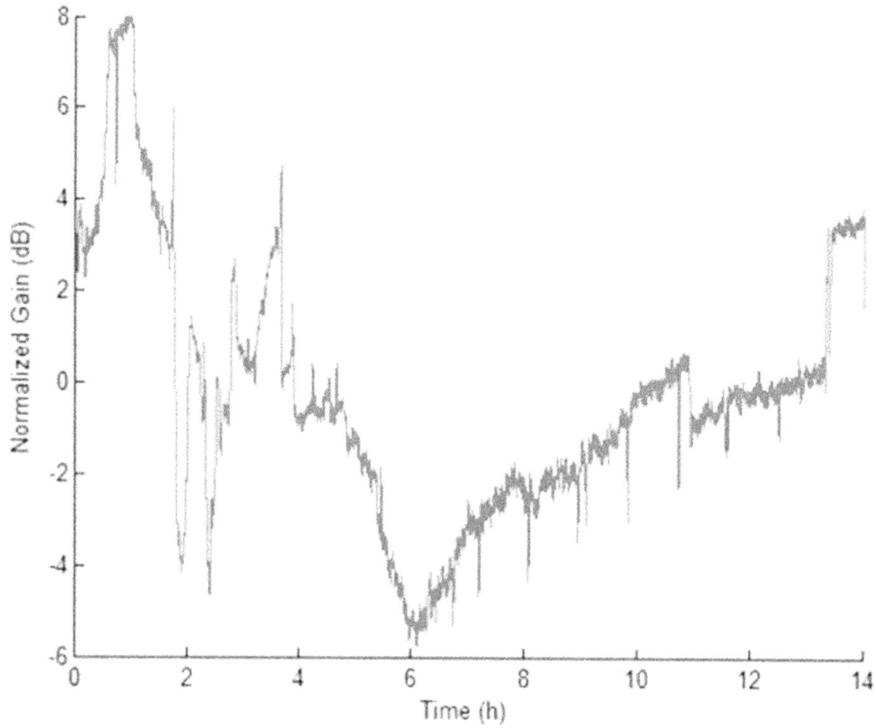

Figure 11. The thickness capacity probability of the standardized channel gain.

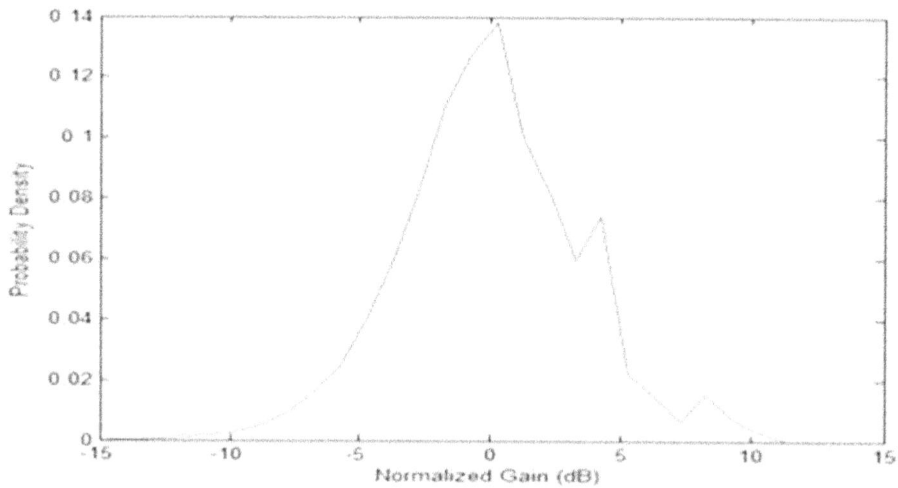

Figure 12. Energy expected utilization for packets effective and loss packet probability is shown, for an AWGN channel.

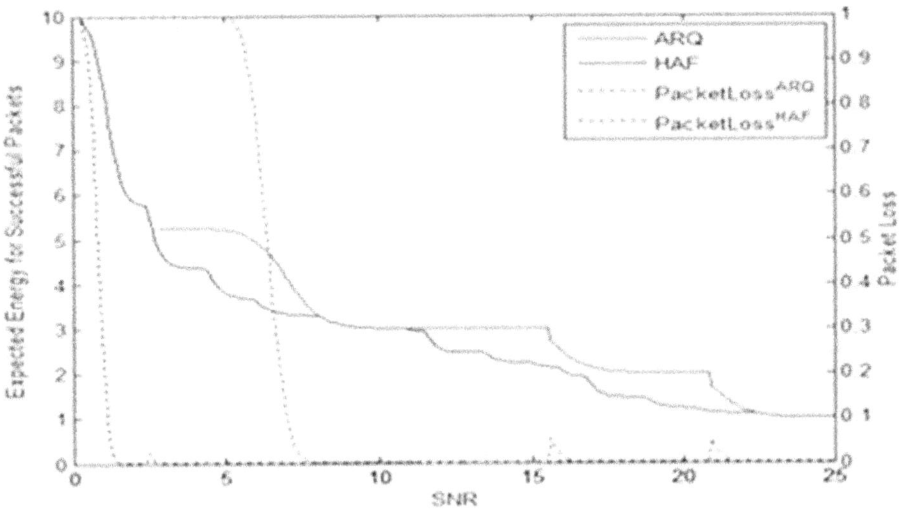

Figure 13 .Energy utilization for packets effective and loss packet probability are shown for a fading channel from an industrial site.

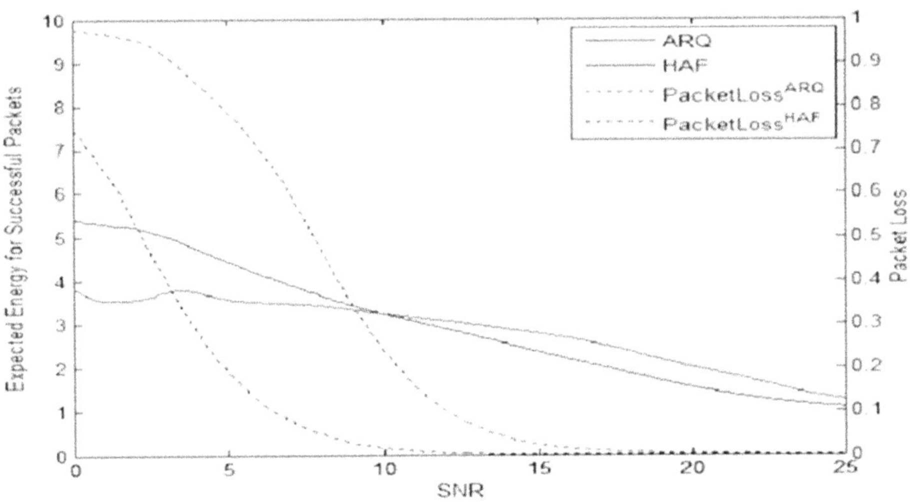

5 12:

due to the fading condition. Furthermore, on the off chance that the researchers look at the display for normal energy in the (transmitting duration) interval [1–7] probabilities of the packet that loss in the transmission in the interval [0.1–0.7]), the researchers have a 6dB gain in SNR.

In comparison to ARQ for AWGN, for Hybrid-ARQ-Adaptive-FEC. In the case of fading, the gain is 5dB. It is also significant that the normal use of energy for Hybrid-ARQ-Adaptive FEC in the fading case interval [0–9] dB is greater or higher than the comparative figures for the ARQ case in the fading case.

This is any path because of the path that ARQ-Hybrid-FEC-Adaptive uses to keep the packet loss of the packets low. See, for example, at $SNR \approx 7\ dB$in the fading case . ARQ-Hybrid-FEC-Adaptive at that stage has a normal energy consumption that is $\sim 15\ dB$ higher than ARQ, but the likelihood of packet loss for ARQ 0.6 is 0.05 for Hybrid-ARQ-Adaptive-FEC. Another focus in the fading case is when ARQ and ARQ-Hybrid-FEC-Adaptive use a similar amount of energy. At that point, the probability of packet loss for ARQ is 0.3 while in the case of ARQ-Hybrid-FEC-Adaptive it is about zero. In contrast to ARQ, this correlation also shows the intensity of Hybrid-ARQ Adaptive-FEC in the fading.

DISCUSSION AND CONCLUSION

The researchers have indicated that in contrast to ARQ, ARQ-Hybrid-FEC-Adaptive shows critical points of interest in anticipated energy consumption, packet loss and latency. The ARQ-Hybrid-FEC-Adaptive typically has less power use as compared to ARQ. The increase in specific areas of SNR (signal to noise ratio) is as much as it is with a comparable use of power for both projects. There is a correlation between the amount of energy consumed and the amount of hops reduced. This exchange is reduced by – comparing with ARQ with Hybrid-ARQ-Adaptive-FEC. At the end of the day, ARQ-Hybrid-FEC-Adaptive can use fewer hops with equivalent energy consumption.

In comparison to the ARQ scenario, the ARQ-Hybrid-FEC-Adaptive plot is also ready to support adequate packet error rates when channel conditions are exacerbated by 6dB. On the off chance, this suggests that CSI be nourished once again by welcoming hubs, at which point it would be extraordinarily appealing to use Hybrid-ARQ-Adaptive-FEC, or any comparable adaptive coding program. For example, integrating CSI in WirelessHART's ACK-packets would be a way to benefit from using ARQ-Hybrid-FEC-Adaptive in a process control situation. This is currently being compared.

REFERENCES

Akyildiz, I. F., & Vuran, M. C. (2010). Error Control. In *Wireless Sensor Networks*. John Wiley & Sons Ltd. doi:10.1002/9780470515181.ch6

Balakrishnan, G., Yang, M., Jiang, Y., & Kim, Y. (2007). Performance analysis of error control codes for wireless sensor networks. *Int. Conf. Information Technology*, 876-879. 10.1109/ITNG.2007.149

Björnemo, E. (2009). Error correcting codes in static channels. Energy Constrained Wireless Sensor Networks.

Björnemo, E., Ahlén, A., & Gidlund, M. (2010). WiComPI: Wireless control for the process industri. *VINNOVA Workshop on Smarter, Faster, and Converging Solutions.*

Cho, S. (2000). Adaptive error control scheme for multimedia applications in integrated terrestrialsatellite wireless networks. *Wireless Communications and Networking Conference*. 10.1109/WCNC.2000.903926

De Biasi, M., Snickars, C., Landernäs, K., & Isaksson, A. J. (2008). Simulation of Process Control with WirelessHART Networks Subject to Packet Losses. In *4th IEEE Conference on Automation Science and Engineering*, (pp. 548-553). Washington, DC: IEEE. 10.1109/COASE.2008.4626509

Goldsmith, A. (2005). *Wireless Communications*. Cambridge University Press. doi:10.1017/CBO9780511841224

Gungor, V. C., & Hancke, G. P. (2009). Industrial Wireless Sensor Networks: Challenges, Design Principles, and Technical Approaches. *IEEE Transactions on Industrial Electronics*, *56*(10), 4258–4265. doi:10.1109/TIE.2009.2015754

HART Communication Foundation. (2007). *2.4GHz DSSS O-QPSK Physical Layer Specification*. HART Communication Foundation.

Howard, S. L., Schlegel, C., Iniewski, K., & Iniewski, K. (2006). Error Control Coding in Low-Power Wireless Sensor Networks: When is ECC Energy-Efficient. *EURASIP Journal on Wireless Communications and Networking*, *2006*(1), 1–14. doi:10.1155/WCN/2006/74812

IEEE. (1964). Table of generators for Bose-Chaudhuri codes. *IEEE Transactions on Information Theory*, 390.

IEEE. (2006). *IEEE Std 802.15.4*. Institute of Electrical and Electronics Engineers, Inc.

Kleinschmidt, J. H., Borelli, W. C., & Pellenz, M. E. (2007). An analytical model for energy efficiency of error control schemes in sensor networks. In *International Conference on Communications 2007*, (pp. 3895-3900). 10.1109/ICC.2007.642

Min, R., Bhardwaj, M., Cho, S.-H., Ickes, N., Shih, E., Sinha, A., & .... (2002). Energy-Centric Enabling Technologies for Wireless Sensor Networks. *IEEE Wireless Communications*, *9*(4), 28–39. doi:10.1109/MWC.2002.1028875

Proakis, J. G. (2001). Block and Convolutional Channel Codes. In *Digital Communications*. McGraw-Hill.

Saifullah, A., Xu, Y., Lu, C., & Chen, Y. (2010). Real-Time Scheduling for WirelessHART Networks. In *2010 31st IEEE Real-Time Systems Symposium*, (pp. 150-159). San Diego, CA: IEEE.

Sankarasubramaniam, Y., Akyildiz, I. F., & McLaughlin, S. W. (2003). Energy efficiency based packet size optimization in wireless sensor networks. *IEEE Int. Workshop on Sensor Network Protocols and Applications*, 1-8. 10.1109/SNPA.2003.1203351

Shih, E., Cho, S., Lee, F. S., Calhoun, B. H., & Chandrakasan, A. (2004). Design Considerations for Energy-Efficient Radios in Wireless Microsensor Networks. *The Journal of VLSI Signal Processing*, *37*(1), 77–94. doi:10.1023/B:VLSI.0000017004.57230.91

Shiozaki, A., Okuno, K., Suzuki, K., & Segawa, T. (1991). A Hybrid ARQ Scheme with Adaptive Forward Error Correction for Satellite Communications. *IEEE Transactions on Communications*, *39*(4), 482–484. doi:10.1109/26.81734

Sklar, B. (2003). Modulation and Coding Trade-Offs. In *Digital Communications*. Prentice Hall PTR.

Song, J., Han, S., & Mok, A. K. (2008). WirelessHART: Applying Wireless Technology in Real-Time Industrial Process Control. *IEEE Real-Time and Embedded Technology and Applications Symposium*, (pp. 377-386). 10.1109/RTAS.2008.15

Spiegel, M. R. (1992). *Theory and Problems of Probability and Statistics*. McGraw-Hill.

Vuran, M. C., & Akyildiz, I. F. (2009). Error Control in Wireless Sensor Networks: A Cross Layer Analysis. *IEEE/ACM Transactions on Networking*, *17*(4), 1186–1199. doi:10.1109/TNET.2008.2009971

Chapter 10
Higher Data Rate Transmission for Underwater Wireless Sensor Communications Using Industrial, Scientific, and Medical (ISM) Bands

Manish Sharma
Chitkara University Institute of Engineering and Technology, Chitkara University, Punjab, India

ABSTRACT

It is known that for underwater communication systems, lower frequency bands are used, which also affects the data rate transmission at lower rate. This is due to different physical phenomena such as reflections, refraction, and energy dispersion. This can be mitigated by installing wireless sensors that are placed close to each other, and hence, for accurate measurements, higher communication bandwidth is required. By using real scenario for measurement at higher data rate, 2.4GHz ISM band is used efficiently for underwater communication system with efficient use of energy.

INTRODUCTION

It is known fact that for underwater communication system, lower frequency bands are used which also effects the data rate transmission at lower rate which are due to physical phenomena such as reflections, refraction and energy dispersion. These above said problems can be mitigated by installing wireless sensors network which are mandatory placed closed to each other and hence for accurate measurements, higher communication bandwidth is required. By using real scenario for measurement at higher data rate, 2.40GHz ISM band is used efficiently for underwater communication system with efficient use of energy. Wireless sensor network has drawn lots of attention for variety of applications. This wireless technology system suffers demerit related to consumption of energy when the number of nodes gets increases. Alternatively, batteries can be powered to run this existing wireless sensor network but us-

DOI: 10.4018/978-1-7998-3640-7.ch010

ing them for underwater wireless communication is also a bigger problem as the issue related to charge them arises. Thus an alternate solution for saving of power becomes necessary (Sendra et al., 2011). A group of cluster cells can be used as underwater local area network for underwater sensor network. By having several nodes in the cluster, data rate can be increased with multiple hops. It is also known that the selection of aggregators, intra-routing (routing from sensor to aggregator) and inter routing (routing from aggregator to sink) are the three main steps involved in data aggregation. By using Fuzzy Logic technique, clustering technique and aggregation can be simplified and optimize the parameters such as residual energy, distance to sink, Node density and link quality (Goyal et al., 2014). One of the major problems faced in underwater wireless communication is that the wireless sensor network gets affected leading to failure of cluster head or cluster member. Fault detection and recovery technique finds the solution for above said problems (Goyal et al., 2018). Electrical property of the wireless system including permeability of the data of liquid water plays vital role in underwater communication system for frequencies upto 1THz which is characterized by double Debye model (Liebe et al., 1991). Underwater communication system requires large amount of energy and also long data transmission of packet thus by effecting cluster head, cluster size and routine scheme discussed earlier. By application of Fuzzy logic in selection of cluster head and cluster size, improvement in the performance in the underwater sensor networks is observed (Goyal et al., 2016). Industrial, Scientific and Radio Band (ISM) working at microwave frequency of 2.40GHz eliminates low frequency demerits for underwater communication such as reflections, refractions, energy dispersion etc. For more accurate measurement in underwater, it becomes necessary that wireless sensors needs t be placed close to each other (Lloret et al., 2012). By using sleep-wake up algorithm, aggregation of data is improved at the cluster head and thus by scheduling of data transmission, power required for underwater communication can be reduced (Goyal et al., 2017a). Cluster based underwater sensor communication networks can be also used for congestion control and load balancing technique in underwater-wireless-sensor-network for cluster. Also the network performance is enhanced by controlling the congestion and balancing the traffic (Goyal et al., 2016). At lower frequencies, communication link can be established for very longer distance which also prevents power losses that are generated for higher frequencies. By keeping frequency as low as 3 KHz can achieve communication of about 40 meters between nodes. Due to the advantage of lower error probability, BPSK modulation system is best suited for underwater wireless communication (Sendra et al., 2011). Path loss of the signal traveling from one end to other is also a major concern in underwater communication system and hence restriction in use of RF signals is limited (Chaitanya et al., 2011). Also, the nodes which are created for sensor networks are combined and monitored for underwater communication for a defined smaller area (Goyal et al., 2017b). Also, the dielectric constant of the water is much higher which effects the propagation parameters of electromagnetic waves used for underwater communication. These few parameters are skin depth, total path loss and frequency function which decides the distance of wireless communication between transmitter and receiver (Chakraborty et al., 2009). Potentially, three different technologies including radio, acoustic and optical form the basis of underwater communication system. It was noted that for small coastal erosions Electromagnetic based communication system for underwater communication provided different advantages when compared with the other two systems (Che et al., 2010). It is also known that aggregate data base is divided into cluster or non-cluster category. Underwater clustering scheme and round based clustering scheme are used generally for underwater communication (Goyal et al., 2019c). There are number of underwater communication system based on sensor network such as collection of data related to graphical structure of ocean, disaster/ prevention management etc. Communication related to underwater involves collection of data from the nodes and then communicate with the sink/load (Goyal et al., 2019a). Underwater communication system suffers drawbacks such as large delay, node mobility and limitation in link capacity (Kumar et al., 2014). An adaptive error control mechanism can overcome demerits related to underwater sensor network and also different architecture for 2-Dimension or 3-Dimension sensor network is proposed (Goyal et al., 2018b: Akyildiz et al., 2004). Also, a PN-sequence (pseudo noise) based underwater communication can be used where a modulated by using binary-shift-keying-technology (Nguyen et al., 2017). Also reactive protocols are generally used for underwater communication network (Garcia et al., 2012: Yadav et al., 2019). By calculating transmission propagation loss of EM penetrating from air to water generalizes the evidence for applications of RF waves in applications for different types of water system (Jiang et al., 2011: Liu et al., 2008). There are various technologies has been proposed for underwater communication system such as CAN (Cognitive Acoustic Networks), SDN (Software Defined Networks), NFV (Network Function Virtualization), Cloud-Computing, Fog-Computing and internet of underwater-communication things. (Huma et al., 2019). It is well known fact that radio waves as well as optical waves suffers huge attenuation for communication in underwater communication networks. Based on the assumption that unknown nodes are moving and with actual speed of light, proposed algorithm reveals that good performance is achieved which is non-sensitive to variations in velocity of sound and movement of node underwater (Shaochen et al., 2019). Considering applications in commercial as well as military fields, positioning technology for underwater communication network is very sensitive and important. Challenges such as fluidity, sparse installation with limitation of energy are to be taken care (En et, al., 2019).

Attempts are being made for increase in the distance and the required bandwidth for underwater wireless sensor network system. On the other side, high frequency RF signals in ISM band at 2.40GHz requires larger energy than conventional acoustic/sound underwater communication. Generally underwater communication system such as optical signals, EM (Electromagnetic) signals and ultrasonic signals are focused. It is also the fact that due to the different environment provided by underwater such as chemical composition, above said technology ha its own advantages and disadvantages. Optical communication link faces drawbacks such as back-scattering which is responsible by suspended particles lying underwater and hence the limit of far distance communication affected by the turbidity caused by water flow and hence system fails for longer distance communication. On the other hand, acoustic systems overcome the problem caused by optical communication and show no hindrance to turbidity of water. They can be used a far as around communication for 20Km but, suffers drawbacks as low data rate transmission due to lower frequency band. Also some more drawbacks such a lower carrier frequency, very strong reflections along with attenuation, limits the use of such system for underwater communication. Alternatively, different modulation techniques can be employed to better increase the data rate transfer. Hence, to overcome all the above needs, radio frequency in underwater communication is the best suited but has the limitations that it can be used for very shorter distance. Also, RF communication has an advantage over acoustic system is that it travels with very high speed and also very less immune to reflection and refraction for suspended particles lying underwater.

Underwater EM Propagation (ISM 2.4GHZ band)

Practical study of electromagnetic spectrum for underwater communication analyzed for IEEE802.11b standard for 2.40GHz is studied. A comparative study is shown in Table I which signifies the different wireless communication bands and their respective data transfer rate. It can be also concluded from Table I that ISM band under different wireless protocol is well suited for underwater communication network.

It becomes necessary to study the fundamentals of electromagnetic propagation in underwater communication for 2.40GHz ISM band of IEEE8.2.11 standard and finds several advantages over lower frequency band such as higher data transmission between the nodes. Also, behavior of EM wave varies for different types of water. For consideration of fresh water communication, propagation of EM wave can be approximately given by

$$\text{Speed} \approx \frac{1}{\sqrt{(1+S_e)\times 8.854\times 10^{-12}\times \mu_r \times 4\pi \times 10^{-7}}} \tag{1.1}$$

Where S_e is the electric susceptibility of the medium where EM wave propagates and μr is the permeability of the medium.

Table I. Comparison of different RF bands

Protocol	Frequency Band	Channel Width	Maximum Data Rate Transfer
8.2.11ax	2.4GHz/5.0GHz	20MHz, 40MHz, 80MHz, 160MHz	2.40Gbps
8.2.11n	2.4GHz/5.0GHz	20MHz, 40MHz	450Mbps
8.2.11g	2.4GHz	20MHz	54Mbps
8.2.11a	5.0GHz	20MHz	54Mbps
8.2.11b	2.4GHz	20MHz	11 Mbps

EM wave propagation is governed by

$$\gamma=\alpha+j\beta \tag{1.2}$$

where γ is the propagation constant, α is attenuation constant and β is phase constant.

Propagation of EM wave is based on Maxwell's equations which are given by

$$\nabla\times H = J + \frac{\partial D}{\partial t} \tag{1.3}$$

$$\nabla\times E = -\frac{\partial B}{\partial t} \tag{1.4}$$

$$\nabla . D = \rho_v \tag{1.5}$$

$$\nabla . H = 0 \tag{1.6}$$

$$\frac{E}{H} = \sqrt{\frac{\mu_o}{\varepsilon_o}} \tag{1.7}$$

Where

D=Electric Flux Density, B=Magnetic Flux Density, ε_o=permittivity in free space (8.854 $\times 10^{-12} F/m$), μo=permeability in free space = $4\pi \times 10^{-7} H/m$.

Also, penetration of EM wave in underwater communication is defined in terms of skin depth which is given as "The skin depth is a measure of depth to which an EM wave can penetrate the medium (underwater)" and is given by

$$\delta = \frac{1}{\sqrt{\pi f \mu \sigma}} = \frac{1}{\alpha} \tag{1.8}$$

Absorption coefficient of the wave is given by

$$\alpha \cong \frac{\sigma}{2} \sqrt{\frac{4\pi \times 10^{-7}}{(1+S_e) \times 8.854 \times 10^{-12}}} \tag{1.9}$$

where σ is electrical conductivity.

Figure 1 shows communication between transmitter and receiver in water medium. It shows that transmitted signals suffer multiple path fading as it reaches the receiver. Path gain for any i^{th} signal from transmitter to receiver can be calculated by

Figure 1. Communication between transmitter and receiver which are receiving signals by multiple path reflection

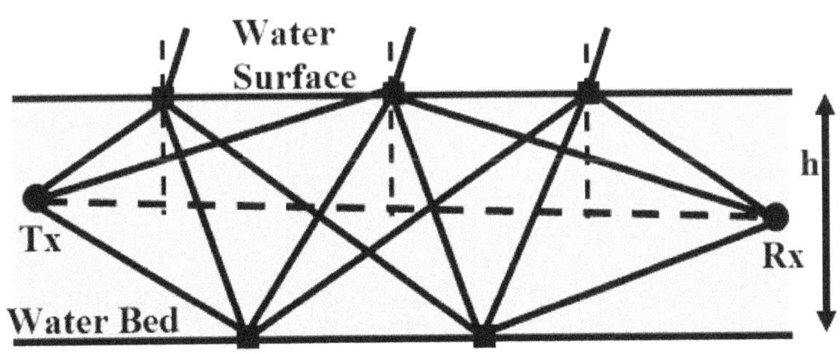

$$P_G = \frac{\Gamma}{\sqrt{G_A}} \tag{2.0}$$

Where Γ is reflection coefficient of ith path, GA is attenuation which is function of (L,f) where L is path length (i^{th}) and f is the frequency. L/c give the path delay.

Table 2. Comparison of acoustic and EM wave used for underwater communication

Parameters	Ground	Underwater
Type of Signal	EM	Sound
Speed	3×10⁸ m/s	1500 m/s
Propagation (Time)	less	high
Power	low	high
Source	EM source	Piezoelectric

Also, propagation of electromagnetic waves in water medium dependent on permeability, permittivity, conductivity and volume charge density. Maxwells equation in time varying field for a underwater medium is given by

$$\nabla^2 E - \mu\sigma \frac{\partial E}{\partial t} - \mu\varepsilon \frac{\partial^2 E}{\partial t^2} = \frac{\nabla \rho_v}{\varepsilon} \tag{2.1}$$

Also magnetic field time-varying equation is given by

$$\nabla^2 H - \mu\sigma \frac{\partial H}{\partial t} - \mu\varepsilon \frac{\partial^2 H}{\partial t^2} = 0 \tag{2.2}$$

Solution for wave equation from Equations 11 and 12 is given by

$$\nabla^2 \phi = \gamma^2 \phi \tag{2.3}$$

ϕ is field density (electric or magnetic) and

$$\gamma = \pm\sqrt{j\omega\mu(\sigma + j\omega\varepsilon} \tag{2.4}$$

It is also known that when wave propagates, it radiates isotropically which means equal radiations in all directions. If z is the direction of propagation of TEM wave, then its respective E and H component is given by

$$E(z,t) = E_o e^{j(\omega t - \beta z)} a_x \quad (2.5)$$

$$H(z,t) = H_o e^{j(\omega t - \beta z)} a_y \quad (2.6)$$

Wavelength of wave in ideal medium with conditions is given by

$$\lambda = \frac{2\pi}{k} = \frac{2\pi}{\omega\sqrt{\mu\varepsilon}} = \frac{c}{f\sqrt{\varepsilon_r}} = \frac{\lambda_o}{\sqrt{\varepsilon_r}} \quad (2.7)$$

$$\alpha = \omega\sqrt{\frac{\mu\varepsilon}{2}[\sqrt{1+(\frac{\sigma}{\omega\varepsilon})^2}]-1} \text{ nepers/m} \quad (2.)$$

$$\beta = \omega\sqrt{\frac{\mu\varepsilon}{2}[\sqrt{1+(\frac{\sigma}{\omega\varepsilon})^2}]+1} \text{ radians/m} \quad (2.9)$$

Electromagnetic waves when passes through one to another medium, interfacing losses also occurs which is governed by following equation

$$Loss(dB) = -20\log\{(7.4586/106) \times \sqrt{f/\sigma} \quad (3.0)$$

It can be also summarized that using ISM band for underwater communication exhibits advantages such as high speed of propagation, can cross boundary between two interfaces (water to air), lower resistance to multipath fading, no impact on sea mammals and is also shows no effect with lower visibility. Few applications regarding underwater communication in ISM 2.40GHz band can be for navigation, sensing and for communication between wireless sensor nodes. There should be proper synchronization between transmitter and receiver including match able modulation wireless communication technique. AUVs (Autonomous-Underwater-Vehicles) are deployed for graphical analysis of underwater ocean with suitable design of transmitter and receiver. This underwater system can be controlled by wireless system to communicate with outside world. Also, this underwater AUVs are to be a robot and have capability of underwater survelliance without any assistance. They are broadly known by unmanned-underwater-vehicles and are controlled non-autonoumsly.

Also, underwater vehicles include remotely controlled vehicles from surface of water and are sometimes are referred as unmanned-undersea-vehicle for military applications. Few of the applications with respect to range is listed in Table 3 given above.

Table 3. Applications of RF communication system for underwater (Range dependent)

Distance	Less than 1m	For 10m	For 50m	For 200m	For 10Km
Applications	AUV docking	Data download (sensor network)	Networking between nodes	Control of AUV	Telemetry (Deep water)

Underwater Performance of EM Communication System

For different types of water like sea water, fresh water, free space and acoustic medium a comparison with respect to parameters is shown in Table 4 while important parameters such as propagation velocity (m/s), wavelength (m) and distance of propagation for 100dB attenuation is studied. It is also recorded that above 10KHz signal frequency for underwater communication signals moves very faster than signals compared to acoustic.

Fresh Water Underwater Experimental Setup

Underwater communication system is experimentally studied by using EM waves for different wireless sensor nodes. The RTT value is estimated/calculated and then the loss of packet for each modulation technique is analyzed.

Table 4. Performance of EM underwater communication

Parameters	Different Conditions	Frequencies (Hz)					
		100	1000	10000	100000	1 Million	10 Million
Propagation Velocity (m/s)	Sea Water	1.78×10^{4}	4.90×10^{4}	1.54×10^{5}	4.84×10^{5}	1.54×10^{6}	4.32×10^{6}
	Fresh Water	3.18×10^{5}	1.05×10^{6}	3.18×10^{6}	1.05×10^{7}	3.18×10^{7}	1.05×10^{8}
	Free Space	3.00×10^{8}	3.00×10^{8}	3.00×10^{8}	3.00×10^{8}	3.00×10^{8}	3.00×10^{8}
	Acoustic	1.50×10^{3}	1.50×10^{3}	1.50×10^{3}	1.50×10^{3}	1.50×10^{3}	1.50×10^{3}
Wavelength (m)	Sea Water	1.76×10^{2}	4.89×10^{1}	1.54×10^{1}	4.84×10^{0}	1.54×10^{0}	4.32×10^{-1}
	Fresh Water	3.18×10^{3}	1.02×10^{3}	3.18×10^{3}	1.05×10^{3}	3.18×10^{3}	1.05×10^{3}
	Free Space	3.00×10^{8}	3.00×10^{8}	3.00×10^{8}	3.00×10^{8}	3.00×10^{8}	3.00×10^{8}
Propagation distance (m) for 100 dB attenuation	Sea Water	3.24×10^{2}	8.94×10^{1}	2.80×10^{1}	8.82×10^{0}	2.90×10^{0}	7.89×10^{-1}
	Fresh Water	5.79×10^{3}	1.85×10^{3}	5.79×10^{2}	1.85×10^{2}	5.79×10^{1}	1.85×10^{1}

Setup for 1Mbps Data Transfer Rate

Comparison of two models for underwater system is shown in Figure 1.2 with respect to distance.

Figure 2. Communication setup for 1Mbps data loss in underwater wireless sensor networks (a) Packet Loss at 1Mbps (b) Average RTT (ms) for 1Mbps.

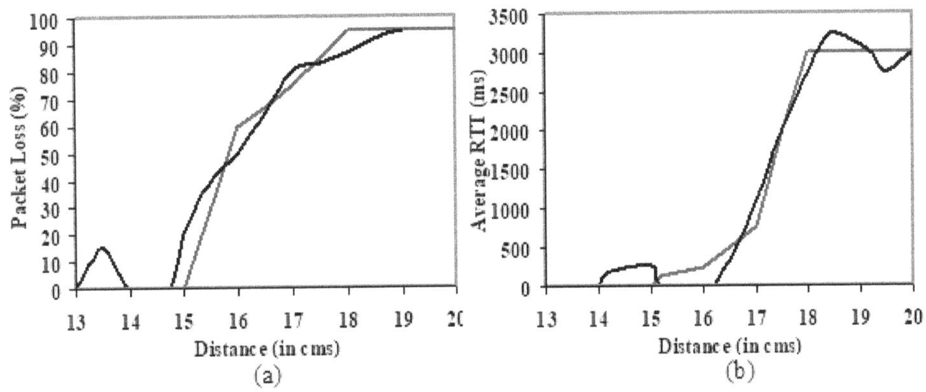

$$Packet\ Loss_{1Mbps}(\%) = \begin{vmatrix} -0.0012x^6 + 0.120x^5 - 5.018x^4 \\ +110.20x^3 - 1.352x^2 + 8.818x^1 - 23.796 \end{vmatrix} \quad (3.1)$$

$$RTT_{Average\_1Mbps}(ms) = \begin{vmatrix} 6.928x^6 - 681.21x^5 + 27.774x^4 \\ -60.072x^3 + 7\times10^7 x^2 - 5\times10^7 x^1 + 10^8 \end{vmatrix} \quad (3.2)$$

Figure 1.2 shows comparison of two different methods for underwater wireless sensor network communication setup. BPSK modulation technique is considered for both the models. As observed in Figure 2(a), for the distance range lesser than 15.5cms, lost data rate also do not exceed more than 20% for almost both the models. For around distance of 15.50cms, lost percent of data rate shoots up to 70% and almost in between 90%-100% for distance above 18.0cms. Black curve signifies following the curve represented by Equation 3.1. Also, Figure 1.2(b) shows the graph of average RTT with respective to distance for BPSK underwater modulation technique at data rate transfer of 1Mbps. It can be observed that the average value of RTT is less than 500ms for distance less than 16cm and is around 1000ms for distance 17cm. This value rises to about 3000ms for distance of 18cm due to the fact that many packets are lost. Also, the correlation coefficient between the two curves comes out to be 0.981 (R^2).

Setup for 2Mbps Data Transfer Rate

$$Packet\ Loss_{2Mbps}(\%) = \begin{vmatrix} -0.0011x^6 + 0.1020x^5 - 4.260x^4 \\ +93.841x^3 - 1.1562x^2 + 7.751x^1 - 20.431 \end{vmatrix} \quad (3.3)$$

Figure 3. Communication setup for 2Mbps data loss in underwater wireless sensor networks (a) Packet Loss at 2Mbps (b) Average RTT (ms) for 2Mbps

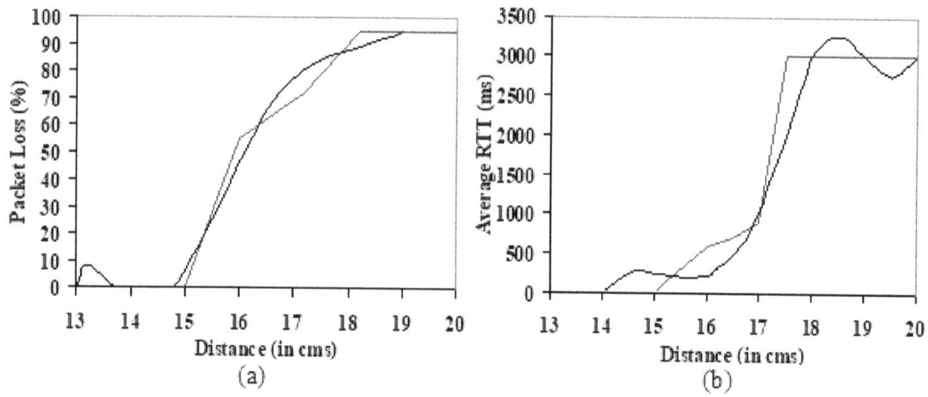

$$RTT_{Average\_2Mbps}(ms) = \begin{vmatrix} 5.091x^6 - 501.41x^5 + 20.478x^4 \\ -44.375x^3 + 5\times10^6 x^2 - 3\times10^7 x^1 + 9\times10^8 \end{vmatrix} \quad (3.4)$$

Figure 3 shows comparison of two different methods for underwater wireless sensor network communication setup. BPSK modulation technique is considered for both the models. As observed in Figure 1.3(a), for the distance range lesser than 15.5cms, lost data rate also do not exceed more than 10% for almost both the models. For around distance of 15.50cms, lost percent of data rate shoots up to 70% and almost in between 95% for distance above 17.0cms. Black curve signifies following the curve represented by Equation 3.3. Also, Figure 1.3(b) shows the graph of average RTT with respective to distance for BPSK underwater modulation technique at data rate transfer of 2Mbps. It can be observed that the average value of RTT is less than 500ms for distance less than 16.5cms and is around 1000ms for distance 17cm. This value rises to about 800ms for distance of 17.2cms and shoots upto 3000ms for distance of

Figure 4. Communication setup for 5.50Mbps data loss in underwater wireless sensor networks (a) Packet Loss at 5.50Mbps (b) Average RTT (ms) for 5.50Mbps

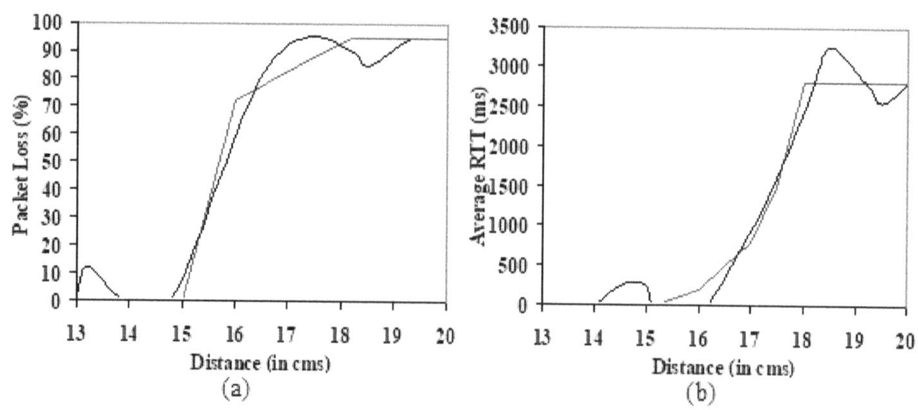

18.5cms due to the fact that many packets are lost. Also, the correlation coefficient between the two curves comes out to be 0.985 (R^2).

Setup for 5.50Mbps Data Transfer Rate

$$Packet\ Loss_{5.5\,Mbps}(\%) = \begin{vmatrix} -0.0020x^6 + 0.2080x^5 - 8.609x^4 \\ +188.4x^3 - 2.3031x^2 + 14.982x^1 - 40.314 \end{vmatrix} \tag{3.5}$$

$$RTT_{Average\_5.5\,Mbps}(ms) = \begin{vmatrix} 6.925x^6 - 681.22x^5 + 27.775x^4 \\ -60.072x^3 + 7\times10^7\,x^2 - 5\times10^7\,x^1 + 10^8 \end{vmatrix} \tag{3.6}$$

Figure 4 shows comparison of two different methods for underwater wireless sensor network communication setup. BPSK modulation technique is considered for both the models. As observed in Figure 1.4(a), for the distance range lesser than 15.5cms, lost data rate also do not exceed more than 10% for almost both the models. For around distance of 15.50cms, lost percent of data rate shoots up to 80% and almost in between 95% for distance above 17.5cms. Black curve signifies following the curve represented by Equation 3.5. Also, Figure 1.4(b) shows the graph of average RTT with respective to distance for BPSK underwater modulation technique at data rate transfer of 5.50Mbps. It can be observed that the average value of RTT is less than 500ms for distance less than 15.20cms and is around 1000ms for distance 17.20cm. This value rises to about 800ms for distance of 17.2cms and shoots upto 2250ms for distance of 18.00cms due to the fact that many packets are lost. Also, the correlation coefficient between the two curves comes out to be 0.992 (R^2).

Figure 5. Communication setup for 11.0Mbps data loss in underwater wireless sensor networks (a) Packet Loss at 11.0Mbps (b) Average RTT (ms) for 11.0Mbps

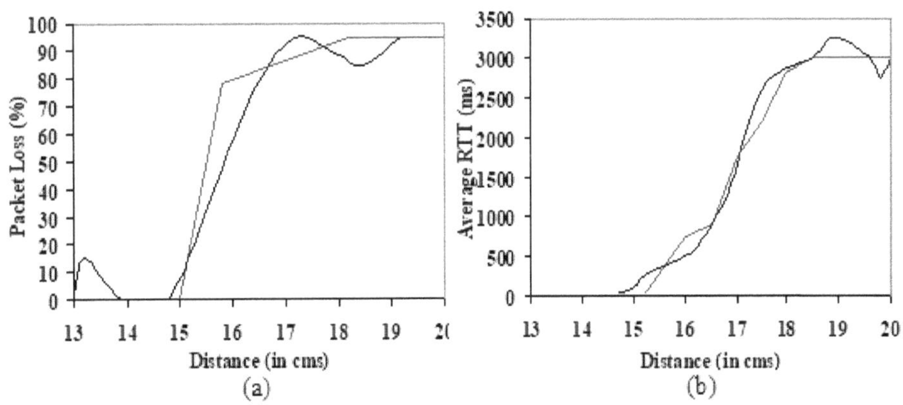

Setup for 11.0Mbps Data Transfer Rate

$$Packet\ Loss_{11Mbps}(\%) = \begin{vmatrix} -0.0020x^6 + 0.2290x^5 - 9.469x^4 \\ +206.97x^3 - 2.532x^2 + 16.426x^1 - 44.15 \end{vmatrix} \tag{3.7}$$

$$RTT_{Average\_11Mbps}(ms) = \begin{vmatrix} 2.1320x^6 - 205x^5 + 8.165x^4 \\ -17.238x^3 + 2\times10^6 x^2 - 1\times10^7 x^1 + 3\times10^7 \end{vmatrix} \tag{3.8}$$

Figure 5 shows comparison of two different methods for underwater wireless sensor network communication setup. BPSK modulation technique is considered for both the models. As observed in Figure 1.5(a), for the distance range lesser than 15.5cms, lost data rate also do not exceed more than 15% for almost both the models. For around distance of 15.50cms, lost percent of data rate shoots up to 80% and almost in between 95% for distance above 17.5cms. Black curve signifies following the curve represented by Equation 3.7. Also, Figure 1.5(b) shows the graph of average RTT with respective to distance for BPSK underwater modulation technique at data rate transfer of 11.0Mbps. It can be observed that the average value of RTT is less than 500ms for distance less than 15.20cms and is around 1000ms for distance 17.25cm. This value rises to about 800ms for distance of 17.20cms and shoots upto 2350ms for distance of 18.00cms due to the fact that many packets are lost. Also, the correlation coefficient between the two curves comes out to be 0.998 (R^2).

Table 5. Lost of packets (minimum amd maximum values)

	a_1	a_2	a_3	a_4	a_5	a_6	a_7
Minimum Value	-0.0021	0.1021	-0.9469	93.85	-2.5325	8.818	-44.151
Maximum Value	-0.0010	0.228	-4.262	206.90	-1.354	16.421	-20.431

When the signals are transmitted from transmitter to receiver for underwater wireless sensor network, the lost packets average and average RTT is calculated by following Equation 3.9 given as

$$Y = a_1x^6 + a_2x^5 + a_3x^4 + a_4x^3 + a_5x^2 + a_6x + a_7 \tag{3.9}$$

Here, a_1, a_2, a_3, a_4, a_5, a_6 and a_7 depend on transfer of data rate and also the distance.

Table 6. Average RTT (Minimum amd Maximum values)

	a_1	a_2	a_3	a_4	a_5	a_6	a_7
Minimum Value	2.134	-681.21	8.164	-60.072	2.0×10^6	-5.0×10^7	10^8
Maximum Value	6.927	-205.0	27.774	-17.237	6.0×10^7	-1.0×10^7	9.0×10^8

Comparison with Other Techniques

An intensive measurements comparison is shown for underwater technologies in Table 7.

Table 7. Comparison of acoustic, electromagnetic and optical waves

Type of Wave	Frequency (Working)	Wave Length	Modulation Technique	Distance Covered	Data Rate Transfer
EM Waves	3.0KHz	N/app	N/av	40.0m	100bps
EM Waves	100.0KHz	N/app	BPSK	6.00m	1.0Kbps
EM Waves	10.0KHz	N/app	BPSK	16.0m	1.0Kbps
EM Waves	1.0KHz	N/app	BPSK	2.00m	1.0Kbps
Optical Waves	N/av	420.00nm	PPM	1.80m	100Kbps
Acoustic Waves	800KHz	N/app	BPSK	1.00m	80Kbps
EM Waves	100.0MHz	N/app	N/av	0.0530m	N/av
Acoustic Waves	12.0KHz	N/app	MIMO(OFDM)	N/av	24.360Kbps
Acoustic Waves	24.0KHz	N/app	QPSK	2.5KM	30.0Kbps
EM Waves	25.0MHz	N/app	N/av	85.0m	N/av
EM Waves	5.0MHz	N/app	N/av	90.0m	500.0Kbps
Optical Waves	N/av	470.0nm	N/av	11.0m	9.690Kbps
Optical Waves	N/av	N/app	N/av	10.0m	10.0Kbps
Acoustic Waves	70.0KHz	N/app	ASK	70.0m	0.20Kbps
EM Waves	2.40GHz	N/app	BPSK	0.170m	1.00Kbps
EM Waves	2.40GHz	N/app	QPSK	0.170m	2Mbps
EM Waves	2.40GHz	N/app	CCK	0.160m	5.50Mbps
EM Waves	2.40GHz	N/app	CCK	0.160m	11.0Mbps

These comparisons are based on acoustic waves, electromagnetic waves and optical waves. As observation from Table VII, it is very clear that 2.40GHz ISM band is more suitable for underwater communication with highest data rate achieved at 2.00, 5.50 and 11.0Mbps speed. Different tests at different frequencies and modulations are compared with respect to minimum depth, distance between the underwater devices and transmission of signal. These above tests are carried out in the range of 2.412GHz-2.442GHz ISM band. BPSK and QPSK modulation techniques provide better communication underwater link. It is also observed that there is loss by 30% of packets for distance as shorter as 16cm for BPSK and by 30% loss of packets for distance 17cm in case of QPSK system. Value of RTT also corresponds to 25ms for 16cm distance where wireless sensor were operated at 2.432GHz ISM band. It can be concluded that for distance of 16cm, both modulation techniques (BPSK, QPSK) offers minimal packet loss at 2.432GHz.

REFERENCES

Akyildiz, I. F., Pompili, D., & Melodia, T. (2004). Challenges for efficient communication in underwater acoustic sensor networks. *Embedded Sensor Networks and Wireless Computing, 1*(2), 3–8. doi:10.1145/1121776.1121779

Chaitanya, D. E., Sridevi, V., & Rao, G. S. B. (2011). Path loss analysis of underwater communications. *Proceeding of the 2011 IEEE Students' Technology Symposium*, 65-70. 10.1109/TECHSYM.2011.5783803

Chakraborty, U., Tewary, T., & Chatterjee, R. P. (2009). Exploiting the Loss-Frequency Relationship using RF Communication in Underwater Communication Networks. *2009 International Conference on Computers and Devices for Communication*, 1-4.

Che, X., Wells, I., Dickers, G., Kear, P., & Gong, X. (2010). Re-Evaluation of RF electromagnetic communication in underwater sensor networks. *IEEE Communications Magazine, 48*(12), 143–151. doi:10.1109/MCOM.2010.5673085

Cheng, E., Wu. L., Yuan, F., Gao, C., & Yi, A.D. (2019). *Node selection algorithm for underwater acoustic sensor network based on particle swarn optimization*. Academic Press.

Garcia, M., Sendra, S., Atenas, M., Lloret, J. (2012). Underwater Wireless Ad-hoc Networks: A Survey. *Mobile Ad Hoc Networks: Current Status and Future Trends Survey*. doi:10.1201/b11447

Ghafoor, H., & Noh, A. Y. (2019). An overview of next generation underwater target detection and tracking: An integrated underwater architecture. *IEEE Access: Practical Innovations, Open Solutions, 7*, 98841–98853. doi:10.1109/ACCESS.2019.2929932

Goyal, N., Dave, M., & Verma, A. K. (2014, February). Fuzzy based clustering and aggregation technique for under water wireless sensor networks. In *2014 International Conference on Electronics and Communication Systems (ICECS)* (pp. 1-5). IEEE. 10.1109/ECS.2014.6892804

Goyal, N., Dave, M., & Verma, A. K. (2016). Energy efficient architecture for intra and inter cluster communication for underwater wireless sensor networks. *Wireless Personal Communications, 89*(2), 687–707. doi:10.100711277-016-3302-0

Goyal, N., Dave, M., & Verma, A. K. (2016, December). Congestion control and load balancing for cluster based underwater wireless sensor networks. In *2016 Fourth International Conference on Parallel, Distributed and Grid Computing (PDGC)* (pp. 462-467). IEEE. 10.1109/PDGC.2016.7913240

Goyal, N., Dave, M., & Verma, A. K. (2017a). Improved data aggregation for cluster based underwater wireless sensor networks. *Proceedings of the National Academy of Sciences, India Section A: Physical Sciences, 87*(2), 235-245. 10.100740010-017-0344-y

Goyal, N., Dave, M., & Verma, A. K. (2017b). Trust model for cluster head validation in underwater wireless sensor networks. *Underwater Technology, 34*(3), 106–113. doi:10.3723/ut.34.107

Goyal, N., Dave, M., & Verma, A. K. (2018a). A novel fault detection and recovery technique for cluster-based underwater wireless sensor networks. *International Journal of Communication Systems, 31*(4), e3485. doi:10.1002/dac.3485

Goyal, N., Dave, M., & Verma, A. K. (2018b). Adaptive error control technique for cluster-based underwater wireless sensor networks. In *International Conference on Wireless Intelligent and Distributed Environment for Communication* (pp. 269-280). Springer. 10.1007/978-3-319-75626-4_20

Goyal, N., Dave, M., & Verma, A. K. (2019a). Protocol stack of underwater wireless sensor network: Classical approaches and new trends. *Wireless Personal Communications, 104*(3), 995–1022. doi:10.100711277-018-6064-z

Goyal, N., Dave, M., & Verma, A. K. (2019c). Data aggregation in underwater wireless sensor network: Recent approaches and issues. *Journal of King Saud University-Computer and Information Sciences, 31*(3), 275–286. doi:10.1016/j.jksuci.2017.04.007

Jiang, S., & Georgakopoulos, S. (2011). Electromagnetic Wave Propagation into Fresh Water. *Journal of Electromagnetic Analysis and Applications, 3*(07), 261–266. doi:10.4236/jemaa.2011.37042

Kumar, M., & Goyal, N. (2014). Reviewing underwater acoustic wireless sensing networks. *International Journal of Computer Science and Technology, 5*(2), 95–98.

Liebe, H. J., Hufford, G. A., & Manabe, T. (1991). A model for the complex permittivity of water at frequencies below 1THz. *International Journal of Infrared and Millimeter Waves, 12*(7), 659–675. doi:10.1007/BF01008897

Liu, L., Zhou, S., & Cui, J. H. (2008). Prospects and problems of wireless communication for underwater sensor networks. *Wireless Communications and Mobile Computing, 8*(8), 977–994. doi:10.1002/wcm.654

Lloret, J., Sendra, S., Ardid, M., & Rodrigues, J. J. P. C. (2012). Underwater wireless sensor communications in the 2.4GHz ISM frequency band. *Sensors (Basel), 12*(4), 4237–4264. doi:10.3390120404237 PMID:22666029

Nguyen, V. D., Luong, N. S., & Patzold, M. (2017). A method to estimate the path gains and propagation delays of underwater acoustic channels using the arrival phase information of the multipath components. *International Journal of Electronics and Communications, 73*, 129–138. doi:10.1016/j.aeue.2017.01.003

Sendra, S., Lamparero, J. V., Lloret, J., & Ardid, M. (2011). Underwater Communications in wireless sensor networks using WLAN at 2.4GHz. *2011 Eighth IEEE International Conference on Mobile Ad-Hoc and Sensor Systems*, 892-897. 10.1109/MASS.2011.106

Sendra, S., Lloret, J., Garcia, M., & Toledo, J. F. (2011). Power saving and energy optimization techniques for Wireless Sensor Networks. *Journal of Communication*, *6*(6), 439–459.

Yadav, S., & Kumar, V. (2019). Hybrid Compressive Sensing Enabled Energy Efficient Transmission of Multi-Hop Clustered UWSNs. *International Journal of Electronics and Communications*. doi:10.1016/j.aeue.2019.152836

Zhang, S., Chen, K., Fan, Z., Cheng, E., & Su, A. W. (2019). The localization algorithm based on symmetry correction for underwater acoustic networks. *IEEE Access: Practical Innovations, Open Solutions*, *7*, 121127–121135. doi:10.1109/ACCESS.2019.2937106

Chapter 11
Secure Communication Techniques for Underwater WSNs

Esha Rani
Institute of Integrated and Honors Studies, Kurukshetra University, India

Vikas Juneja
Seth Jai Parkash Mukand Lal Institute of Engineering and Technology, India

ABSTRACT

Earth is solitary among the rocky planets revolving in the solar system. It is the only planet that is 3/4 covered with oceans of liquid water. In today's era, wireless sensor network is used in almost places and also is an interesting topic for researchers too. But with the advancement in technology, a network similar to wireless sensor network can be deployed under the water with acoustic signals for data transmission. Because of harsh environment and some more challenges, underwater communication becomes tricky. Underwater wireless sensor networks (UWSNs) is an advanced technology for underwater explorations and have demonstrated their quality in underwater applications. UWSN is a blend of wireless technology with sensor technology that has some extraordinary features like smart sensing, intelligent computing, and communication capabilities. Further, to maintain the secure communication underwater, it becomes more and more tricky. This chapter presents an expansive analysis of challenges, attacks, and security techniques of UWSNs.

INTRODUCTION

Wireless sensor network is a type of network in which many sensors are attached at different locations to make the communication possible from different locations wirelessly e.g. to sense anything, a sensor is situated at any place and information or data can be collected from that sensing device through a centralized network.

Wireless sensor network can be categorized in two sections:

DOI: 10.4018/978-1-7998-3640-7.ch011

1. Organized, Systematic and well regulated WSNs
2. De-organized and unregulated WSNs

In Organized WSNs, sensor nodes are distributed in pre-planned manner and less deployed that's why maintenance cost is very low. On another side, in de-organized and unregulated WSNs; all the sensor nodes are densely collected and use Ad-hoc type of deployment that's why maintenance of these type of network is a tough task and maintenance cost is also high.

1. Underwater wireless sensor networks (UWSNs) is an advanced technology for underwater explorations and have demonstrated their quality in underwater applications e.g. Surveillance, resource exploration, ocean monitoring and military use in harsh underwater environments (Cui et al.,2005, Akyildiz et al.,2005).

UWSN is a wireless technology with micromechanical sensor technology having features like sharp sensing, intelligent computing, and communication capabilities.

Figure 1. Underwater wireless sensor network (UWSN) environment

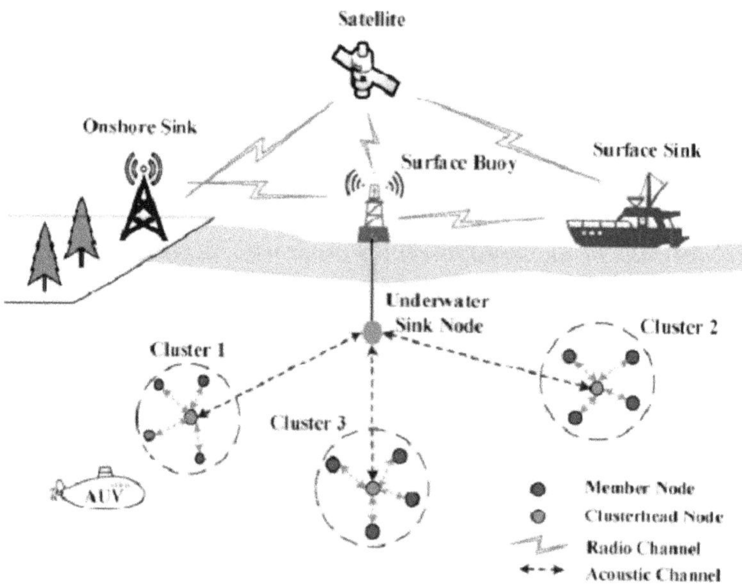

2. Figure1 (Yang et al., 2019) demonstrates the working environment of UWSNs. Data communication is performed in UWSNs by several components like underwater sensor nodes, underwater sink nodes, onshore sink, surface buoy station and surface sink which are elaborated as follows:

   a. Initially, all the sensor nodes are dropped at the ocean floor. Every sensor nodes is in itself a standalone network which is responsible for sensing, processing and transmitting the information to the remote station.

b. Underwater sensor nodes monitor or sense water related physical or environmental conditions, such as pressure, sound, temperature, etc. and mutually transfer data to the underwater sink node. Different types of underwater sensor nodes are used as moored sensors, cabled seafloor sensors, acoustically connected sensors.
c. Underwater sink nodes with horizontal transceivers collects data from sensor nodes and relays data to surface buoy with vertical transceiver.

Besides this, autonomous underwater vehicles (AUVs), satellite and vessel helps to grow the sense power and increase the communication range. The AUVs work as Power nodes in UWSN. With high energy level, AUVs move independently, and behaves like routers between fixed sensors. AUVs manage to reconfigure the network.

Figure1 shows the cluster-based network architecture. This architecture works not only to manage the nodes in an easy manner but also to extend the lifetime of the whole network.

Process: Among all the cluster member nodes, it is tricky to find any node as sink node. So, cluster-head (CH) node is assumed as the sink-node of the cluster which accumulates, broadcast packets within cluster and in-between clusters. Hence, the energy consumption of CH is greater than member nodes. To balance energy consumption as well as to extend the life-time of the whole network, sensor nodes are elected to be the CH. The protocol used behind the scene may be based on the best energy consumption (Wang et al., 2007), residual energy (Fang et al., 2007) and the optimal number of CHs (McGlynn et al., 2001).

UWSN is almost similar to WSN only by the difference of environment under which it works. Due to harsh environment, challenges faced in underwater wireless sensor network are discussed in section II. In section III, threats and attacks in UWSN are introduced. Security issues and techniques to maintain security will be discussed in section IV. In section V, conclusion and future prospects will be discussed.

UNDERWATER ACOUSTIC ENVIRONMENT WITH CHALLENGES

Due to harsh working environment underwater, some challenges are faced in UWSN which are discussed as follows:

1. **Power Control:** Since underwater no any other energy resources exist like solar energy etc. except battery i.e. every underground sensor node requires battery power to communicate. As attenuation is more, underground sensor node require more energy. So, energy consumption rate is higher. To change batteries time to time and sometimes, replacement of batteries become a difficult task. That's the reason Power Conservation becomes necessary under water and it is possible by implementing Energy Efficient Protocols.
2. **Finite Bandwidth and Low Data Rates:** Underwater communication is done through acoustic channels. The bandwidth range of underwater acoustic channels is finite and limited and depends on both transmission range and depth. The bandwidth is a range of frequencies to send signals from one location to another. Because underwater acoustic communication is possible only over limited bandwidths, proportionately the data rate of the underwater acoustic channel is much slower than radio channel in the air. That's why an extended version of communication range is more required than high data transfer rate.

3. **Physical Vulnerability:** Generally, UWSNs nodes are compressed, impermeable and precise in nature and may be located at smooth and hard sea regions. Underwater so many physical factors exist which can damage the nodes, make them invalid. That's why it is not possible to save each node from physical damage.
4. **Topology Design:** Network topology can-not be static as underwater sensor nodes are mobile most of the time due to water flow. Highly dynamic network topology is used underwater. Figure1 shows the Cluster based network architecture which is the best example of dynamic environment as AUVs are used which is of ad-hoc type. They join and exit the cluster only on demand. AUVs are used to enhance the communication range. Highly dynamic network topology lacks the collaboration among the nodes as no any node remains at its position for a long time.
5. **Environmental Hazards:** In sea or deep water, not only water exists but marine life also exists. Protection from insects, animals or water is needed. Underground environment is not ideal for electronic devices.
6. **Transmission Loss:** Transmission Loss is measured in two terms:

Spreading Loss Attenuation Cost

At the time of spreading, the attenuation cost is high due to energy loss which is absorbed by the medium. Spreading loss is nothing it is the power loss in between the path in the meanwhile of spreading period from source to destination. Spreading Loss can be shown by two models: Spherical as well as Cylindrical.

Attenuation Cost (AC): During the data transfer from one location to another i.e. propagation period, conversion of energy of an acoustic wave into many forms (e.g. heat) is done and immersed into transmission medium. In general, the attenuation cost can be calculated by frequency.

7. **Noise:** Noise also affects the communication on earth as well as underwater acoustic channel. As discussed above, in underwater not only water exist but also marine life. So noise can be due to water as well as marine animals. In water, various categories of noises exist like noise by waves caused by movement of water, self-noises from vessels, the radiated noises and different surface motions; turbulence noise, thermal noise and noise from marine animals. Acoustic communication is influenced by noises of different frequency band which can be described as follows:
    a. Turbulence noise (Nt, <10 Hz),
    b. Steam noise (Ns, 10–100 Hz),
    c. wafting noise (Nw, 100 Hz–100 kHz), and
    d. Thermic noise (Nth, over 100 kHz) (Zhou et al., 2015)
8. **Multipath:** It will always be not possible that a signal transmitted will follow one path from Source to Destination. Some signals follow two or more paths to travel. This is called Multipath effect. At any point, signals arriving from the two (or more) paths may strike at the target node. In UWSNs, Multipath effect has more significant than in WSNs due to different water environment. In underwater environment, water level is different at different places. In deep water, extreme upper and lower reflections are considered as negligible and medium is homogeneous. On the other side, in shallow water environment, depth of water is not so high even transmission distance is more than water depth. That's why division of shallow water environment is done in many layers from surface to bottom level. Acoustic wave is transmitted and reflected through many layers and when multiple arrivals of the same signal is received at receiver, it shows multipath effect.

THREATS AND ATTACK IN UNDERWATER WIRELESS SENSOR NETWORK (UWSN)

Due to harsh working environment, UWSN are vulnerable to various threats and attacks. Attacks are categorized into two sections according to the interruption of communication act which are described as follows:

1. Active Attacks
2. Passive Attacks

Figure 2. Threats and challenges in UWSNs

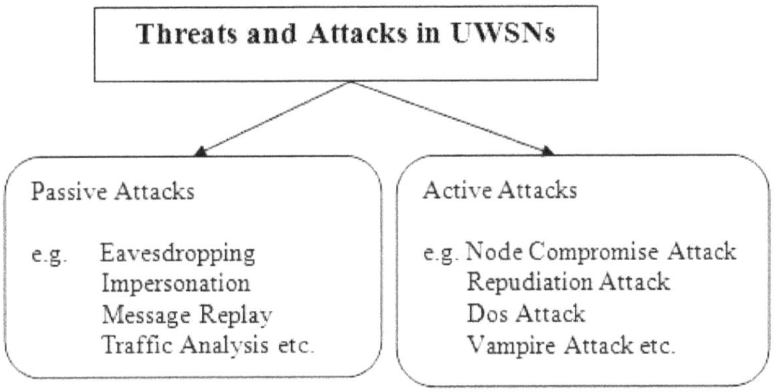

1. **Passive Attack:** In this attack, the Malicious node M attempts to break the system and obtain the information without interact or disrupt any resources or parties involved in communication, Data transmitted in network is obtained without interrupting the communication e.g. eavesdropping, impersonation, interfering, traffic analysis, secret information leakage, message replay etc. Underwater acoustic channel is an open channel that makes it more vulnerable to various types of attacks. Malicious node can easily detect or capture the disseminated packets in the channel using resources like underwater microphone, hydrophone etc. and find the location of nodes. Detection of passive attack is not so easy since neither any system resources nor any network operation is affected. In WSN, encryption techniques are used to detect passive attacks but these mechanisms are directly not implemented in UWSN.
2. **Active Attack:** Active attack attempts to disrupt the communication and modify or drop the packets by various methods like injection, alteration and deletion etc. Not only internal nodes but external nodes can also be executed this attack. Active attacks can be performed by internal as well as external nodes. To detect external node as malicious node is easy to detect and defend. But if the network is damaged from internal node, it will be difficult to detect. The reason is Malicious nodes behaves as normal node in the network (Juneja et al., 2018).

MISCELLANEOUS ATTACKS

Eavesdropping

An eavesdropping attack is a variety of passive DOS attack which grasps the packets transmitted in the channel after the continuous monitoring of the acoustic transmissions. Without the encryption scheme over packets, data and private information can be easily read by malicious nodes. As normal communication is not disrupted, to detect the occurrence of eavesdropping is difficult. In (Wang et al.,2016), the authors presented an analytical model which investigates the probability of eavesdropping attacks with consideration of underwater acoustic channel conditions, including signal attenuation and ambient noise i.e. eavesdropping success condition is directly related to underwater acoustic signal channel. Eavesdropping probability is derived by considering both array eavesdropper and isotropic eavesdropper. Some other factors also help to calculate probability like wind speed, spreading factor, node density and acoustic signal frequency.

Jamming Attack

Malicious nodes interfere in communication by emit useless signals in regular mode. As the whole network is dependent on some special nodes for data collection, packet transmission as well as maintain the network such as base station or cluster head. So, instead of affecting all the nodes, network can be paralyzed by jamming only these special nodes.

In UWSN, underwater sink nodes collect data packets from sensor nodes and transfer these packets to on-shore base station. Malicious node in form of jamming attackers continuously send useless data packets or requests to underwater sink nodes to divert them from accepting the response of legitimate requests.

Many solutions exist to defend against jamming attack in WSNs like Spread-spectrum technique, a multi-point acoustic medium access method but no feasible solution exist for UWSNs.

We can say, in an underwater environment localization protocols are effectively disrupted by jamming attack.

Figure 3. Jamming Attack

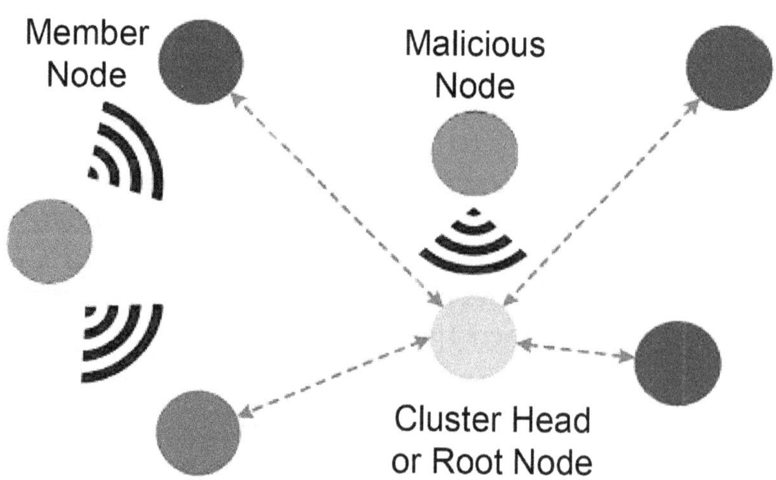

Replay Attack

This attack is also known as Playback attack. In this attack, data packet transmission is repeated or delayed. In a replay attack, malicious node M intercepts or traps the message sent by Sender S for receiver R, modify it and resend it after a short delay. The packet is received by the same receiver for which sender has sent only by the difference of arrival time. As a consequence, malicious node can be detected by the estimation of attenuation loss and signal arriving time. Security against replay attack can be provided by Authentication as well as anti-replay techniques.

In (Ibragimov et al., 2016), to resolve security issues like authenticity, confidentiality and protection against replay attack, an underwater media access control protocol is proposed which is based on the RTS/CTS mechanism with a cipher block chaining-message authentication code.

In (Xu et al., 2015), A secure MAC protocol is proposed to ensure the security of data transmission for cluster-based UWSNs. A random number is generated to prevent against replay attack after each data transmission. Therefore, only the authorized parties can accept or send the message.

DOS Attack

DOS attack is called Denial-of Service attack which can be active or passive. In this attack, an announcement of the unavailability of the services and resources to the authorized nodes is made and preventing them to access these services offered by the network. These attacks are fatal, complicated and not easily available.

Even though encryption technology protects UWSNs, but DOS attacks can still disrupt the coordination and computation among nodes and make the entire network less available. Besides this, DOS attacks are economic, deadly, and even worse, tough to detect and defend. In DOS attack, malicious node performs tasks like reduce the life span of legitimate nodes by imposing more power tasks; severe damage at very low cost; act as an authorized node to dupe neighbor nodes.

Selective Packet Forwarding Attack

As name suggests, in this attack, selective packets are forwarded from one node to another, some packets are dropped time to time. In (Pandarinath, 2011), malicious nodes behave as authorized nodes and refuse to forward the received packets to other nodes by declaring that there is no valid path to the destination. But more refusals in time duration can create risk to being detected as malicious node. To avoid this, selective packets are forwarded and certain packets are dropped. The malicious node takes keen interest in interruption and then modification of a packet to limit suspicion of its malpractice.

In (Zhou et al., 2015), a watchdog mechanism is used to raise intrusion alert by pre-defined rules in WSNs which is not suitable for UWSN. Its implementation is restricted in some cases of collision like indeterminate receiver, incomplete packet dropping, fraud misbehavior, inadequate transmission power.

Behavior Assessment mechanisms like Trust management and Reliable Reputation mechanism is used to detect malicious node and disconnected them from the network (Jiang et al., 2017; Prasan et al., 2012).

Homing Attack

In this attack, different areas of the network are analyzed by malicious node and on the basis of volume of traffic, location of special nodes like base station or cluster head is identified. Moreover, in hard situations, malicious attacker may take help of other DOS attacks to shatter these special nodes. Even though Data packet header encryption is a simple corrective measure but traffic inspection is not entirely prevented.

In (Chen et al., 2015), A scheme is proposed to divert eavesdroppers form special node locations by reducing the volume of traffic at base station and increase the volume of traffic at some places by "dummy packets". The burden of dummy packets relatively reduces node's strength in UWSNs. That's why; this technique is rarely used and used specially in those cases where prevention of traffic analyzing is utmost required.

Decentralization Attack

In this attack, spurious packets with fake sequence numbers are transmitted by an adversary node to disrupt the active or current communication (synchronization) among the nodes. By this, accuracy of the synchronized clocks and efficiency of scheduled operations is affected. In UWSNs, for systematic communication, synchronization is mandatory. Likewise, GPS is not suitable since synchronization maintenance is very difficult. GPS free synchronization mechanisms are used. Even though, this method is suitable but it is unsafe against wormhole attack. If de-synchronization attack is combined with a wormhole attack, cooperation and communication of the entire network is disrupted severely.

For example, wormhole, Sybil, or replay attacks may cause fake measurement results on ranges or round-trip time (RTT) between legitimate nodes, which are important parameters for time reference alignment. Header or full packet authentication can defeat such an attack.

SECURITY ISSUES IN UWSN

As underwater wireless sensor network is also of ad hoc type, the security goals are similar for the terrestrial ad hoc sensor networks or even underwater sensor nodes. Setting security goals for sensor networks will depend on knowing what needs to be protected. Underwater sensor networks share basic features of terrestrial networks. Therefore the security goals encompass terrestrial networks as well as underwater sensor networks. The security goals can be primary and secondary. The primary goals are also called standard security goals such as Confidentiality, Integrity, Authentication and Availability (CIAA). The secondary goals are Fresh Data, Organization of nodes by themselves, Time Sync and Secure Localization. The four security goals for underwater sensor networks are determined as Confidentiality, Integrity, Authentication and Availability (CIAA).

1. **Confidentiality:** This is the most important issue in network security. Confidentiality is the strength to wrap sensitive data (e.g. secret keys and security credentials etc.) in a way to protect from unauthorized passive attacker. Data wrapping is done in a way that message or data packet during communication from one location to another will not be opened. Confidentiality is not only confined to user information, it extends its limit to routing information, MAC etc. Confidentiality should be maintained in a way that any attacker if try to gain access to information precluded from

the network. Distribution of keys and ID proof to every sensor node in the network helps to establish a secure communication among the sensor nodes. That's why Encryption method is best suited to WSN as well as UWSN. The cipher text stealing technique is a lightweight encryption technique used in UWSNs.

2. **Integrity:** Data integrity in underwater sensor networks is required to ensure the reliability of the data i.e. confirmation of the corruption of the data in form of modification, deletion or any alteration while on the network by unauthorized nodes either by malicious node or radio failure. Even though the network has confidentiality measures in place, but still a possibility that the data's integrity has been compromised by alterations.

Queries arise regarding integrity of the network if:

- Injection of fake messages in the network by malicious node.
- Unstable conditions due to underwater acoustic channel cause damage or loss of data.

Message Authentication Code can be used for data authentication can be applied in UWSNs for reliability and good scalability.

3. **Authentication:** Authentication focuses on the Source from where message comes and ensures the reliability of the message by its origin. Attacks are not only judged by the alteration of packets but attackers can inject fake packets to divert. Every receiving node should have the capability to confirm the stem from the node claiming to have sent it.

As discussed above, underwater acoustic channel is open. Without any encryption scheme, data packets can be easily modified by the malicious node. Hence, filtration process is required at every receiving node to identify the malicious node by knowing the source of the packet. Authorized users can only share and access the channel or even modify the data. Symmetric or asymmetric mechanisms can be applied for data authentication where secret keys are shared between sending and receiving nodes to compute the message authentication code (MAC). Intrusion Detection scheme and a trust management mechanism is used to detect and eliminate malicious node from the network with the condition that only the unauthorized users can perform this operation.

4. **Availability:** It determines the availability of services even in case of failure or attack. Network must be so powerful that a short failure in the network should not stop the operation or services. Even this is very challenging and complex security measures impose a higher consumption of energy and computation power. UWSN use self-adaptive approach and proper redundancy to provide availability.
5. **Data Freshness**: Even though confidentiality and data integrity are most important issues in network security but still there is a need to ensure the freshness of each message. Freshness should ensure that data is recent and not the retransmission of legacy data. Route update notifications should be forwarded time to time to ensure the real time environment. Any delay in updates can lead to the wrong status of the nodes and even information can be lost.
6. **Self-Organization**: Underwater wireless sensor network is an ad hoc network requires every sensor node be independent flexible enough to be self-organizing and self-healing. No fixed Infrastructure

is available for the purpose of network management in a sensor network. Self-healing is a great challenge to wireless sensor network security. Lack of self-organization in a sensor network leads to risky environment which can cause devastating results.

7. **Time-Synchronization**: This feature is used by sensor nodes for tracking applications. End-to-end delay of a packet is measured by the sensor nodes when data packet travels between two sensor nodes.
8. **Secure Localization**: Sensor nodes should be capable to locate other sensors in the network and also locate the faults and pin point the location of the fault. Unfortunately, unsecured locations can be easily manipulated by malicious node by Replaying signals or Reporting false signal strengths.

SECURITY TECHNIQUES IN UWSN

As security incidents are increasing day by day. To secure the network from malicious attackers becomes a very big concern in underwater sensor network. Some security techniques are used to resolve security issues discussed in the previous section, arise in the network. The security methods applied to tackle with security issues of UWSNs are: Intrusion Detection Scheme, Encryption based technique, Trust value based technique, location information security, temporal security, Routing security.

Intrusion Detection Technique: Intrusion detection method is used to detect, identify and isolate malicious attackers (including internal or external intruders) from the network. It is not easy to detect malicious intruders. On the other hand, an efficient security mechanism named as Intrusion Tolerance mechanism can be used to protect network in the presence of malicious intruders. UWSNs security has been improved using IDS. Some examples of IDS are discussed as follows:

Autonomous IDS: No exchange of data among nodes; every node implements IDS according to their own need and detects the attack by passive monitoring of not only node routing behavior but also monitor the neighborhood behavior concurrently.

Shared IDS: Every node participates in intrusion detection by sharing local and global detection decision-making.

Ordered IDS: This is suitable for multilayered UWSNs. In every cluster, Cluster head node takes the charge to implement the task of IDS and perform like checkpoints such as routers in wired networks. Since hierarchical IDS is based on cluster structure which is suitable for UWSNs.

Encryption Based Technique

To resolve the security issues like Authentication, Confidentiality, integrity and non-repudiation, Cryptography and key management techniques are used. As we know, in UWSNs, underwater acoustic channel is an unsecured network. Confidential matter should be beyond the reach of unauthorized users or should not be read and modify by them. So, Cryptography enables the storage and delivery of important information in a secured manner. Encryption as well as key management mechanisms used in WSN is not directly applicable for UWSNs. In public key cryptography (PKC) schemes, certificate authority is required to manage certificate and key distribution for public key infrastructure (PKI) based methods results in high computation and communication cost which is not economical for UWSNs (Teguig et al., 2017, Simplicio et al., 2017). Unlike PKC, no PKI is required for identity based encryption (IBE) scheme. In this scheme, the public key computation is done according to the node ID. Hence resource cost for storing and granting certificates is saved which makes it a feasible scheme. On the other hand, private key distribution in PKC will raise communication cost, overloads UWSNs.

Recent years research has been presented for UWSNs in (Jiang et al., 2018; Jiang et al., 2015; Luo et al., 2016; Peng et al., 2016; Yuan et al., 2017; Hamid et al., 2018; Ateniese et al., 2015; Capossele et al., 2017).In Worst case, the existing key management and cryptography mechanisms have issues like computational complexity and Expansion of cipher text.

In cryptography, message size expansion is done by codes and message padding causes more energy consumption on computation and transmission. In AES encryption, message blocks size 128 bits can be expanded to message of 720 bits due to 18% padding in UWSNs (Dini et al., 2012).

Trust Value Based Technique

In Intrusion detection, a trust management mechanism based on cryptography plays an important role. But some challenges are experienced due to some constraints in UWSNs (Goyal et al., 2017).

Three schemes of trust management mechanisms are: centralized scheme, distributed scheme, and hierarchical scheme.

Centralized Scheme: in a centralized scheme, a centralized node also called a parent node or a base station or hub where trust management is done for each node in the network. The centralized schemes are not appropriate for UWSNs, because the exchange of trust values between sensor nodes and the base station requires more energy consumption which is more expansive.

Distributed Scheme: In this scheme, each and every node calculate the trust values of the entire network independently and maintain them. But as resources are limited in underwater sensor nodes, so this scheme is almost impossible for UWSNs. UWSN neither compatible with pure centralized nor pure distributed scheme.

Hierarchical Scheme: The public key infrastructure (PKI) uses this scheme. Certificate authority (CA) is at top level and trustworthy. It may itself issue certificates as well as issues the certificates to them which further may issue certificates just to complete the chain to bottom level. The bottom level certificate's validity is verified by moving backwards from its issuer to other issuers, until a directly trusted root (CA) certificate is found. The hierarchical scheme is represented as a tree structure. Trust value implementation in form of computation as well as transmission is not only in computation but also in a hierarchical manner. The trust values are calculated and evaluated in a tree like manner. The trust values are moved in a systematic manner i.e. from leaf node to root node. That's why it is considered more suitable for cluster- based topology used in UWSNs. In UWSNs, in some applications, sink nodes act as an intermediate between UWSN and land network which must be authorized to deliver control information as well as retrieving the reading from underwater sensor nodes.

Location Information Security

In tracking applications, Location estimation technique plays an important role in source detection. In UWSNs, location estimation is done by underwater sensor nodes during the localization phase to re-trieve the location as well as speed of mobile nodes. Localization rules made for WSNs can't be directly implemented in underwater environment due to some reasons like (Finite bandwidth, long propagation delay, and multi-path and node mobility).

In UWSNs, location estimation schemes are: Scope free scheme, scope based scheme and hierarchical scheme (Luo et al., 2018).

In scope free scheme, energy consumption rate is higher. Energy consumption and hardware problems make the unavailability of information about the neighbor distances and angle. In scope based scheme, neighbor distance and angle information is available. So Node location can be calculated easily. Hierarchical scheme is different from both schemes, it includes four types of nodes i.e. Anchor nodes, ordinary nodes, surface buoys and Unreachable Generator Transceivers (UGTs). Firstly, A surface buoy is installed on water surface with GPS. Then UGT is connected to surface buoy to broadcast its position by moving up and down. UGTs position information is used to calculate the position of anchor nodes. Examples of localization specific attacks are: Black Hole Attack, Sybil Attack, Worm hole Attack. In these attacks, location information about nodes is utilized or modified to cause great damage in the network.

Above discussed location estimation schemes are used to estimate the location of sensors, even in UWSNs. Now the issue arises, whether this location information is true or false. To protect against false injection of location information, some efficient cryptographic algorithms need to be created (Zhang et al., 2013; Han et al., 2016; Das et al., 2017; Varadharajan et al., 2018).

Temporal Security

In UWSNs, Synchronization is very important factor. But to maintain it in underwater environment among the nodes is not an easy task. Synchronization security protocols used in WSNs are unsuitable for UWSNs applications.

In (Liu et al., 2013 & 2014; Mortazavi et al., 2017), security issues were not discussed in the existing time synchronization schemes. Time synchronization attacks can be defended by attacks (e.g. Replay attack, masquerade and manipulation attacks) by Cryptographic techniques. Protection metrics discussed in (Vaseghi et al., 2017; Cheng et al., 2014; Vaseghi et al., 2017) for WSNs are not directly implemented in UWSNs.

Routing Security

Security techniques are used to secure routing protocols as well as individual nodes. Information must be exchanged with their neighbors by applying any of the routing protocols (reactive, proactive and hybrid) to build up network topology.

Two aspects to secure route are as: 1. Data packets forwarding in a secure way 2. Secure Routing.

Secure routing requires the nodes to maintain coordination among them to share accurate routing information keeping the network efficient and reliable whereas in other, data packets must be forwarded while protected from altering, dropping or tampering by illegal party. In (Dini et al., 2011; Porkodi et al., 2018; Peng et al., 2017; Aman et al., 2018; Shen et al.,2015; Han et al., 2017) many research papers have been presented to secure routing in UWSNs.

CONCLUSION AND FUTURE WORK

UWSN constraints restrict the direct implementation of some technologies used in WSN. Even though some technologies were tried to secure UWSN, but only theory stage is applicable, pragmatic research is under work. To resolve security issues in UWSNs, few proposals are given, but still some security issues have not been solved.

In addition, UWSN security is a complicated matter, and to reduce the consumption of extra resources, an optimal utilization of various security technologies is given.

In UWSN, sensor nodes are deployed underwater, protection of every underwater node and detection of threatened nodes is not as easy task. Prior installation of the security methods in the sensor nodes is done. If any node is threatened or rejected in the network, then there is a risk of more and severe damage. So, reconfiguration of security system is required time to time while it is not an easy task. Though, it is a challenging issue, but an important issue.

To secure UWSN is a complicated task due to network deployment cost and maintenance cost. However, the UWANs specific facts also impose challenges, and should be exploited in the design of security schemes. Many applications require different security schemes but implementation of these will put a heavy burden on the network in terms of power utilization. Therefore security of nodes in UWSNs will be a very challenging issue in future research.

REFERENCES

Akyildiz, I. F., Pompili, D., & Melodia, T. (2005). Underwater acoustic sensor networks: Research challenges. *Ad Hoc Networks*, *3*(3), 257–279. doi:10.1016/j.adhoc.2005.01.004

Aman, W., Rahman, M. M. U., & Qadir, J. (2018). *Secure Routing in OFDM based Multi-Hop Underwater Acoustic Sensor Networks*. arXiv; arXiv:1807.01556

Ateniese, G., Capossele, A., Gjanci, P., Petrioli, C., & Spaccini, D. (2015). Security framework for underwater acoustic sensor networks. *Proceedings of the MTS/IEEE OCEANS*, 1–9.

Capossele, A., Petrioli, C., Saturni, G., Spaccini, D., & Venturi, D. (2017). Securing Underwater Communications: Key Agreement based on Fully Hashed MQV. *Proceedings of the ACM International Conference on Underwater Networks & Systems*, 12. 10.1145/3148675.3152760

Chen, H., & Lou, W. (2015). on protecting end-to-end location privacy against local eavesdropper in wireless sensor networks. *Pervasive and Mobile Computing*, *16*, 36–50. doi:10.1016/j.pmcj.2014.01.006

Cui, J.H., Kong, J., Gerla, M. Zhou, S. (2005) *Challenges: Building scalable and distributed Underwater Wireless Sensor Networks (UWSNs) for aquatic applications*. Academic Press.

Das, A. P., & Thampi, S. M. (2017). Fault-resilient localization for underwater sensor networks. *Ad Hoc Networks*, *2017*(55), 132–142. doi:10.1016/j.adhoc.2016.09.003

Dini, G., & Duca, A. L. (2011). A secure network discovery protocol for Underwater Acoustic Networks. *Proceedings of the 2011 IEEE Symposium on Computers and Communications (ISCC)*, 636–638. 10.1109/ISCC.2011.5983910

Dini, G., & Lo Duca, A. (2012). *A secure communication suite for underwater acoustic sensor networks*. Academic Press.

Goyal, N., Dave, M., & Verma, A. K. (2017). Trust model for cluster head validation in underwater wireless sensor networks. *Underwater Technology*, *34*(3), 107–114. doi:10.3723/ut.34.107

Hamid, M. A., Abdullah-Al-Wadud, M., Hassan, M. M., Almogren, A., Alamri, A., Kamal, A. R. M., & Mamun-Or-Rashid, M. (2018). A key distribution scheme for secure communication in acoustic sensor networks. *Future Generation Computer Systems*, *86*, 1209–1217. doi:10.1016/j.future.2017.07.025

Han, G., Jiang, J., Sun, N., & Shu, L. (2015). Secure communication for underwater acoustic sensor networks. *IEEE Communications Magazine*, *53*(8), 54–60. doi:10.1109/MCOM.2015.7180508

Han, G., Liu, L., Jiang, J., Shu, L., & Rodrigues, J. J. (2016). A collaborative secure localization algorithm based on trust model in underwater wireless sensor networks. *Sensors (Basel)*, *16*(2), 229. doi:10.339016020229 PMID:26891300

Han, S. Y., Chen, Y. H., & Tang, G. Y. (2017). Fault diagnosis and fault-tolerant tracking control for discrete-time systems with faults and delays in actuator and measurement. *Journal of the Franklin Institute*, *354*(12), 4719–4738. doi:10.1016/j.jfranklin.2017.05.027

He, J., Chen, J., Cheng, P., & Cao, X. (2014). Secure time synchronization in wireless sensor networks: A maximum consensus-based approach. *IEEE Transactions on Parallel and Distributed Systems*, *25*(4), 1055–1065. doi:10.1109/TPDS.2013.150

Ibragimov, M., Lee, J. H., Kalyani, M., Namgung, J. I., Park, S. H., Yi, O., & Lim, Y. K. (2016). CCM-UW security modes for low-band underwater acoustic sensor networks. *Wireless Personal Communications*, *2016*(89), 479–499. doi:10.100711277-016-3283-z

Jiang, J., Han, G., Zhu, C., Chan, S., & Rodrigues, J. J. (2017). A trust cloud model for underwater wireless sensor networks. *IEEE Communications Magazine*, *55*(3), 110–116. doi:10.1109/MCOM.2017.1600502CM

Jiang, S.M. (2018). On securing underwater acoustic networks: A survey. *IEEE Commun. Surv. Tutor.*

Law, Y. W., Doumen, J., & Hartel, P. (2006). Survey and benchmark of block ciphers for wireless sensor networks. *ACM Transactions on Sensor Networks*, *2*(1), 65–93. doi:10.1145/1138127.1138130

Li, X., Fang, S., & Zhang, Y. (2007). The study on clustering algorithm of the underwater acoustic sensor networks. *Proceedings of the 14th IEEE International Conference on Mechatronics and Machine Vision in Practice (M2VIP 2007)*, 78–81.

Liu, J., Wang, Z., Zuba, M., Peng, Z., Cui, J. H., & Zhou, S. (2014). DA-Sync: A Doppler-assisted time-synchronization scheme for mobile underwater sensor networks. *IEEE Transactions on Mobile Computing*, *13*(3), 582–595. doi:10.1109/TMC.2013.13

Liu, J., Zhou, Z., Peng, Z., Cui, J. H., Zuba, M., & Fiondella, L. (2013). Efficient time synchronization for mobile underwater sensor networks. *IEEE Transactions on Parallel and Distributed Systems*, *24*(2), 406–416. doi:10.1109/TPDS.2012.71

Luo, J., Fan, L., Wu, S., & Yan, X. (2018). Research on localization algorithms based on acoustic communication for underwater sensor networks. *Sensors (Basel)*, *18*(2), 67. doi:10.339018010067 PMID:29301369

Luo, Y., Pu, L., Peng, Z., & Shi, Z. (2016). RSS-based secret key generation in underwater acoustic networks: Advantages, challenges, and performance improvements. *IEEE Communications Magazine*, *54*(2), 32–38. doi:10.1109/MCOM.2016.7402258

McGlynn, M. J., & Borbash, S. A. (2001). Birthday protocols for low energy deployment and flexible neighbor discovery in ad hoc wireless networks. *Proceedings of the 2nd ACM International Symposium on Mobile Ad Hoc Networking & Computing*, 137–145. 10.1145/501416.501435

Mortazavi, E., Javidan, R., Dehghani, M. J., & Kavoosi, V. (2017). A robust method for underwater wireless sensor joint localization and synchronization. *Ocean Engineering*, *137*, 276–286. doi:10.1016/j.oceaneng.2017.04.006

Pandarinath, P. (2011) Secure localization with defense against selective forwarding attacks in wireless sensor networks. *Proceedings of the 2011 IEEE 3rd International Conference on Electronics Computer Technology (ICECT)*, 5, 112–117. 10.1109/ICECTECH.2011.5941968

Peng, C., & Du, X. (2017). SDBR: A Secure Depth-Based Anonymous Routing Protocol in Underwater Acoustic Networks. *International Journal of Performability Engineering*, *13*, 731–741. doi:10.23940/ijpe.17.05.p16.731741

Peng, C., Du, X., Li, K., & Li, M. (2016). An ultra-lightweight encryption scheme in underwater acoustic networks. *Journal of Sensors*, *2016*, 8763528. doi:10.1155/2016/8763528

Porkodi, K.P., & Zubair Rahman, A.M.J.M. (2018). *Enhanced Underwater Wireless Sensor Networks Security with Tic-Tac-Toe Ai-Minimax Algorithm in Game Theory*. Academic Press.

Prasan, U. D., & Murugappan, D. S. (2012). S.A. Combinatorial Multi-Objective Trust Model for Efficient and Secured Routing in UWSN. *Int. J. Eng. Trends Technol.*, *22*(6), 259–264. doi:10.14445/22315381/IJETT-V22P255

Shen, J., Tan, H. W., Wang, J., Wang, J. W., & Lee, S. Y. (2015). A novel routing protocol providing good transmission reliability in underwater sensor networks. *J. Internet Technol.*, *16*, 171–178.

Simplicio, M. A. Jr, Silva, M. V., Alves, R. C. A., & Shibata, T. K. C. (2017). Lightweight and escrow-less authenticated key agreement for the internet of things. *Computer Communications*, *98*, 43–51. doi:10.1016/j.comcom.2016.05.002

Teguig, E. H., Touati, Y., & Ali-Cherif, A. (2017). ECC Based-Approach for Keys Authentication and Security in WSN. *Proceedings of the 2017 9th IEEE-GCC Conference and Exhibition (GCCCE)*, 1–4. 10.1109/IEEEGCC.2017.8447901

Varadharajan, K. (2018). Secure localization using coordinated gradient descent technique for underwater wireless sensor networks. *ICTACT J. Commun. Technol.*, *9*(1), 1716–1720. doi:10.21917/ijct.2018.0252

Vaseghi, B., Pourmina, M.A., & Mobayen, S. (2017). Finite-time chaos synchronization and its application in wireless sensor networks. *Trans. Inst. Meas. Control.*

Vaseghi, B., Pourmina, M. A., & Mobayen, S. (2017). Secure communication in wireless sensor networks based on chaos synchronization using adaptive sliding mode control. *Nonlinear Dynamics*, *89*(3), 1689–1704. doi:10.100711071-017-3543-9

Vikas Juneja & Gupta. (2018). *Security against Vampire Attack in ADHOC Wireless Sensor Network: Detection and Prevention Techniques*. Academic Press.

Wang, P., Li, C., & Zheng, J. (2007). Distributed minimum-cost clustering protocol for underwater sensor networks (UWSNs). *Proceedings of the IEEE International Conference on Communications (ICC'07)*, 3510–3515. 10.1109/ICC.2007.580

Wang, Q., Dai, H. N., Li, X., Wang, H., & Xiao, H. (2016). On modeling eavesdropping attacks in underwater acoustic sensor networks. *Sensors (Basel)*, *16*(5), 721. doi:10.339016050721 PMID:27213379

Xu, M., Liu, G., & Guan, J. (2015). Towards a secure medium access control protocol for cluster-based underwater wireless sensor networks. *International Journal of Distributed Sensor Networks*, *11*(5), 325474. doi:10.1155/2015/325474

Yang, G., Dai, L., & Wei, Z. (2019). Challenges, Threats. *Security Issues and New Trends of Underwater Wireless Sensor Networks*, *147*, 210–216.

Yuan, C., Chen, W., & Li, D. (2017). A Hierarchical Identity-Based Signcryption Scheme in Underwater Wireless Sensor Network. *Proceedings of the China Conference on Wireless Sensor Networks*, 44–54.

Zhang, Y., Jin, Z. G., Luo, Y. M., & Du, X. (2013). Node secure localization algorithm in underwater sensor network based on trust mechanism. *Jisuanji Yingyong*, *33*(5), 1208–1211. doi:10.3724/SP.J.1087.2013.01208

Zhou, P., Jiang, S., Irissappane, A., Zhang, J., Zhou, J., & Teo, J. C. M. (2015). Toward energy-efficient trust system through watchdog optimization for WSNs. *IEEE Transactions on Information Forensics and Security*, *10*(3), 613–625. doi:10.1109/TIFS.2015.2389145

Chapter 12
Cyber Attacks and Cyber Security in Underwater Communication

Sachin Umrao
Yale University, USA

ABSTRACT

This chapter is structured around the concepts of risk analysis due to underwater deployment of the cables for data transfer. Most of the organizations have deployed their networks below the water for shortening the distances between peers and also to reduce the physical destruction cost of cables. Furthermore, some organizations like Google kept their servers below the water because it reduced the cost of getting it cool, which in turn increases the efficiency. However, security consultants around the world in recent past expressed their considerations that a cyber-attack on these servers or cables might result in miserable economic collision. This might be overstated but there are infrequent situations in which cable breakage could be riotous. Although organizations cannot rule the threat of attacks on these apparatuses, there are fewer check measures that could reduce the possible attack chances in underwater communication.

INTRODUCTION

Previously, cybersecurity remained measured as the work of IT people only. But now days, cybersecurity is the responsibility of one and all. In 2019 cybersecurity is at the highest primacies for an organization. Detectable correspondence system are the good channel and errand well over less separations, still they are unprotected to assault and can undoubtedly be compacted. The organization NATO in 2016 for Underwater Communications, ebb and flow condition of-workmanship and network gathering talked about the close term difficulties, as work by a group of Italian who have created secure submerged remote interchanges and marketed it with the goal that it can achieve 10Mb/s at a scope of 32 feet, in harbour aquatic circumstances(Zhong et al., 2019; Salama and Abdelrahman, 2019).In 1977, the examination vessel Alvin dove 2500 meters underneath the sea waves empowering geologists to see legitimately, in the absolute first time, the peculiar lifeforms that run about the hot outlets at the planet's mid-sea edges.

DOI: 10.4018/978-1-7998-3640-7.ch012

After forty years, new types of exceptional beasts watch these profundities. Utilized in mine chasing, frame investigate, ecological observing, port watch, and different capacities, submerged unmanned underwater vehicles (UUV's) are currently so unique and countless in their plan and strategic the US has built up an innovation guide which is directly in its third form for coordinating the innovation into standard peacetime and war fighting missions (Lehto et al., 2019).

Various telecommunication industries had undergone through deep transformation resulting into exponential growth in recent past. Cable deployment technology is still the primary network deployment strategy for infrastructure development and communication medium, resulting in higher bandwidth for current and evolving technologies. With the increasing demand according to new technology and services, cable deployment also promise a scalable and flexible network platform with higher capacity. But many times cable cuts results in failure and network outages. This chapter investigates the categories of cyber-attacks during underwater communication, with emphasis on maritime security recommending the appropriate solutions as well. Fiber deployment management in Ghana is threatened by the Lack of regulatory guidelines and policies. Various security officials around the world have expressed concern regarding cyber-attack on these deployed cables and can result in great economic impact too. As shown in figure 1 the submarine cables are deployed globally and covering vast area below the water (Umrao, 2020).

Figure 1. Global map of submarine cables

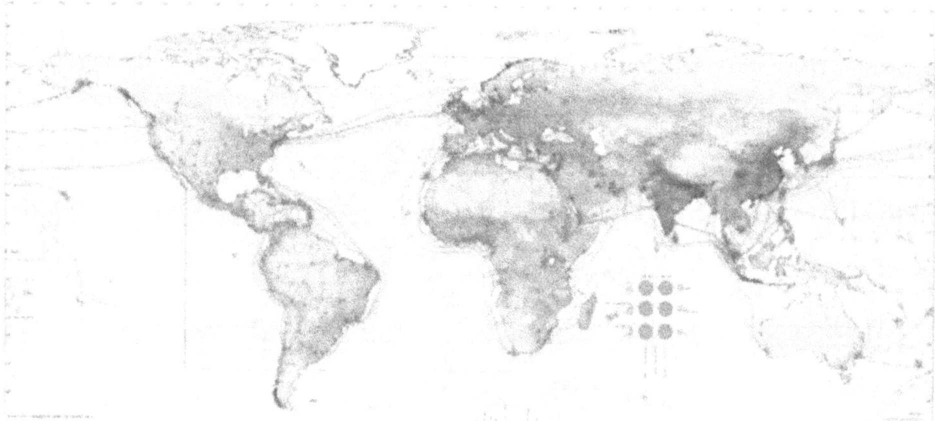

This chapter consists of 4 sections mainly with section II to categorize the cyber-attacks during underwater communication. Further in section III cyber risks along with cyber security measures are discussed. Furthermore the chapter leads to the conclusion of these parameters.

CATEGORIES OF CYBER ATTACKS DURING UNDERWATER COMMUNICATION

Some of the most common categories of attacks that can happen underwater also are as (Hummelholm, 2019; Wanglai 2018; Davidovic et al., 2019; Silverajan, 2018):

1. **Malware:** You may have ever observed an antivirus ready discourse box on your screen, or in the event that you have clicked a vindictive email connection unintentionally, at that point you may have a narrow escape with malware. Assailants generally use malware to increase a base in clients' PC and just as in the workplaces they have worked in on the grounds that it tends to be so useful. It alludes to different sorts of hurtful programming, for example, ransomware and infections. When infection comes in your framework, it can wreck a wide range of devastation; even it can assume responsibility for your framework, to screen the keystrokes, activities and quietly sends classified information from your PC to the aggressor's command post. Assailants will utilize an assortment of strategies to get infection into your framework, yet at some stage it is commonly required by the client to make a move to introduce the malware. In this incorporates click on a download connection to download a record, or opening a connection that may look innocuous, however really has a malware installer covered up inside (channels et al., 2012).
2. **SQL Injection Attack:** SQL represents organized query linguistic and it is a programming language used to connect with databases to oversee database. Numerous servers that store the significant profound information for sites and administrations utilize Structured Query Language (SQL) for dealing with the information into the database. An SQL injection attack aims this type of server, using hateful code to disclose info it usually would not. This is especially challenging if the server supplies sensitive customer data from the website credentials, or another personal identified data, which are attractive and profitable targets of the attacker. The attack of SQL Injection affect the whole thing by utilizes any of the known SQL vulnerabilities that allow the SQL server to implement the malicious code. For instance, if a SQL server is perilous to an Infusion Attack, an aggressor may be able to go to a site's internet searcher and type in code that could compel the webpage's SQL server to dispose of all the put away usernames and passwords for the site (Goyal et al., 2019; Li et al., 2012).
3. **Cross-Site Scripting (XSS):** As discussed above in SQL Injection Attack, an assailant drives next to a susceptible website to mark its critical deposited data, like identifications or further significant data. Nonetheless if the assailant unswervingly targets the user, they may choose for XSS attack. This attack also consist of putting harmful code, in this the website is not being attacked but the damaging code runs in browser and when user visits this, it goes after the visitor directly. The most common ways an attacker uses is by injecting malicious data into any comment/script that can run repeatedly. XSS attacks harm a website's status knowingly through putting users' important data such as their username or password at danger without any sign that anything had occurred (Awan et al., 2017).
4. **Denial-of-Service (DoS):** In this attack if you have a communication link overloaded with more traffic in comparison to that it was thought of to control, the server gets overload and it becomes tough to serve for its data to users. It may occur due to harmless facts surely. This type of traffic overloading is very damaging, as an assailant overflows a server with a huge quantity of traffic to essentially shut it down for all users. In some cases, these Denial of Service attacks are performed by many computer systems at the same time. This kind of attack is known as a (DDoS) Distributed Denial-of-Service Attack and can be even more difficult to overcome due to the assailant appearing from many different IP addresses around the world simultaneously (Dong et al., 2014).
5. **Session Hijacking and Man-in-the-Middle Attacks:** In UWSN there is network connection between every sensor nodes present in the water, and that network is connected to parent node which transfers data to the main base station. When communication gets established between client and

server deployed below the water, certain transactions takes place which helps the server to identify basic information of the client e.g. client id and what type of client is. Also the server requests the client for services and act on that after manipulating the service required by the client. These things should only happen when you are browsing on internet or trying to access some another website by using some login credentials. When there is a suitable connectivity between the server and client/computer, a session is establish between them and also at same time a session id is generated between two parties which must be always private. During the attack, attacker uses different techniques to get that session id, after getting it the attacker acts as new system that is unknown to the server and can access sensitive data stored on the server like, to read full history of the session or to fetch all the unauthorized data present on the server. It also make attacker capable to cut the communication link between the two parties which generally known as man-in-the-middle attack (Hu et al., 2017).

6. **Credential Reuse:** In this growing world as technology is growing day by day it is also a fact that this advanced technology can be harmful for our technological systems because attackers are also using those technologies for hacking at more advanced level. Today everyone around world uses different websites for different purposes, some are using for business or some for education. For the security, reliable and efficient use of these sites every user has to pass through some type of login credentials in the form of username and password. But only for their own ease many people are using same credentials for different sites that make it very easy for the attackers to get access to it. Many security professional recommends having unique username and password for the application and the websites they are working on. Once an attacker gets credential it might be very easy for them to access different sites. So it is essential to have different credentials for different purposes, even by using credentials managing systems (Connell et al., 2011).
7. **Phishing Attacks:** These attacks are generally activated when a user performs some task on mail or some links sent by the unknown person. The unknown client is/can be an attacker trying to get your sensitive data in the form of fake links and mails. The link contains a virus or malware which when clicked, spreads out into the computer system to affect that system. That affected system pretends to do that task from user whereas user didn't perform that. Human curiosity is the major factor on which this attack depends as the attacker gets all your personal information and uses it in erroneous mode. In order to conflict this attempt, significance of confirming email sources and supplement materials is very important (Arachchilage, 2015).
8. **Spear Phishing Attack:** These attacks are generally performed in the form of emails sent to organization which is working for government or any other secret services to fetch the sensitive data. Sometimes these attacks are also performed by government officials to get sensitive data for various fields like education, healthcare, military secret information, submarine data below water etc. These attacks are also performed by cybercriminals for earning purposes. They fetch the sensitive data from various organizations and sell it to the government and various multinational companies in exchange of big amount. These attackers uses various social engineering toolkits and methods to make such a effectively personalize website or messages (Jones, 2015).
9. **Whale Phishing Attack** These attacks generally aims to the people who are considered to be the important and are buildings block for the company like CFO, CEO, MD etc. Being an administrator for a particular organization they have full access to sensitive data. The word whale symbolizes the size of the attack to be performed on a particular system. These attacks are aimed, that's why they are almost negligible to identify them. Security administrator plays a major rule in any organization

as they are the people who are mostly responsible for security of every system in the company. So, there must be some programs related to security which helps to spread some awareness in each and every employee of the company (Garon, 2018).

These are just some common types of attacks and methods. It is not predetermined to be comprehensive and attackers advance new approaches as desired. However one must be aware of these kinds of attacks and reduce these attacks will surely improve your system security.

MARITIME CYBER SECURITY

King college in London recently briefed that there are around thousands of companies which are at higher level of risk, having low security in their network and also the employee working in that companies are also not aware of the security systems. These companies can become more vulnerable to the different attacks (Goyal et al., 2018).

1. **Risk Barometer:** During the study it is observed that cyber-attacks mainly interrupt the business and is at the top of the Allianz risk barometer. In broad losses mainly cyber-crime, privacy breaches, ransomwares and distributed denial of service attack are considered to be the major factors in today's time. According to Allianz services the data loss due to these attacks are more as compared to the losses during some technical failure or some glitches (Paltrinieri et al., 2017).
2. **Increasing Cyber Risks:** Every organization has to pay huge amount, around millions of dollars when their services have been disrupted by some type ransomwares or attacks. Due to advancement in technology criminals are using advances level of methods and techniques to steal data of different organizations and sell them at higher prices. These types of services can also put an adverse effect on the economy of the country like sometimes a ransom was spread to a system of some government companies which ask for around millions of dollars to continue the services. In those cases, the government has to fulfil the demand of the ransom and these types of practices can lead to huge change in the economy level of the country (Babcock, 2015).
3. **Biggest Problems:** Whenever we talked about business interruptions and large scale attacks the scenario behind this is the small scale attacks like phishing for example attacker mainly sent fake mails to the employee working in companies this is the first and utmost step of the beginning of the attack, from this many attacker can fetch very sensitive data of the company resulting in the use of the data in an inappropriate manner. According to the search by EY majorly most of the attacks are done through phishing by which they can have an unauthorized access to the system and data (Sovacool et al., 2011).

CONCLUSION

This chapter deals with the cyber security risk analysis and prevention methods while underwater communication that can happen in both wired and wireless fashion. To the best of authors' knowledge it is the only chapter well described about various possible attacks and their effects on various networks or computer system for underwater signal transfer. Underwater sensor networks communication is the most sensitive network communication among different countries by different companies deploying underwater servers. It will help to understand how to manage and secure the sensor nodes from various attacks performed by the attacker like phishing, advanced engineering tools and how we can get rid of these attacks. The overall dependency is upon the public community that how to secure their networks and organization.

REFERENCES

Arachchilage, N. A. G. (2015). *User-Centred Security Education: A Game Design to Thwart Phishing Attacks*. arXiv preprint arXiv:1511.03459

Awan, J. H., Memon, S., Pathan, S. M., Usman, M., Khan, R. A., Abbasi, S., ... Hussain, Z. (2017). A user friendly security framework for the protection of confidential information. *Int. J. Comput. Sci. Netw. Secur*, *17*(04), 215–223.

Babcock, C. (2015). *Preparing for the Cyber Battleground of the Future*. 50th Space Communications Squadron Schriever AFB United States.

Channels, M. O. F., & Abdelzaher, T. (2012). *Mining Diverse Opinions Abo, Gavin High-Efficiency Ferrite Meander Antenna (HEMA) for LTE Applications*. Academic Press.

Connell, I. T. W., Connell, T. W., & Amand, J. P. S. (2011). *U.S. Patent No. 8,063,737*. Washington, DC: U.S. Patent and Trademark Office.

Davidovic, S., Loukoianova, M. E., Sullivan, C., & Tourpe, H. (2019). *Strategy for Fintech Applications in the Pacific Island Countries*. International Monetary Fund. doi:10.5089/9781498326735.087

Dong, Y., Dong, H., & Zhang, G. (2014). Study on denial of service against underwater acoustic networks. *Journal of Communication*, *9*(2), 135–143. doi:10.12720/jcm.9.2.135-143

Garon, J. M. (2018). Cyber-World War III. *Journal of Law & Cyber Warfare*, *7*(1), 1–60.

Goyal, N., Dave, M., & Verma, A. K. (2018). A novel fault detection and recovery technique for cluster-based underwater wireless sensor networks. *International Journal of Communication Systems*, *31*(4), e3485. doi:10.1002/dac.3485

Goyal, N., Sandhu, J. K., &Verma, L. (2019). *Machine Learning based Data Agglomeration in Underwater Wireless Sensor Networks*. Vol. 9, Issue 6, pp. 240-245. Academic Press.

Hu, Q., & Hancke, G. P. (2017, March). A session hijacking attack on physical layer key generation agreement. In *2017 IEEE International Conference on Industrial Technology (ICIT)* (pp. 1418-1423). IEEE. 10.1109/ICIT.2017.7915573

Hummelholm, A. (2019, July). Undersea Optical Cable Network and Cyber Threats. In *European Conference on Cyber Warfare and Security*. Academic Conferences International Limited.

Jones, M. (2015). *The effects of conformity and training in a phishing context: Conforming to the school of phish*. The University of Alabama in Huntsville.

Lehto, M., Hummelholm, A., Iida, K., Jakstas, T., Kari, M. J., Minami, H., & Saunavaara, J. (2019). Arctic Connect Project and Cyber Security Control, ARCY. *Informaatioteknologiantiedekunnanjulkaisuja*, (78/2019).

Li, C. S., & Hu, Y. (2012). Study on State-based Remote Detection Technology for SQL Injection Vulnerability. *Information Security and Communications Privacy, 5.*

Paltrinieri, N., Landucci, G., & Salvo Rossi, P. (2017, June). Real-Time Data for Risk Assessment in the Offshore Oil and Gas Industry. In *ASME 2017 36th International Conference on Ocean, Offshore and Arctic Engineering*. American Society of Mechanical Engineers Digital Collection. 10.1115/OMAE2017-61486

Salama, E. M., & Abdelrahman, A. (2019). *Security of Critical Cyber-Physical Systems: Fundamentals and Optimization* (Doctoral dissertation). Virginia Tech.

Silverajan, B., Ocak, M., & Nagel, B. (2018, July). Cybersecurity Attacks and Defences for Unmanned Smart Ships. In *2018 IEEE International Conference on Internet of Things (iThings) and IEEE Green Computing and Communications (GreenCom) and IEEE Cyber, Physical and Social Computing (CPSCom) and IEEE Smart Data (SmartData)* (pp. 15-20). IEEE. 10.1109/Cybermatics_2018.2018.00037

Sovacool, B. K., & Bulan, L. C. (2011). Behind an ambitious megaproject in Asia: The history and implications of the Bakun hydroelectric dam in Borneo. *Energy Policy*, *39*(9), 4842–4859. doi:10.1016/j.enpol.2011.06.035

Umrao. G. (2020). *Undersea Cables and the Challenges of Protecting Seabed Lines of Communication.* Center for International Maritime Security. Retrieved 9 May 2020, from http://cimsec.org/undersea-cables-challenges-protecting-seabed-lines-communication/35889

Wanglai, G. (2018). BRICS cybersecurity cooperation: Achievements and deepening paths. *China Int'l Stud.*, *68*, 124.

Zhong, S., Zhong, H., Huang, X., Yang, P., Shi, J., Xie, L., & Wang, K. (2019). Networking Cyber-Physical Systems: System Fundamentals of Security and Privacy for Next-Generation Wireless Networks. In *Security and Privacy for Next-Generation Wireless Networks* (pp. 1–32). Springer. doi:10.1007/978-3-030-01150-5_1

Chapter 13
Machine Learning Techniques for Underwater Wireless Sensor Networks:
A Comprehensive Study

Deepti Rani
Chandigarh University, India

Anju Sangwan
https://orcid.org/0000-0002-5433-1484
Guru Jambheshwar University of Science and Technology, India

Anupma Sangwan
Guru Jambheshwar University of Science and Technology, India

Tajinder Singh
University of Information Science and Technology, Macedonia

ABSTRACT

With the enormous growth of sensor networks, information seeking from such networks has become an invaluable source of knowledge for various organizations to enhance the comprehension of people interests. Not only wireless sensor networks (WSNs) but its various classes also remain the hot topics of research. In this chapter, the primary focus is to understand the concept of sensor network in underwater scenario. Various mechanisms are used to recognize the activities underwater using sensor which examines the real-time events. With these features, a few challenges are also associated with sensor networks, which are addressed here. Machine learning (ML) techniques are the perfect key of success to resolve such issues due to their feasibility and adaption in complex problem environment. Therefore, various ML techniques have been explained to enhance the operational performance of WSNs, especially in underwater WSNs (UWSNs). The main objective of this chapter is to understand the concepts of UWSNs and role of ML to address the performance issues of UWSNs.

DOI: 10.4018/978-1-7998-3640-7.ch013

INTRODUCTION

Advances in the area of hardware and wireless network technologies have taken us to the verge of a new age where little wireless devices give access and participate actively in smart environments anytime, anywhere. WSN has achieved great popularity universally due to its potential applications in variety of fields. Wireless sensor network refers to a network of geographical distributed devices, known as nodes, having dedicated sensors. These can sense the habitat and communicate the environment's physical conditions collected from the field monitored via wireless links to base stations (Yick et al., 2008). The major objective of such networks is information collection by environment sensing and placing the collected data to relevant location. The number of sensor nodes depends on various applications and can vary significantly. Consequently, the management of so many nodes needs an algorithm that is scalable and efficient. Furthermore, the WSNs can change dynamically as desired by system designers or due to other external factors. Such situations can be handled by reprogramming and network reconfiguration (Kumar et al., 2019).

Underwater Wireless Sensor Networks (UWSNs)

Majority of the earth surface, approximately 70 percent, is covered with water. Out of the total water covered area, more than 95 percent area is oceanic. Rest of the water subsists in seas, rivers, streams, ponds, and lakes. Hence, a lot of activities take place under water which could neither be supervised by human beings nor by devices that are used for above earth surveillance. Underwater sensor network (UWSN) is a key technology for the surveillance of various under-water activities. Technologies used underwater are very much different from technologies used above water. UWSN is a technique that facilitates users and devices to send and receive messages in underwater scenario. Wireless sensor-based technology is globally adopted by various users as per need and requirement in different domains of WSNs. Architecture of Underwater WSNs is vulnerable to various challenges. UWSN provide solution for many area related problems.

The UWSNs consist of several kinds of sensors like acoustic sensors used to track underwater physical information such as noise, temperature, pressure, disaster symptoms, attacks, pollution. Underwater Acoustic Sensor Networks (UASNs) are the network of sensor nodes that are used for monitoring the aquatic activities. UWSN is very frequently used for underwater object exploration, navigation of underwater vehicles, monitoring the behaviour of aquatic living and non-living bodies. Data from underwater sensors is gathered through some autonomous vehicles specially designed for collaborative monitoring of sea bottom (Awan et al., 2019). Sensors are the backbone of every kind of smart network which extract the data associated with physical and environmental conditions. The information based on collected data is diffused and shared among connected nodes through wireless links. Sharing of information is based on network topologies (bus, star, ring, tree, or mesh), which can be decided as per need and requirement as WSNs are deployed in different kinds of environments. Characteristics of UWSNs are also different from other traditional networks. Before the emergence of UWSNs, there were some flaws in the traditional approaches used for data monitoring and other operations used for underwater environments. It was very difficult to monitor the area covered with water in depth such as oceans, Marianas Trench, and sea. Acoustic Sensors are efficient in such circumstances (Singh et al., 2013).

Figure 1. A typical underwater sensor network (Tuna et al., 2017)

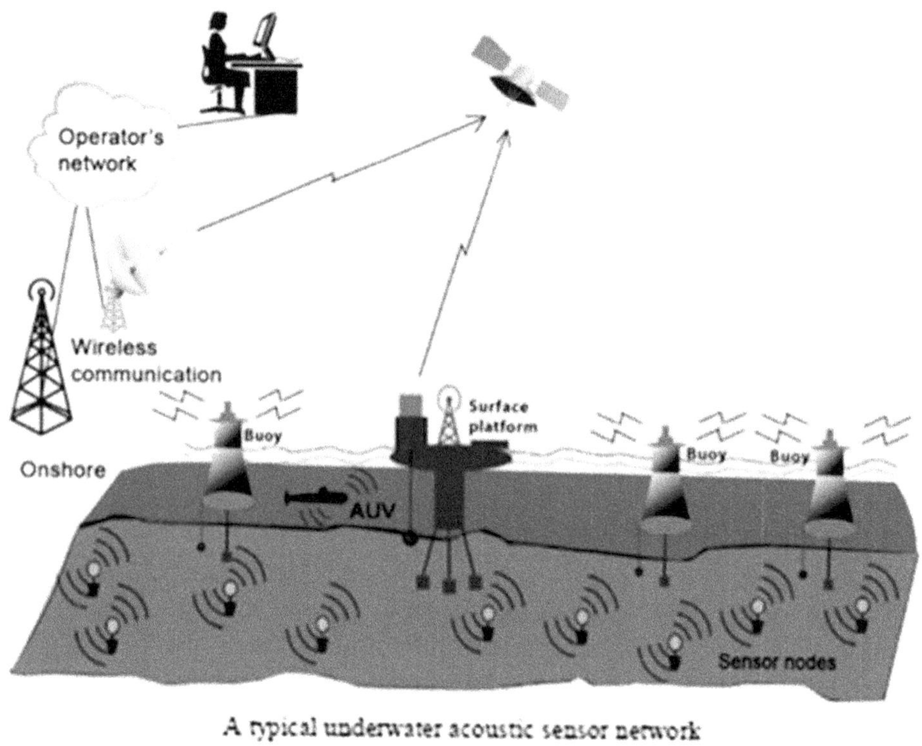

Limitations in Traditional Ocean Monitoring Approach Before Emergence of Underwater WSNs (UWSNs) (Freitag et al., 2002)

- Online processing of some kinds of acquired information is not allowed.
- Bi-directional or closed loop communication is almost impractical.
- Identification of breakdown and failures in real time is impossible.

In last some decades, UWSN has been deployed and finely tuned with high interest in many ocean related application areas like: seismic waves monitoring, oceanographic data collection, sea water pollution measurement, on board data recording recovery, assessment of water quality, underwater waste monitoring, biological monitoring, underwater robotic missions, and security monitoring (Gkikopouli et al., 2012).

Applications of UWSNs

The water covered area also contains rich valued assets. Underwater Sensor Networks have a large spectrum of applications covering a wide range of different fields such as scientific analysis, commercial advantage, sports facilitation, disaster prevention etc. With recent technological advancements, this application spectrum of UWSNs have not only fascinated the attraction of research community but have an orientation towards industrial and defence lobby. Defence or other services can also be appointed for the protection and monitoring of asset sources such as oil, natural gas and coal etc. Every country appoints their own underwater military forces and secret agents for the safety purposes. Some major application areas of Underwater WSNs are (Singh et al., 2013; Kumar et al., 2014; Felemban et al., 2015):

- Oceanographic Information Gathering
- Disaster Prevention and Management
- Tactical Surveillance
- Under-water and Over-water Transportation Management (e.g. ships, boats, submarines, and water crafts)
- Secret Agencies
- Undersea Exploration
- Aquatic Study
- Mine Reconnaissance
- Tactical Surveillance
- Water Pollution and Environment Monitoring
- Earth Temperature Monitoring
- Physical Monitoring (e.g. Pressure and Temperature)
- Monitoring Underwater Pipelines
- Seismic Monitoring
- Mine Detection
- Offshore Exploration
- Habitat Monitoring
- Classification of Underwater Objects
- Tracking Radar and Sonar

In this chapter, we are addressing various aspects of Underwater Wireless Sensor Networks. There are several challenges in UWSNs such as limited bandwidth, energy consumption, long propagation delay, sensor failures, link reliability, noise, security, traffic, memory, and lifetime of battery used in sensors etc. Data transmission in UWSNs is also very challenging. Optical and radio signals are not suitable for data transmission in underwater environment (Ayaz et al., 2009). Acoustic channels showcase good performance while underwater communications with characteristics like dynamic topology, high latency, and low bandwidth (Zhu et al., 2019). Now, UWSNs have emerged as significant interest area for numerous researchers in different application domains such as military, industry, academy, commerce, and science (Lu et al, 2020). Researchers working on UWSNs have drawn great anticipation for applications in aquatic environment protection, marine exploration, offshore functioning, and underwater military and undersea observations (Felemban et al., 2015).

Underwater environment is more harsh than terrestrial environments. Here, communication is much more difficult as compared to land. Hence, enhancing the energy efficiency is one of the most significant concern while designing any framework for underwater sensor networks. Protocols used in underwater scenario should be more energy efficient for successful data transmission. This chapter highlights various machine learning techniques and its algorithms which are much feasible to address and resolve various issues in different application areas of UWSNs. Machine Learning algorithms provide promising solutions for performance enhancement of network communication in various range of environments like static, dynamic, centralized, decentralized, and grid. The network performance can be improved with the enhancement of network lifetime, functional accuracy and reducing the time complexity of measuring devices.

MACHINE LEARNING

Machine Learning is an automated development or learning mechanism that operates without having been deliberately programmed (Mitchell, 1997; Langley et al., 1995; & Ayodele, 2010). ML has improved our computational processes in cost effectiveness, efficiency and reliability. ML manufactures models automatically, quickly, and accurately by analysing even more complex data. Machine learning techniques are extensively used in Search Engine Result Refining, Email Spam and Malware Filtering, Product Recommendations, Predictions while Commuting, Online Customer Support, Virtual Personal Assistants, Social Media Services like face recognition, Videos Surveillance and Online Fraud Detection etc.

For the following principal reasons, machine learning is important in sensor network applications:

- Sensor networks typically monitor rapidly-changing dynamic environments. For instance, because of sea turbulence or soil erosion, the position of the nodes can change. Sensor networks that can be adapted to these environments and operated efficiently need to be developed.
- WSNs may be used in exploratory projects to collect additional knowledge on unattainable and dangerous sites (Paradis et al., 2007) (e.g. volcanic activity or drainage monitoring). Due to the unpredictable behaviour trends in such scenarios, system designers may develop solutions, which may not function, as expected at initial stage. System designers would prefer reliable algorithms in machine learning which are capable of testing themselves against newly established information.
- For system model, for example, routing problems, machine learning offers low-complexity estimates, where otherwise complicated algorithms are needed (Krishnamachari et al., 2002; Al-Karaki et al., 2004).
- Designers of sensor network generally have exposure to vast volumes of data, yet they cannot draw significant correlations. The WSN application, for instance, often has minimal specifications for coverage of data that must be met with limited sensor hardware resources, not to mention ensuring communications connectedness and energy sustainability (Romer et al., 2004). Machine learning approaches may be used to explore significant associations within the sensor data and recommend optimal usage of the sensor for utmost data coverage.
- Novel applications and integration of WSNs have been introduced with the objective of promoting intelligent decision-making and self-control, for example in Cyber-Physic Systems (CPS), Machine-to-Machine (M2M) and Internet of Things (IoT) technologies (Wan et al., 2013). In this context, ML is essential to extract the numerous levels of abstraction, necessary to carry out AI tasks with minimal human intervention (Bengio, 2009).

The strength of the Machine Learning is its ability through a framework that can learn to enhance its performance to provide general solutions. ML plays a crucial role in numerous applications comprising medical, computing and engineering, due to its interdisciplinary nature. In order to solve various problems on WSN, recent ML advances have been applied (Alsheikh et al., 2014).

Machine Learning is a subset of Artificial Intelligence (AI) and also deals in statistics and optimization. It facilitates computer systems to learn automatically. Learning and building of algorithms is based on previous patterns and prediction of data sets (Simon et al., 2016). For every training pattern with input variable '*p*' there is a corresponding output say variable '*q*'. Here, *q* contains the predetermined value.

Application of Machine Learning in WSN

Certain ML applications in WSN are given below (Kumar et al, 2019):

- The optimum number of sensor nodes required for region coverage, can be conveniently determined by the ML approach in target area coverage problem.
- ML is used to distinguish normal sensor nodes from damaged sensor nodes thereby increasing network performance.
- Data routing plays a significant role in extending network's life. In order to enhance system performance, dynamic routing schemes are required for dynamic behaviour of sensor networks.
- Sending all information to the base station increases network overhead. ML also help in reducing data movement at the cluster head level or sensor.
- Sensor nodes might be relocated through internal or external factors. The ML algorithm makes accurate localization smooth and fast.
- The long-lasting and self-powered maintenance of WSNs deployed in harsh environments can be achieved through energy harvesting. Machine Learning algorithm ameliorates the performance of WSNs in predicting energy that can be harvested within a certain timeline.

Application of Machine Learning in UWSN

Machine learning motivates towards optimization of resources to prolong the network lifetime. The usage of machine learning techniques is also profitable in the field of UWSN. Being in its initial stage with a wide scope of expansion, some of ML applications in UWSN are as follows:

- Machine learning classification and regression algorithms are useful to improve localization accuracy, tracking accuracy as well as error reduction.
- The clustering algorithms in machine learning are helpful in reducing energy consumption and load balancing. Thus, provides an expansion to the network lifespan.
- The clustering algorithms are also fruitful in order to gain higher throughput, improved packet delivery while maintaining the energy consumption in the network.
- The deep learning techniques are appropriate to improve diagnosis accuracy and prevention of over-fitting while training of network.

Classification of Machine Learning Approaches

Machine learning approaches can be mainly categorized into 3 types of higher-level learning which are further divided into sub-categories.

Supervised Machine Learning: Maximum practical machines follow the concept of supervised learning. In supervised learning approach, a system is trained on a pre-determined set of training samples in order to provide a precise solution to new dataset problems (Simon et al., 2016). Training set contains

Table 1. Machine learning applications for sensor networks

ML Techniques	Applications in Sensor Networks	Basic Procedure
	Supervised	
Regression (Lefort et al., 2017)	Localization, Connectivity, Data aggregation, Energy harvesting	Predicts accurate results with minimum error
Decision Tree (Chander et al., 2018; Kumari et al., 2019)	Classification, Solving the design related problems to: improve reliability, reduce failure mean, time of link loss rate, and corruption rate; Reliability identification of link in the network	Collected data is compared and classified in the form of decision trees on the basis of decision conditions
KNN (Yu et al., 2005; Winter et al., 2005; Branch et al., 2009; Jayaraman et al., 2010; Chander et al., 2018)	Preparation of function for query processing subsystem; N/w outlier detection; Energy efficient processing; Query processing in 2D and 3-D; Target detection and tracking	Regression and classification on the basis of near data sample's values
Neural Networks (Shareef et al., 2008; Zou, 2014; Harinath et al., 2019)	Localization of sensor nodes using received signal strength; Measuring distance and propagation angle; I-matrix to control light, Measuring access dependability on WSN, Broadcast scheduling; Prevent false message flooding	Derived from concept of biological neural network. One neuron sends signal to another connected neuron(s)
Bayesian (Rao et al., 2018)	Tracking and navigation Underwater, Port-starboard ambiguity	Relationship is detected among datasets using various statistical and testing methods
	Unsupervised	
Clustering (Sandeep et al., 2017)	Data aggregation, LEACH is widely used for routing	Grouping of most similar set of nodes
Hierarchical clustering	Temperature monitoring and many spatial and temporal applications	Hierarchy of clusters is built, produces decomposed set of objects.
K-means clustering (Zhang et al., 2015)	Energy conservation, routing	Reasonably cluster the network and choose primary heads for energy sharing
	Reinforcement	
Q-Learning (Shah et al., 2007; Lu et al., 2020)	Improving the lifetime of UWSN, Resource management, Quality of Service, Energy efficient adaptive routing	Learns from actions and guide an agent to find the best

pairs of input and expected output (Kumar et al., 2019). Trained data is labeled and categorized into one or more groups. Input dataset is split up into two parts: one is called 'training dataset' and another is 'test dataset'. Training dataset consists of output variable that is required to be predicted (Dey, 2016). Mapping between input and output is learnt by the training process.

For example, there is an input variable x and output variable y. A learning algorithm is designed to learn the input to obtain the predicted output. After completing the training process, we can obtain a function from the input '*p*' with finest estimation of output '*q*' i.e. *y (f: p®q)*. On the basis of training points, the system learns how 'test' data can be categorized and labeled (The Royal Society, 2017). Supervised learning is capable to cope with various challenges in sensor networks. Supervised learning is classified into two parts: regression and classification.

Two phases of supervised machine learning are:

- **Training Phase:** Training samples are used to develop algorithms to learn correlation among training samples.
- **Prediction Phase:** Estimation of sensor node location coordinates using training models.

Unsupervised Machine Learning: In unsupervised learning approach, the learning takes place without labeled input. Training data set is not needed for learning. The goal of unsupervised learning is to discover the characteristics which are used to form data points to classify the introduced data sets (Simeone, 2018). For example, clusters can be created for given data. Data are assigned to clusters on the basis or their similarity (more or less) or dissimilarity. Most similar data are assigned to a common cluster near to the origin of a cluster. With the reduction of similarity, data points are kept away from the origin (Royal Society, 2017). When any new data is given, it is assigned to a cluster according to its similarity or dissimilarity with other data present in a cluster. Hence, in unsupervised learning the method learns some features from the data. Previously learnt features are used to recognize the class of new introduced data (Dey, 2016).

Reinforcement Learning: Reinforcement learning lies between supervised and unsupervised learning. The learner has no knowledge about the actions which have to be taken, until it is given any situation. The reinforcement learning is based on two criteria: trial and explore the error (Dey, 2016). Reinforcement learning algorithms continue learning from surrounding environment interaction and collect information to perform certain actions. This learning approach improves the performance by determining the most ideal results for decision making (Kumar et al., 2019; Chander et al, 2018). Fig. 1 shows classification of machine learning approaches.

Machine Learning Based Performance Enhancement Algorithms for Underwater Sensor Networks

There are numerous issues in UWSNs which need to be reduced using some advance methodologies and algorithms. ML has been proved as an efficient technique to reduce these challenges up-to certain level. This section states the benefits of using the ML-specific solution to address UWSN issues. Following are some issues in UWSNs which need to be addressed carefully.

- Localization
- Coverage & connectivity
- Fault detection
- Routing
- Data aggregation

SML-Based Localization Approaches in UWSNs

Localization is the procedure to identify the position of sensor nodes in Wireless sensor network (Bhatti, 2018). It is characterized by differentiating geographical/ physical area of a sensor node. While placing the sensor nodes in UWSNs, the positions are not exactly known. Numerous provisions have been discovered over WSNs for penetrating location and accessing the hidden points. Underwater localization has limited schemes in comparison to terrestrial environmental network localization. Localization methods developed for UWSNs make location discovery functionally independent to GPS in contrast to WSNs. Several algorithms have been developed so far based on different parameters. Some methods developed for range-based localization are: Angle of Arrival (AoA), Time Difference of Arrival (TDoA) and Time of Arrival (ToA). Received Signal Strength Indicator (RSSI) is a localization method that has been developed for distance estimation (Srinivasan et al., 2007; Srbinovska et al., 2008; Gkikopouli et

Figure 2. Classification of machine learning approaches

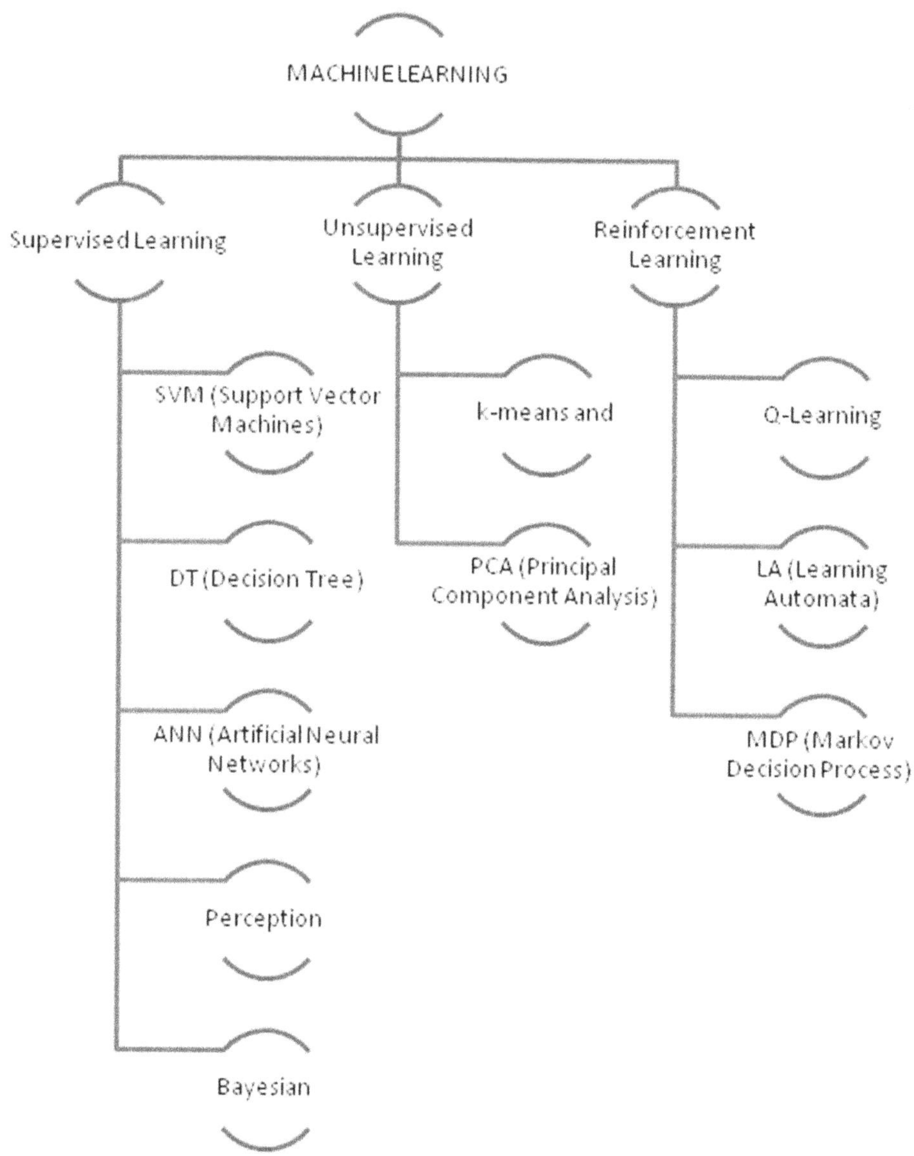

al., 2012; & Bhatti, 2018). For many years, ML techniques have been applied to improve the accuracy of the node location (Benras et al., 2015; Cottone et al., 2016).

Authors have proposed non-linear regression of source localization in fluctuating underwater environment (Lefort et al., 2017). Traditional inversion approaches, MUSIC algorithm and the Matched Field Beamforming are compared with local kernel regression and linear regression. Machine learning (ML) only needs training datasets. This environmental feature underlying the regression model makes ML a suitable candidate for fluctuating channels. Simulation results have shown that machine learning techniques outperform classical inversion approach by reducing the source localization error.

Authors have applied Deep Neural Network (DNN) in a range-based localization scheme (Dong et al., 2017). In this scheme, range-based approaches have been used instead of range-free due to their relatively high precision. Proposed DNN localization scheme is compared with Generalized Least Squares (GLS) and Least Squares Support Vector Machines (LS-SVM). Numerical results have confirmed that the Deep Neural Network localization algorithm achieves higher efficiency and localization accuracy than conventional approaches regardless of variance of distance measurement error, size of training set and number of anchor nodes.

Authors have developed a calibration (Classification and Regression) model for Time Difference of Arrival (TDoA) data outputs in a controlled field environment (Rauchenstein et al., 2018).In Washington (USA), on the Snake River, classification and regression algorithms of machine learning (ML) have been applied to measure localization errors of TDoA. There it has been used to track hydro-electric dam. The calibration process has two steps classification and regression. Classification algorithms and classification tree ensembles are used to find and eliminate outliers from potentially relevant input data vectors. The accuracy of the three-dimensional localization can be improved by this model in any TDoA acoustic sensor array in a stable scenario. This model described here is confined to the site and time domain for which it was created.

ML-Based Coverage and Connectivity Approaches in UWSNs

Besides efficiency of energy, other most challenging issues in UWSNs are coverage and connectivity. Coverage refers to the monitoring efficiency of deployed sensor in the area of interest. Deployment of sensor node depends vary according to application. Coverage is mainly categorized into two classes, namely; partial coverage and full coverage (Elhoseny et al., 2017). Further sub-categories of partial coverage are: focused, sweep, target, and barrier coverage (shown in Fig. 3).

Connectivity demonstrates that there is no isolated sensor node in the network. Data in the network is either sent directly via single node or via relay nodes. Suppose there are two nodes, node A and node

Figure 3. Hierarchical structure of coverage

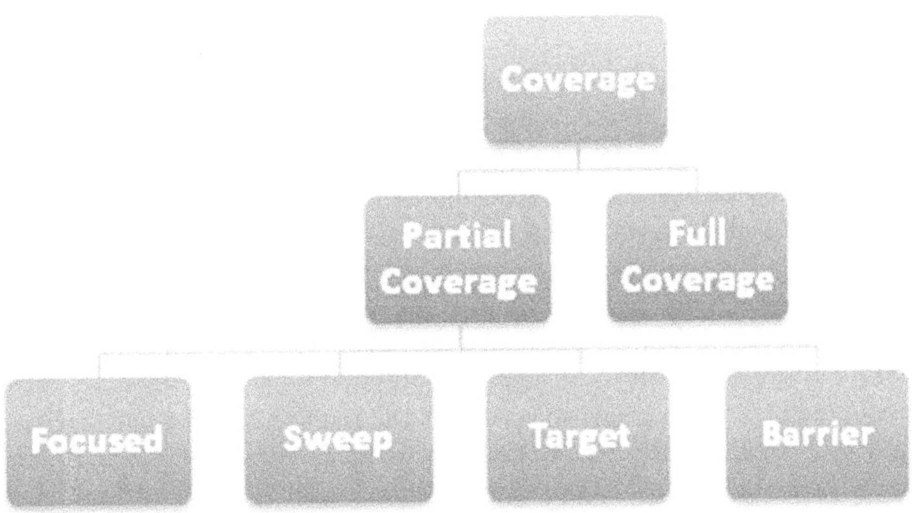

B. These two nodes are said to be isolated nodes if these are not connected to any other node or to each other too. These nodes are also called disconnected nodes. The nodes in WSNs are placed very densely. Several algorithms have been introduced so far to solve the issues related to coverage and connectivity (Farhat et al., 2019). After exhaustive literature survey, it has been observed that machine learning techniques are widely used to enhance coverage and connectivity in WSNs but to the best of our belief, no work has been reported for underwater sensor networks so far. The strategies explained ahead are specifically related with wireless sensor networks (WSNs).

There are negligible energy resources of recharge-ability for underwater sensor devices. Clustering is widely used technique to increase the scalability of system. Authors have suggested clustering-based approach for non-acoustic techniques of communication in underwater areas (Sandeep et al., 2017). They have explored various clustering, connectivity, and coverage issues and related solutions for UWSN. Researchers have used regression ML technique along with end-to-end delivery reliability algorithm for controlling the link quality in WSNs (Sun et al., 2018). Using this algorithm, a mapping function is captured among the packet reception ration, background noise, and RSS to optimize the quality of network and reliability in sensor nodes. Shu and others (2017) have explained SVM technique with decision tree algorithm. The combination also facilitates link quality estimation in WSNs. SVM uses the classification for quality of communication. Random forest-based algorithm is presented by Elghazel and co-authors (2015) for the monitoring of WSNs. RF changes the random quality of features in a better way. The authors (Yang et al., 2016) have performed location tracking of human in sensor networks using Naïve Bayes classification-based algorithm.

ML-Based Fault Detection Approaches in UWSNs

Under the water, a sensor network consists of several nodes/sensors of different sizes. Many hardware devices and software systems are used to operate these sensor nodes like memory, battery, processor, radio, device drivers, and datasets. These devices are very sensitive towards various kinds of malicious attacks. The sensor network can become hostile to cyber-attacks. A reliable underwater sensor network can be established by detecting of sensor faults at earlier stage. Sensors under the water work in complex environments and fault detection at earlier stage can save network break down. Use of machine learning based technologies in detecting faults in sensors can proved to be one of important step to move forward in this direction.

Authors (Sun et al., 2018) have proposed a fault diagnosis method for Autonomous Underwater Vehicle (AUV). The approach is mainly based on Deep Neural Network (DNN) and Denoising Autoencoder (DAE). In order to define DNN based fault diagnosis method, initially, the authors have considered the difference value of theoretical and measured state of AUV. Three steps have been taken for pre-training and tune. In first step, DAE is used to add a disturbance between input and hidden layers. By using original distributed data, maximum output of DAE is restored to improve network robustness. In next step, Back Propagation (BP) algorithm is used to provide a fine tune for network weights. Then the dropout method is used to stop overfitting within a small sized sample network. And, the experimental results show that the proposed DNN based methodology outperforms in terms of diagnostic accuracy in comparison to traditional fault diagnosis methods, that are; SVM- based and BP-based diagnosis methods. It provides classification accuracy of 98.21% where as SVM- based and BP-based diagnosis methods gives an accuracy level of 93.45% and 73.58% respectively.

Authors (Li et al., 2017) have elaborated a machine learning based fault detection method for the sensors. Generally, Autonomous Underwater Vehicles are used for sensing purpose under the water. The faults such as constant deviation, signal jump, invalidation and gain faults can be associated with these AUVs. The present method is based on the combination of Phase Space Reconstruction (PSR) and Extreme Learning Machine (ELM). The phase space reconstruction provides input for ELM model. Then, ELM model predicts faults in the sensors. A dataset collected from a fault free sensor is used to train ELM model. The steps taken to complete the whole procedure are as follows: i) data from sensors is collected, ii) it is pretreated by phase space reconstruction to obtain residual errors, iii) the threshold is fixed in accordance to the change of AUV state and iv) in the last, a check is placed to identify sensor fault as it is not possible to determine a faulty sensor immediately. If there exist a fault, the outputs received from ELM model are used to compensate with these failures. The dataset for simulation purpose is taken from sea test data. And, the outputs reveal that the proposed ELM model is lead over the BP- neural network in terms of mean square error and mean absolute error. Due to few training steps and fast training speed of ELM model, the proposed method can recognize faulty sensors efficiently.

ML-Based Routing Approaches in UWSNs

Routing is a major challenge for WSNs. It determines the most optimal route from the source node to the destination among the available paths. Underwater wireless sensor network (UWSN) is an autonomous and controlled system that can collect, process, classify, compress and transmit data automatically. Sensor nodes can be randomly or permanently placed in the monitoring area to collect real-time information and transmit collected sensed data through a specific route. It is difficult to develop networking protocols for UWSN because of its characteristics like dynamic structure, limited bandwidth, high latency and high energy consumption. Owing to several unique features such as adaptability to changing environments, distributed autonomous behaviour, invaluable decision-making tool and complex distributed problems like routing can be efficiently solved using machine learning techniques. The literature is rich in approaches tackling the routing problem in UWSNs using ML.

Authors have presented (Hu et al., 2010) a distributed Q-Network-based Energy and Latency-Aware Routing protocol (QELAR) based on Reinforcement Learning (RL) techniques This protocol is adaptive, lifetime-aware and energy efficient. In QELAR, Q-learning algorithm is used to transmit packets with lowest cost or highest rewards. In order to compute reward function, the Q-learning algorithm employs an action-value function (Q-value) that provides the expected reward by performing an action in a particular state. Sensor node with the highest reward function can be selected as an appropriate packet forwarder. QELAR routing protocol can balance workloads among sensor nodes, reduce overhead and extend network life. However, QELAR cannot be implemented on a large network because every sensor node has to record large amount of data for the Q-value calculation, that demands huge storage space.

Authors (Zhang et al., 2013) have proposed a cluster-based routing protocol using the modified K-means algorithm, to address the unbalanced energy consumption problem thereby extending the network lifetime. Maximum distance method, is adopted to select initial cluster centers by considering the depth of location of nodes and the density of the surrounding nodes. Factors such as the distance from base station to nodes, distance from the cluster center to nodes, and residual energy of nodes play significant role in electing primary and aided cluster heads.

Authors (Su et al., 2019) have developed an adaptive Deep Q-Network-based Energy and Latency-aware Routing protocol (DQELR) to extend the network lifetime. A deep Q-network algorithm is used in the DQELR to make optimal routing choices globally, using both off- and on-policy approaches. Depending on nodes' depth state and energy, nodes having maximum Q-value can be selected as forwarder by taking energy and latency into account. The DQELR is able to extend network life with lower energy consumption and strict latency limits.

ML-Based Data Aggregation Approaches in UWSNs

Data aggregation refers to the process of data collection from sensor nodes and performing aggregation. Many routing algorithms have been proposed so far to improve the efficiency of sensors used in UWSNs.

Figure 4. Cluster based Data Aggregation in WSN
(Source: Nasridinov et al., 2013)

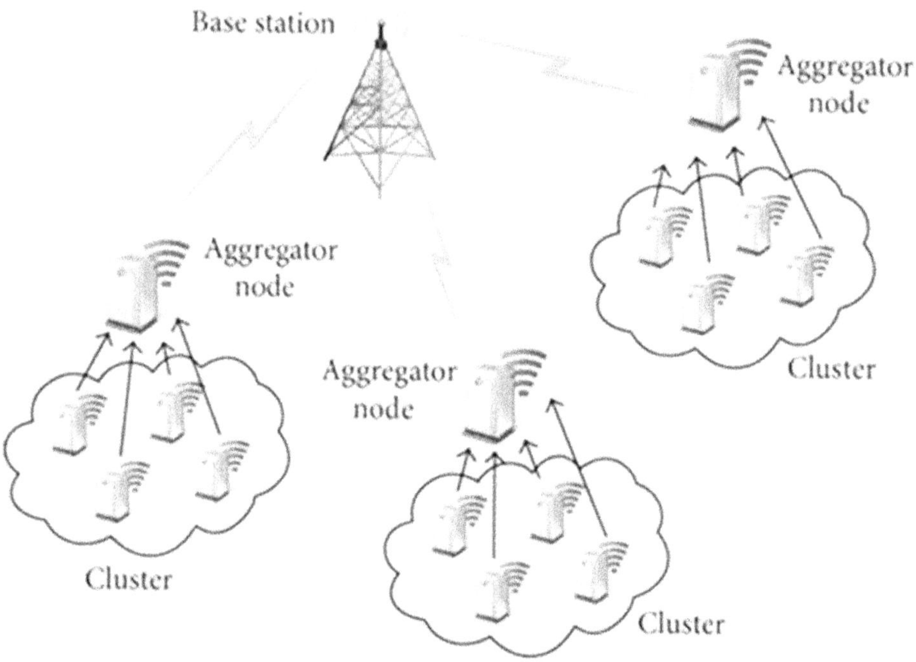

The main target of aggregation is to ignore one to one communication toward sink and to create an aggregate path over the network. UWSN contain a set of underwater LANs. These LANs are also called clusters. It is very difficult to gather data for long time using same sensors due to reserved lifetime. Sensors are very much vulnerable to faults and failures.

Authors (Goyal et al., 2014), have proposed a fuzzy based clustering and aggregation technique for underwater sensor network. Fuzzy logic technique is used for cluster size determination and cluster head selection. The proposed technique has considered node density, distance to sink, residual energy, and link and load quality as parametric inputs for fuzzy function. After using NS-2 simulator, it has been observed that proposed approach reduces the delay and average energy consumption and improves the packet-delivery ratio.

For three-dimensional underwater acoustic sensor networks (UASNs), authors (Lin et al., 2015) have proposed an energy-efficient compressed data aggregation framework, that uses data sampling and compressing mechanism in two consecutive layers. The proposed data aggregation mechanism tries to minimize the amount of energy consumed while transmitting the sensed data. Compressed sampling layer is the bottom layer, where nodes are grouped together to form the cluster. Sampling is done by randomly selecting the nodes and data transmission takes place through random access channels to cluster heads. Data aggregation layer is the top layer, where full sampling is performed. A method to identify the number of clusters and sampling probability is also proposed. Simulation results have revealed that the proposed data aggregation framework can minimize the energy consumption as well as the number of sampling nodes. This further provide guidance for the development of long-term deep-water monitoring networks. Following table represents a brief of different machine learning based approaches for UWSNs.

Table 2. Different machine learning based approaches for UWSNs

ML Strategy	Issue Addressed	Benefits
Regression (Lefort et al., 2017)	Localization	Improved localization, Error reduction
Deep NN (Dong et al., 2017)	Localization	Higher accuracy, localization accuracy, Minimize cost, No chance of over-fitting
SVM (Rauchenstein et al., 2018)	Localization	Improved localization accuracy, tracking accuracy, error reduction in depth
Deep Learning (Sun et al., 2018)	Fault Detection	High diagnosis efficiency, Improved diagnosis accuracy, Prevent over-fitting while training of network
Extreme Learning Machine and Phase Space Reconstruction (Li et al., 2017)	Fault Detection	Improved accuracy
QELAR (Hu et al., 2010)	Routing	Reduce overload and extend lifetime
K-means Clustering (Zhang et al., 2013)	Routing	Balanced energy consumption and prolong network lifetime
DQELR (Su et al., 2019)	Routing	Reduce energy consumption and communication latency, enhance lifetime
Fuzzy Based Clustering (Goyal et al., 2014)	Data Aggregation	Reduce energy consumption, higher throughput, improved packet delivery
Clustering (Lin et al., 2015)	Data Aggregation	Improved energy efficiency, Reduce data transmission cost

CONCLUSION

In this chapter, machine learning mechanisms have been explored in many ways which can be applicable in real time scenarios. It can play a leading role in the development and execution of numerous applications in the underwater scenarios. This chapter presents conceptual framework and basic ideas behind how machine learning and sensing technologies can work together in order to reach new goals in the vast area of UWSNs. Classification of machine learning approaches has been described for addressing various issues like localization, routing, fault detection, coverage, connectivity, data aggregation etc. This chapter can be viewed as a complete package of introductory knowledge about UWSNs, ML and their related strategies. Recently, researchers from interdisciplinary areas are working together to achieve new heights so as to lead the way of substantial inventions. This chapter provides a simple roadmap to the researchers and can help them in further advancements. For future research, they can consider working in this dynamic and heterogeneous environment by linking machine learning and sensing technologies under the water. Majority of existing positioning algorithms are only tested for two-dimensional space while most real-time implementations deploy sensor nodes in three dimensions. Therefore, localization technologies for static and mobile UWSN spaces need to be created. After exhaustive literature survey, it has been found that machine learning techniques are widely used to enhance coverage and connectivity in WSNs but to the best of our belief, no work has been reported for UWSNs so far. Therefore, machine learning based solutions can be worked out so as to address the issues of coverage and connectivity under the water. Hence, it can be used as one of the best alternatives that can be practiced.

REFERENCES

Al-Karaki, J. N., & Kamal, A. E. (2004). Routing techniques in wireless sensor networks: A survey. *IEEE Wireless Communications, 11*(6), 6–28. doi:10.1109/MWC.2004.1368893

Alsheikh, M. A., Lin, S., Niyato, D., & Tan, H. (2014). Machine learning in wireless sensor networks: Algorithms, strategies, and applications. *IEEE Communications Surveys and Tutorials, 16*(4), 1996–2018. doi:10.1109/COMST.2014.2320099

Awan, K. M., Shah, P. A., Iqbal, K., Gillani, S., Ahmad, W., & Nam, Y. (2019). Underwater wireless sensor networks: A review of recent issues and challenges. *Wireless Communications and Mobile Computing, 2019*, 1–21. doi:10.1155/2019/6470359

Ayaz, M., & Abdullah, A. (2009). Underwater wireless sensor networks: Routing issues and future challenges. *Proceedings of 7th International Conference on Advances in Mobile Computing and Multimedia*, 370–375. 10.1145/1821748.1821819

Ayodele, T.O. (2010). *Introduction to Machine Learning*. Academic Press.

Bengio, Y. (2009). Learning deep architectures for AI. *Foundations and Trends in Machine Learning, 2*(1), 1–127. doi:10.1561/2200000006

Bhatti, G. (2018). Machine learning based localization in large-scale wireless sensor networks. *Sensors (Basel), 18*(12), 1–20. doi:10.339018124179 PMID:30487457

Branch, J. W., Szymanski, B., Giannella, C., Wolff, R., & Kargupta, H. (2009). In-network outlier detection in wireless sensor networks. *Knowledge and Information Systems, 34*(1), 1–28.

Chander, B., Kumar, P., & Kumaravelan. (2018). A analysis of machine learning in wireless sensor network. *International Journal of Engineering & Technology, 7*(4), 185-192.

Dey, A. (2016). Machine learning algorithms: A review. *International Journal of Computer Science and Information Technologies*, *7*(3), 1174–1179.

Dong, Y., Li, Z., Wang, R., & Zhang, K. (2017). Range-based localization in underwater wireless sensor networks using deep neural network: poster abstract. *Proceedings of the 16th ACM/IEEE International Conference on Information Processing in Sensor Networks*, 321-322. 10.1145/3055031.3055069

Elghazel, W., Medjaher, K., Zerhouni, N., Bahi, J., Farhat, A., Guyeux, C., & Hakem, M. (2015). Random forests for industrial device functioning diagnostics using wireless sensor networks. *IEEE Aerospace Conference*, 1–9. 10.1109/AERO.2015.7119275

Elhoseny, M., Tharwat, A., Farouk, A., & Hassanien, A. E. (2017). K-Coverage model based on genetic algorithm to extend WSN lifetime. *IEEE Sensors Letters*, *1*(4), 1–4. doi:10.1109/LSENS.2017.2724846

Farhat, A., Guyeux, C., Makhoul, A., Jaber, A., Tawil, R., & Hijazi, A. (2019). Impacts of wireless sensor networks strategies and topologies on prognostics and health management. *Journal of Intelligent Manufacturing*, *30*(5), 2129–2155. doi:10.100710845-017-1377-4

Felemban, E., Shaikh, F. K., Qureshi, U. M., Sheikh, A. A., & Qaisar, S. B. (2015). Underwater sensor network applications: A comprehensive survey. *International Journal of Distributed Sensor Networks*, *2015*(11), 1–14. doi:10.1155/2015/896832

Freitag, L., Stojanovic, M., Grund, M., & Singh, S. (2002). Acoustic communications for regional undersea observatories. Proceedings of Oceanology Inti, 1-10.

Gkikopouli, A., Nikolakopoulos, G., & Manesis, S. (2012). A survey on underwater wireless sensor networks and applications. *Proceedings of 20th Mediterranean Conference on Control & Automation (MED)*, 1147-1154. 10.1109/MED.2012.6265793

Goyal, N., Dave, M., & Verma, A. K. (2014). Fuzzy based clustering and aggregation technique for under water wireless sensor networks. *International Conference on Electronics and Communication System*, 1-5. 10.1109/ECS.2014.6892804

Harinath, K., Vishalakshi, R., Kumar, R. R., & Shanmukhi, M. (2019). Exploring machine learning techniques for localization in wireless sensor networks. *Journal of Information and Computational Science*, *9*(11), 568–575.

Hu, T., & Fei, Y. (2010). QELAR: A machine-learning-based adaptive routing protocol for energy-efficient and lifetime- extended underwater sensor networks. *IEEE Transactions on Mobile Computing*, *9*(6), 796–809. doi:10.1109/TMC.2010.28

Jayaraman, P. P., Zaslavsky, A., & Delsing, J. (2010). Intelligent processing of k-nearest neighbors queries using mobile data collectors in a location aware 3D wireless sensor network. *Proceeding of International conference on Industrial, Engineering and other Applications of Applied Intelligent Systems, Trends in Applied Intelligent Systems*, 6098, 260–270. 10.1007/978-3-642-13033-5_27

Krishnamachari, B., Estrin, D., & Wicker, S. B. (2002). The impact of data aggregation in wireless sensor networks. *Proceeding of 22nd International Conference on Distributed Computing Systems Workshops*, 575– 578. 10.1109/ICDCSW.2002.1030829

Kumar, D. P., Amgoth, T., & Annavarapu, C. S. R. (2019). Machine learning algorithms for wireless sensor networks: A survey. *INFORM FUSION*, *49*, 1–25. doi:10.1016/j.inffus.2018.09.013

Kumar, R., & Singh, N. (2014). A survey on data aggregation and clustering schemes in underwater sensor networks. *International Journal of Grid and Distributed Computing*, *7*(6), 29–52. doi:10.14257/ijgdc.2014.7.6.04

Kumari, V. S., & Dondeti, V. (2019). Survey: Wireless sensor networks in node localization by techniques of machine learning. *International Journal of Research and Analytical Reviews*, *6*(1), 728–737.

Langley, P., & Simon, H. A. (1995). Applications of machine learning and rule induction. *Communications of the ACM*, *38*(11), 54–64. doi:10.1145/219717.219768

Lefort, R., Real, G., & Dremeau, A. (2017). Direct regressions for underwater acoustic source localization in fluctuating oceans. *Applied Acoustics*, *116*, 303–310. doi:10.1016/j.apacoust.2016.10.005

Li, X., Song, Y., Guo, J., Feng, C., Li, G., Yan, T., & He, B. (2017). Sensor fault diagnosis of autonomous underwater vehicle based on extreme learning machine. *IEEE International Symposium on Underwater Technology*, 1-5.

Lin, H., Wei, W., Zhao, P., Ma, X., Zhang, R., Liu, W., Deng, T., & Peng, K. (2015). Energy-efficient compressed data aggregation in underwater acoustic sensor networks. *Wireless Networks*, *22*(06), 1–13.

Lu, Y., He, R., Chen, X., Lin, B., & Yu, C. (2020). Energy-efficient depth-based opportunistic routing with q-learning for underwater wireless sensor networks. *Sensors (Basel)*, *20*(4), 1–25. doi:10.339020041025 PMID:32074989

Mitchell, T. M. (1997). *Machine Learning* (1st ed.). McGraw-Hill, Inc.

Nasridinov, A., Ihm, S.-Y., & Park, Y.-H. (2013). Skyline-based aggregator node selection in wireless sensor networks. *International Journal of Distributed Sensor Networks*, *9*(9), 1–7. doi:10.1155/2013/356194

Paradis, L., & Han, Q. (2007). A survey of fault management in wireless sensor networks. *Journal of Network and Systems Management*, *15*(2), 171–190. doi:10.100710922-007-9062-0

Rao, M., & Kamila, N. K. (2018). Bayesian network based energy efficient ship motion monitoring. *Karbala International Journal of Modern Science*, *4*(1), 69–85. doi:10.1016/j.kijoms.2017.11.001

Rauchenstein, L. T., Vishnu, A., Li, X., & Deng, Z. D. (2018). Improving underwater localization accuracy with machine learning. *The Review of Scientific Instruments*, *89*(7), 1–13. doi:10.1063/1.5012687 PMID:30068124

Romer, K., & Mattern, F. (2004). The design space of wireless sensor networks. *IEEE Wireless Communications*, *11*(6), 54–61. doi:10.1109/MWC.2004.1368897

Sandeep, D. N., & Kumar, V. (2017). Review on clustering, coverage and connectivity in underwater wireless sensor networks: A communication techniques perspective. *IEEE Access: Practical Innovations, Open Solutions*, *5*, 11176–11199. doi:10.1109/ACCESS.2017.2713640

Shah, K., & Kumar, M. (2007). Distributed independent reinforcement learning (DIRL) approach to resource management in wireless sensor networks. *Proceeding of International Conference on Mobile Adhoc and Sensor Systems*, 1-9. 10.1109/MOBHOC.2007.4428658

Shareef, A., Zhu, Y., & Musavi, M. (2008). Localization using neural networks in wireless sensor networks. *Proceedings of the 1st International Conference on Mobile Wireless Middleware, Operating Systems, and Applications, 4*, 1-7.

Shu, J., Liu, S., Liu, L., Zhan, L., & Hu, G. (2017). Research on link quality estimation mechanism for wireless sensor networks based on support vector machine. *Chinese Journal of Electronics, 26*(2), 377–384. doi:10.1049/cje.2017.01.013

Simeone, O. (2018). A very brief introduction to machine learning with applications to communication systems. *IEEE Transactions on Cognitive Communications and Networking, 4*(4), 648–664. doi:10.1109/TCCN.2018.2881442

Simon, A., Deo, M., Selvam, V., & Babu, R. (2016). An overview of machine learning and its applications. *International Journal of Electrical Sciences & Engineering, 1*(1), 22–24.

Singh, T., Bedi, R., & Marwaha, M. (2013). Need and role of underwater acoustic sensor networks. *International Journal of Electronics Communication and Computer Engineering, 4*(4), 1155–1163.

Sun, Y., Wang, Z., & China, P. R. (2018). Fault diagnosis method of autonomous underwater vehicle based on deep learning. *Proceeding of IOP Conf. Series: Materials Science and Engineering*, 1-6.

The Royal Society. (2017). *Machine learning: the power and promise of computers that learn by example*. Author.

Wan, J., Chen, M., Xia, F., Di, L., & Zhou, K. (2013). From machine-to-machine communications towards cyber-physical systems. *Computer Science and Information Systems, 10*(3), 1105–1128. doi:10.2298/CSIS120326018W

Winter, J., Xu, Y., & Lee, W.-C. (2005). Energy efficient processing of k nearest neighbor queries in location-aware sensor networks. *Proceeding of 2nd Annual International Conference on Mobile and Ubiquitous Systems: Networking and Services*, 281-292. 10.1109/MOBIQUITOUS.2005.28

Yang, B., Lei, Y., & Yan, B. (2016). Distributed multi-human location algorithm using Naive Bayes classifier for a binary pyroelectric infrared sensor tracking system. *IEEE Sensors Journal, 16*(1), 216–223. doi:10.1109/JSEN.2015.2477540

Yick, J., Mukherjee, B., & Ghosal, D. (2008). Wireless sensor network survey. *Computer Networks, 52*(12), 2292–2330. doi:10.1016/j.comnet.2008.04.002

Yu, L., Wang, N., & Meng, X. (2005). Real-time forest fire detection with wireless sensor networks. *Proceeding of International Conference on Wireless Communications, Networking and Mobile Computing*, 2, 1214–1217.

Zhang, Y., Sun, H., & Yu, J. (2015). Clustered routing protocol based on improved K-means algorithm for underwater wireless sensor networks. *Proceeding of IEEE International Conference on CYBER Technology in Automation, Control, and Intelligent Systems (CYBER)*, 1304-1309. 10.1109/CYBER.2015.7288132

Zhu, J., Du, X., Han, D., Wang, L., & Li, M. (2019). LEER: layer-based and energy-efficient routing protocol for underwater sensor networks. *Proceeding of International Conference on Ad Hoc Networks*, 306, 26-39. 10.1007/978-3-030-37262-0_3

Chapter 14
Deep Reinforcement Learning Methods for Energy-Efficient Underwater Wireless Networking

Ahmed Ali Saihood
https://orcid.org/0000-0002-0147-7121
University of Thi-Qar, Iraq

Laith Alzubaidi
University of Information Technology and Communications, Iraq

ABSTRACT

The wireless sensor networks have been developed and extended to more expanded environments, and the underwater environment needs to develop more applications in different fields, such as sea animals monitoring, predict the natural disasters, and data exchanging between underwater and ground environments. The underwater environment has almost the same infrastructure and functions with ground environment with some limitations, such as processing, communications, and battery limits. In terms of battery limits, many techniques have been proposed; in this chapter, the authors will focus in deep reinforcement learning techniques.

INTRODUCTION

During the recent years, Underwater Wireless Sensor Networks (UWSNs) have gained a lot of attention. These networks provide a large number of applications including, for example in studies on marine mammals as well as investigating commercial exploitations strategies. The main driving force for investigating the design process for underwater wireless sensor networks involves the need for collecting oceanography data in order to provide the opportunity for human beings to study large areas underwater in various ways (Jin, Ma, Su, Li, & Fu, 2017). The UWSN constitutes a number of sensors and vehicles

DOI: 10.4018/978-1-7998-3640-7.ch014

specifically designed for the target underwater domain in order to collect the required data and provide the opportunity for collaborative monitoring. In a general sense, in order to proceed with monitoring the ocean bed, there have been a number of oceanographic sensors developed and proposed that are capable of recording data in a stable manner, while also providing the opportunity to recover the utilized instruments once the task of collecting the required data is completed. One of the problems with using conventional methods involves the fact that there is no interactive communication between the two ends, increasing the likelihood of losing the recorded data during the transmission stage. Moreover, in such methods, the recorded data will be damaged in case of failure (Awan, et al., 2019). On the other hand, some of the major concerns and problems influencing underwater routing involve the issue of the mobility of the nodes, the limited available bandwidth, the low speed of sound, major possibilities of failure, and the limitations in energy consumption. In general, when developing routing algorithms, keeping a complete picture of the problem in mind is a highly difficult task. Therefore, there is a need for achieving a tradeoff among various limitations. Similar to any other network, in UWSNs, it is highly important and integral to set up appropriate paths and provide a secure and efficient way for transmitting the collected data (Dosaranian-Moghadam & Amo-Rahimi, 2018). The schematic architectural configuration of UWSNs is presented in Figure 1 below. There is a strong similarity between Underwater Wireless Sensor Networks (UWSNs) and Terrestrial Wireless Sensor Networks (terrestrial WSNs). However, the challenges facing the UWSNs differ from those facing the terrestrial WSNs due to difference in the technology utilized in the physical layers along with the difference in the propagation medium. In clearer terms, radio signals are used in terrestrial WSNs to transmit the data; however, such a technology is not appropriate for underwater settings since it requires a lot of energy and radio signals dissipate rapidly in

Figure 1. The architecture for the underwater sensor network (Zhu & Wei, 2018)

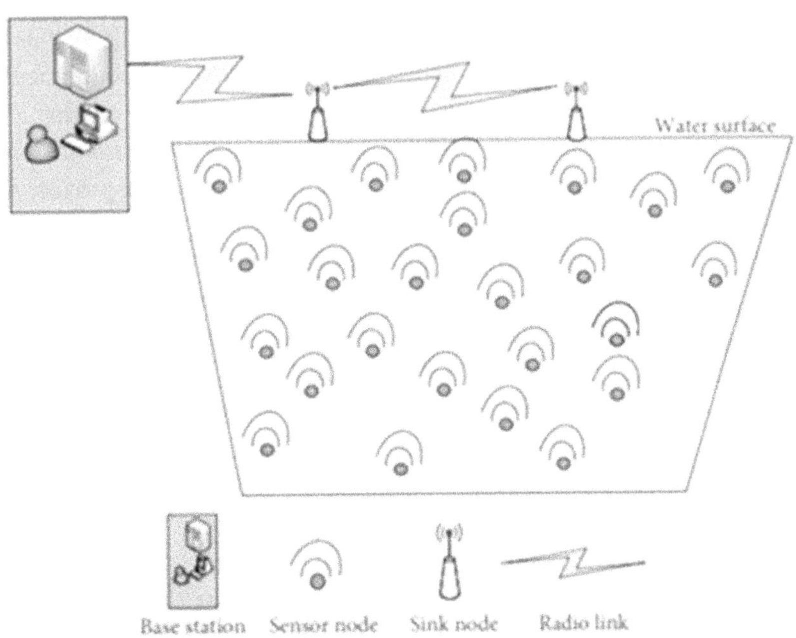

water. Therefore, acoustic signals can constitute a much better choice as a communication medium when working underwater. However, the main issue with the acoustic signal is that its low speed compared to radio signals results in a high level of propagation delay. Moreover, the multipath fading channel results in slightly higher error ratios and low bandwidth. These issues cause a number of problems including low levels of reliability, high energy consumption, and the high rate of retransmission. Therefore, designing appropriate routing protocols for UWSNs is one of the main challenges in this field (Zhu & Wei, 2018).

In recent years, a number of protocols for efficient energy consumption have been proposed. Among these protocols, the Q-learning-based energy-efficient and lifetime-aware routing (Hu & Fei, 2010) and the VBF protocol (Xie, Cui, & Lao, 2006) can be mentioned. The VBF protocol reduces energy consumption by limiting the number of senders. In this protocol, suitable senders are the ones which have the shortest geographical distance to the destination node, with the shortest distance being referred to as the 'routing vector'. In this protocol, the routing pipe is shaped around the routing vector. Some of the nodes might be located outside the routing pipe, and thus they will not be considered as appropriate senders. Moreover, they are not allowed to send any received packages. In this way, the system can save on energy consumption by reducing the number of transmission at unsuitable nodes.

On the other hand, the QELAR protocol increases the lifetime of the Underwater Acoustic Wireless Sensor Networks (UA-WSNs) by achieving an evenly-shared residual power among the nodes. In QELAR, which is a specific version of SARL, the residual energy of the nodes and the energy distribution in a group of nodes will be employed for obtaining a reward function to select appropriate senders for the packages. The results obtained from simulations indicate that compared to the VBF protocol, the QELAR protocol increases the lifetime of the network by 20 percent on average (Xie, Cui, & Lao, 2006).

Moreover, Basagni, Valerio, Gjanci, & Petrioli (2017) developed and proposed a forwarding method called the MARLIN protocol (*Multi – modAl Reinforcement Learning – based RoutINg*). This protocol proceeds by selecting the best relay node and the best communication channel. Moreover, MARLIN can be configured in a way to search for reliable routes to the final destination, or in a way to accelerate thedelivery of packets. On the other hand, Gjanci, et al. (2018) presentd the GAAP (*Greedy and Adaptive AUV*) path-finding heuristic. In the GAAP method, a heuristic algorithm is used for finding a path for an autonomous underwater vehicle (AUV) in a way that it can maximize the informational value of the data delivered to sink nodes. Compared to random selection (RS), lawn mower (LM), and traveling salesman (TSP) methods, th GAAP method can provide much higher levels of informational value. However, as the size of the network increases, the advantage of GAAP over TSP reduces. This is because larger networks allow the TSP method to collect more packets, while the GAAP strategy hasa higher average end-to-end delay compared to the TSP method. In addition, there have been a large number of energy-efficient forwarding methods proposed for boosting the lifetime of these networks. For inatance, Hu & Fei (2010) the QELAR (Q-Learning-based Energy-Efficient and Lifetime-Aware Routing) protocol for UWSNs, which utilizes the Q-learning algorithm to define the residual energy of sensor nodes as the reward function. As a result, in the QELAR protocol, the node with the highest residual energy is selected by the sensor nodes as the relay node, increasing the lifetime of the network. Nonetheless, the QELAR does not limit the end-to-end delay. Therefore, when the number of nodes incrases, a longer delay is encountered. In addition, Coutinho, Boukerche, Vieira, & Loureiro (2017) developed an energy valancing routing (*EnOR*) protocol for UWSNs. The *EnOR* protocol is based on the notion of balancing the consumption of energy among neighboring nodes in the forward set through reversing their priority in order to increase the lifetime of the network. Nonetheless, having a large candidate set may lead to high levels of delay due to the fact that because of the long distance between the sender and the nodes with high priority, the link quality of these high priority nodes is generally very low. Moreover, Jin, Ma, Su, Li, & Fu (2017) developed a QDAR (Q-Learning-based Delay-Aware Routing) algorithm for increasing the lifetime of UWSNs. This algorithm defines the costs related to delay and energy consumption in order to account for timeliness and energy efficiency.

Based on the above-mentioned considerations, the current study discusses the role played by reinforcement learning in the energy efficient routing protocols for UA-WSNs. Moreover, the study will assess the methods for utilizing such protocols as a means for increasing the lifetime of UA-WSNs. When using energy-efficient protocols, the sensor nodes can evaluate the environment and adjust or alter the relevant network parameters.

FUNDAMENTALS OF REINFORCEMENT LEARNING

Reinforcement learning is a famous artificial intelligence algorithm which can train intelligent elements to interact with the environment as a means to maximize the cumulative reward (Lu, 2012). This section provides an introduction for the reinforcement learning algorithm,

Markov Decision Process

The Markov Decision Process (MDP) is usually utilized in the reinforcement learning algorithm to formulate factor interaction and the environment. The MDP assumes that at any time step, the environment is in a specific state (s), where an actor performs a specific action (a) to gain reward, $r(s,a)$. As the state of the environment changes to a new one, the probability of the stochastic transition of the environment is affected by the previous action taken by the actor, and it is given as $p(s' \mid s,a)$.

Definition 1: the Markov Decision Process is expressed in the form of a tuple, $[S,A,r,p]$ where S is the different space of the state, A signifies a different space of action, $r: SxA \to R$ signifies a factor reward function (where R is the set of real numbers), and $p: SxAxS \to [0,1]$ denotes the function of transition, which maps the space of state-action to the probability distributions in the space of state s.

Moreover, the policy π along with the $MDP[S,A,r,p]$ are used to show that the actor selects the action as the function of the states. For instance, $\pi: SxA \to [0,1]$ shows the probability of taking an action in state. In addition, the value of state s under policy π is denoted by $v(s,\pi)$, and it is a factor-based reward in state s with policy π. This is calculated as follows:

$$V(s,\pi) = \sum_{t=0}^{\infty} E^{\pi}(r(s_t,a_t) \mid \pi, s_0 = s) \tag{1}$$

In this equation, s is the specific state showing the initial state. Moreover, $r(s_t,a_t)$ is the reward for taking action a at time t, while $\gamma \in [0,1)$ is the discount factor. In addition, $E(.)$ indicates the $(.)$

probability. Afterward, the MDP solution can be considered as the general policy π^*, which increases the long-term reward (Lu, 2012).

Reinforcement Learning

During the recent decades, a large number of solutions have been proposed for MDP. For instance, RL uses an iterative search process to find the stable point of the following equation, which is known as the Bellman equation:

$$v\left(s,\pi^*\right) = \max_a \left\{ r\left(s,a\right) + \gamma \Sigma_{s'} p\left(s' \mid s,a\right) v\left(s',\pi^*\right) \right\} \tag{2}$$

Solution a, satisfying the above equation, is guaranteed to be the general policy for the MDP.

Value of Q

The basic question is if the objective of reinforcement learning is to find π^*, then how the agent learns through interacting with the environment. When using reinforcement learning, it is much simpler to use the Q value instead of finding π^*:

$$\pi^* = argmax_a Q\left(s,a\right) \tag{3}$$

$$V^*\left(s\right) = \max_a Q\left(s,a\right) \tag{4}$$

However, instead of identifying a policy which can increase the value of all states, Equation (3) above seeks to find an action that increases the quality (Q) of all the states. In this case, the values of Q, V^* and π^* will be evaluated based on Equations (4) and (5). If the reward for an action is the ability to observe the next state, it is possible to create iterative error algorithms for finding Q (Gulli & Pal, 2017):

$$Q\left(s.a\right) = r + \gamma max_{a'} Q\left(s'.a'\right) \tag{5}$$

In order to simplify the notations, the next action and the next state are denoted by a' and s', respectively. It should be noted that Equation (5) is known as the Bellman equation, which is based on a Q-learning algorithm. Q-learning is based on estimating the value of the first-order expansion (i.e., Equation (3)) as the present state as well as the action function. When there is no knowledge about the environment, the agent tries action a to see the reward (r) as well as the resulting state (s'). On the other hand, $\max_{a'} Q\left(s',a'\right)$ selects another reasonable action and outputs the maximum Q value for the next state. Afterward, considering all the cases in Equation (5), the value of Q for the current state-action pair will be updated. This iterative process of updating will result in the final value of Q. Q-learning is an off-policy reinforcement learning method, which tries to improve the policy by indirectly sampling the agent's experiences. However, the value of Q can independently be learned based on the policy used by the agent. Once the Q function converges, the policy can be evaluated using Equation (3). It should be noted that the agent must constantly evaluate its environment to use what it has learned so far. This means there is a need in reinforcement learning to strike a balance between utilization and exploration (of the environment). In general, at the beginning of the algorithm, the action is selected in random (exploration). As learning goes on, the agent will use the value of Q (utilization). As an example, at the initial stage, 90 percent of the actions are random while 10 percent are based on the value of the Q function. The randomness will decrease after each iteration, reaching a state where random actions are only 10 percent, while actions based on the value of the Q function account for 90 percent of the actions.

DEEP REINFORCEMENT LEARNING METHODS FOR UNDERWATER SENSOR NETWORKS

Delay-Aware Routing Based on Q-Learning (QDAR)

The QDAR process can take care of the entire routing process. This process has five main steps, including the $data\_ready$ step, the routing decision, the step of interest, the forwarding acknowledgement, and the package. The $data\_ready$ process collects data to provide to the sink node to decide on routing. Then, the sink node will assess the route based on QDAR. At the step of interest, the sink node sends the package for the source node in order to determine the optimal route. Packages will be sent over the route provided that the acknowledgement step is successful; otherwise, the process is repeated. The main communication process in this algorithm is presented in Figure 2 below.

In this algorithm, the following assumptions are considered (Jin, Ma, Su, Li, & Fu, 2017):

- The nodes maintain information about their own depth;
- The nodes in the underwater sensor networks performs the $query\ of\ source-initiated$;
- The sink node keeps information on failed and successful communication attempts;

At the $query\ of\ source-initiated$ step, if the source node wants to send a package to the sink node, it will first broadcast a $DATA-READY$ request to ask for information and communication. This step is discussed in the section on $data-ready$ above. At the $data-ready$ step, the node will send a package only if its depth is smaller than the last one. The first time the sink node receives a package, it can decide if it is new or old data. If it is new, the sink node will chose a route based on the QDAR algorithm, and will send the package of interest to the source node through that route. This step is known as the step of interest, in which every node will record its last hop and takes another hop to forward the package. The nodes selected by the QDAR algorithm will connect to the package forwarding of interest, while the other nodes will sleep, perform other tasks to improve the capacity of the network, or store energy. Then, when the package of interest arrives at the source node, the route will be created. Afterward, the source node will send the data package through the constructed route. Finally, the sink node sends a package of acknowledgement (ACK) once it receives the data successfully. If the source node receives the ACK successfully, the communication session can be terminated. Future packages from the

Figure 2. The mechanism used in delay-aware Q-learning routing (Jin, Ma, Su, Li, & Fu, 2017)

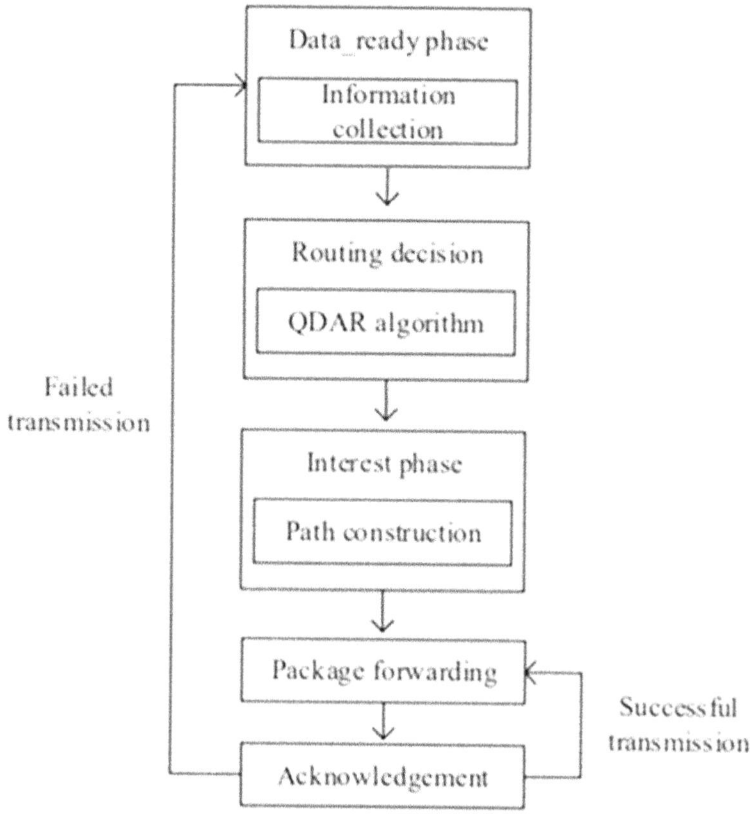

source node will be sent through the same route, and there will be no repeated data. The same routing decision making steps will not be repeated until the transmission route fails.

The QELAR Algorithm

The QELAR algorithm is another routing algorithm recently created for underwater sensor networks, and it is based on the Q-learning approach proposed in (Plate & Wakayama, 2015). In this algorithm, the energy-based reward is estimated by the quality value for sending packages from neighboring nodes to the destination node. In this method, each node evaluates the Q value of its neighboring node, and each node will choose the neighboring nodes with the highest quality value. The reward function will account for parameters such as the basic energy of the node, the energy distribution among neighboring nodes, and the residual energy in order to select the shortest route to maximize the lifetime of the network. Any time the node selects the routing, there will be two possible scenarios: successful action (changing the state to S_{t+1}), or failed action (the state remains the same, s). The reward function in the first scenario will be expressed as:

$$R^{at}_{s_t s_{t+1}} = -g - \alpha_1 \left[c(s_t) + c(s_{t+1}) \right] + \alpha_2 \left[d(s_t) + d(s_{t+1}) \right] \quad (6)$$

In this equation, g is a constant, which is utilized for calculating the energy while the transfer is underway. Moreover, $c(s)$ denotes the penalty of transmission to a node since the energy ratio of the node can be used for reducing the basic energy. In addition, $d(s)$ is based on the second penalty of the amount of energy the node has compared to its neighboring nodes. Finally, α_1, α_2 signify the weighting parameters. In case of failure, the reward function will be expressed as follows:

$$R_{s_t s_t}^{at} = -g\eta - \beta_1 c(s_t) + \beta_2 d(s_t) \tag{7}$$

In this equation, β_1, β_2 signify discrete weighting parameters. In (Lu, 2012), an extra weight $\eta > 1$ is defined for the scale g in order to account for the fact that in case the transmission fails, the cost will not be limited to the transmission cost of the single node; rather, the transmission must be repeated, so there will be extra costs. One of the benefits of using QELAR for improving the routing performance involves the fact that each node evaluates all of its neighboring nodes to find inefficient nodes with lower values of V, while the other node hops because of inefficient estimates of V. When $\eta = 1$, the node tries to select an neighboring node in a way to guarantee successful transmission by selecting the nodes which will create the lowest possibility of transmission failure. Moreover, η indicates the penalty for failing which is higher than the penalty for remaining at the current state. Using every possible case a reward function for failure as well as success can be created to reward each type of performance. Then, the Q function can be expressed as follows:

The probabilities of transmitting to neighboring nodes will be calculated based on the history of the nodes failure or success in its forwarding actions. When the nodes receive information about the forwarding transmissions of neighboring nodes, they use that information to update the V values of the neighboring nodes. During changes to the topology of the network, the nodes will only send header packages in a periodic manner along with the V values of their neighboring nodes if they have not sent transmissions during a specific timeframe. Therefore, virtual experiments using Equation (9), and real observations by listening for the transmission of the neighboring nodes will be employed for performing the routing process.

Q-Learning using Kinematic Sweeping (QKS)

The QKS was developed to be used with mobile ad hoc distributed underwater networks. This method usually focuses on RL-based routing algorithms which can explore the network's environment in order to constantly adapt the routing decisions upon alteration in the topology of the network based on the energy usage in the movement of nodes. The QKS adds the feature of kinematic sweeping to the routing algorithms which are based on Q-learning. The concerns about the slow speed of this algorithm and its relation with pure RL algorithms are discussed in the studies by the authors who presented this method (Plate & Wakayama, 2015; Walikar, 2018). Moreover, in order to accurately estimate the probability of successful transmission based on the history of failed transmissions, a significant number of forwarding actions is requird. This can result in slow convergence, which will be challenging in case the network's topology changes quickly. Therefore, a method for resolving this issue and improving the ability to predict the probability of changes in the state involves increasing the convergence speed of the evaluation process for the V value.

CHALLENGES FACING UNDERWATER SENSOR NETWORKS

Throughput efficiency and energy per bit are employed in order to evaluate the effects of packet size on two protocols and the end-to-end delay. The optimal package size depends on the characteristics of each protocol, including the provided load rate and the bit error rate. If the packet size is not selected accurately, the performance of the network will be affected negatively. In recent years, multi-hop networks have gained a lot of attention because it can cover a much larger area compared to the single hop network. In such networks, selecting the packet size in an optimal manner will boost the efficiency of the network. For instance, Stojanovic has studied the optimization of the packet length in order to maximize the throughput efficiency using the point-to-point scenario at the data link layer. Results show that the performance of the network is significantly affected by the size of the packet (Awan, et al., 2019).

On the other hand, parameters such as throughput efficiency, energy consumption, and latency in multi-hop UWNs can be significantly improved using a proper packet size. Basagni, Petrioli, Petroccia, & Stojanovic (2010) evaluated the performance of CDMA and DACAP by using different predefined packet sizes. Refraining from changing the packet size based on a point-to-point sscenario implies that using this method can improve the performance and channrel utilization based on environmental factors and channel capacity at different locations under the sea. The size of the control packets is much smaller than the size of the data packets (Basagni, Petrioli, Petroccia, & Stojanovic, 2010). If the data packet and control packet collide, the data packet must be discarded. In order to reduce the likelihood of such collissions, long data packets can be divided into smaller parts. In this case, the effects of the collission will be limited to a number of smaller fragments instead of the entire data packet. Moreover, this partitioning method can decrease the amount of traffic as well as the overhead and number of retransmissions. This method is applied to the DACAP, which improves all the abovementioned parameters, except for the overhead, which increases in the network. Therefore, nased on throughput efficiency, there is an optimal number of partictions. It should be noted that increasing the traffic load will also increase the likelihood of collissions.

This partitioning method will not only reduce the likelihood of collission, but it will also decrease the overall energy consumption. In contrast, it will cause a higher overhead for individual data transmissions. Increased trffic and bit error rate will reduce the throughput efficiency. Nonetheless, compared to the cases where such partitioning is not used, the level of wasting energy is much lower. Table 1 shows that partitioning the packat will result in reduced packet latency as well. When an optimal number of partitions is considered, throughput efficiency is increased, energy consumption is lowered, and packet latency decreases.

Table 1. Comparing partitioning of the packet with not partitioning the packet

Factors	Partitioning	No partitioning
Delay	Low	High
Energy consumption	Low	High
Traffic load	High	Low
Channel utilization	Low	High
Collission	Low	High

Using optimized selection of packet sizes, the stop and wait mechanism of UWSNs, which is used in the half duplex mode, can be avoided. The optimized selection of packet sizes positively affects the multi-hop communication.

Moreover, selecting the size of packages based on environmental factors will also boost acoustic communication, which necessitates further investigation. Compared to communication methods that do not account for changes in the underwater environmental factors related to the acoustic channel, those communication method that consider such factors will provide improved performance in terms of highly usable acoustic channel, which is very challenging due to issues such as multipath propagation, underwater noise, the channel cpaacity's dependence on distance, and the Doppler effect. Since ocean currents move the nodes, there is an issue related to localization. Localization and large number of alterations in the topology of the network result in serious challenges for routing protocols. Moreover, selecting the packet sizes in UWSNs is highly important since it can improve energy consumption, channel utilization, and the throughput efficiency of the network. Therefore, these issues and challenges must be considered in future studies.

CONCLUSION

Compared to terrestrial wireless sensor networks, Underwater Acoustic Wireless Sensor Networks (UA-WSNs) require a much higher transmission power. On the other hand, changing or charging the bateries of underwater sensors is both difficult and costly due to the harshness of the deployment environment. Therefore, an effective energy-efficient routing protocol is necessaruy in such networks to conserve energy. Because of the abovementioned issues with the sensors' batteries, improving the efficiency of energy consumption becomes a much more important issue. A number of deep reinforcement learning methods were discussed in this study in relation to Underwater Wireless Sensor Networks (UWSNs). The main goal of these methods is to mitigate the issues encountered underwater, while also providing directions for future studies. Since the action-utility function of the Q-learning method accounts for discounted long-term rewards along with the direct reward, methods which are based on Q-learning are capable of evlauating the global best hop instead of greedy ones. QDAR provides the following benefits:

- Defining steps for data collection and structuring the packages before selecting the best routing as a way to quickly adapt to the changes in the underwater environment;
- Considering residual energy and dely by utilizing two cost functions: delay-based cost and energy-based cost;
- Using an adaptive mechanism to increase the lifetime of the network and reduce delays. When there is enough residual energy, the end-to-end delay will be low. However, when the residual energy of some of the nodes is below a threshold value, the path will include nodes with longer delays, while accessing more residual energy;
- It can simply be expanded. The action-utility function in the QDAR prrotocol can account for a large number of factors including channel capacity, energy consumption, and communication reliability. These can be evaluated in future studies for various applications.

ACKNOWLEDGMENT

Hossein Karshenas.

REFERENCES

Awan, K., Shah, P., Iqbal, K., Gillani, S., Ahmad, W., & Nam, Y. (2019, January 1). Underwater wireless sensor networks: A review of recent issues and challenges. *Wireless Communications and Mobile Computing*, *2019*, 1–20. Advance online publication. doi:10.1155/2019/6470359

Basagni, S., Petrioli, C., Petroccia, R., & Stojanovic, M. (2010). *Optimizing network performance through packet fragmentation in multi-hop underwater communications. In IEEE Sydney, OCEANSSYD 2010*. IEEE.

Basagni, S., Valerio, D., Gjanci, P., & Petrioli, C. (2017). Finding MARLIN: Exploiting multi-modal communications for reliable and low-latency underwater networking. In *IEEE Conference on Computer Communications (INFOCOM)* (pp. 1701-1709). Atlanta, GA: IEEE. 10.1109/INFOCOM.2017.8057132

Coutinho, R., Boukerche, A., Vieira, L., & Loureiro, A. (2017). EnRO: Energy balancing routing protocol for underwater sensor networks. In *IEEE International Conference on Communications (ICC)*, (pp. 1-6). Paris: IEEE.

Dosaranian-Moghadam, M., & Amo-Rahimi, Z. (2018). Energy efficiency and reliability in underwater wireless sensor networks using Cuckoo Optimizer Algorithm. *AUT Journal of Electrical Engineering*, *50*(1), 93–100. doi:10.22060/eej.2018.13347.5154

Gjanci, P., Petrioli, C., Basagni, S., Phillips, C., Boloni, L., & Turgut, D. (2018). Path finding for maximum value of information in multi-modal underwater wireless sensor networks. *IEEE Transactions on Mobile Computing*, *17*(2), 404–418. doi:10.1109/TMC.2017.2706689

Gulli, A., & Pal, S. (2017). *Deep learning with Keras*. Packt Publishing Limited.

Hu, T., & Fei, Y. (2010, February). QELAR: A machine-learning-based adaptive routing protocol for energy-efficient and lifetime-extended underwater sensor networks. *IEEE Transactions on Mobile Computing*, *9*(6), 796–809. doi:10.1109/TMC.2010.28

Hu, T., & Fei, Y. (2010). QELAR: Q-learning-based energy-efficient and lifetime-aware routing protocol for underwater sensor networks. *IEEE Transactions on Mobile Computing*, *9*, 796–809. doi:10.1109/TMC.2010.28

Jin, Z., Ma, Y., Su, Y., Li, S., & Fu, X. (2017). A Q-learning-based delay-aware routing algorithm to extend the lifetime of underwater sensor networks. *Sensors (Basel)*, *17*(7), 1660. doi:10.339017071660 PMID:28753951

Lu, J. (2012). *Reinforcement learning based energy efficient routing protocols for underwater acoustic wireless sensor networks*. University of Western Australia.

Plate, R., & Wakayama, C. (2015). Utilizing kinematics and selective sweeping in reinforcement learning-based routing algorithms for underwater networks. *Ad Hoc Networks*, *34*, 105–120. doi:10.1016/j.adhoc.2014.09.012

Walikar, G. A. (2018, February). Reliable and stable route construction in mobile ad hoc networks using R- learning techniques. *International Journal of Engineering Science Invention*, *2*(2), 58–63.

Xie, P., Cui, J.-H., & Lao, L. (2006). VBF: Vector-based forwarding protocol for underwater sensor networks. In *International Conference on Research in Networking*. Berlin: Springer. 10.1007/11753810_111

Zhu, F., & Wei, J. (2018, December 16). An energy efficient routing protocol based on layers and unequal clusters in underwater wireless sensor networks. *Journal of Sensors*, *2018*, 1–10. Advance online publication. doi:10.1155/2018/5835730

Chapter 15
Relaibility in Underwater Wireless Sensor Networks

Gaurav Sharma
Seth Jai Parkash Mukand Lal Insitute of Engineering and Technology, India

Shilpi Harnal
Seth Jai Parkash Mukand Lal Insitute of Engineering and Technology, India

Neha Miglani
National Institute of Technology, Kurukshetra, India

Savita Khurana
Seth Jai Parkash Mukand Lal Insitute of Engineering and Technology, India

ABSTRACT

Underwater wireless sensor networks (UWSNs) have gained importance as well as diverted attention of many researchers, domain experts to a great extent in recent past. The devices used for UWSN deployment are resource-constrained like storage issue, low processing speed, as well are vulnerable to a wide class of security threats and malicious attacks, which affect reliable communication. For reliable data delivery, a system should include packet delivery ratio, battery life, delays incurred, and energy consumption, etc. Numerous reliability models for underwater networks have been designed to incorporate the parameters and performance metrics in optimized manner. The chapter deals with focusing on such models and their efficiency in terms of battery life, packet loss, error handling mechanism, and network delays. Further, it is also explained how and why the error-controlled schemes should be designed and implemented in order to incorporate reliable data delivery in limited resources-constraints of UWSN along with the consideration of efficiency and performance concerns.

DOI: 10.4018/978-1-7998-3640-7.ch015

INTRODUCTION

Nowadays the wireless networking has become the one of the foremost domains in the research world (Rawan et al.,2019). Further categorization of wireless sensor networks can be done as Terrestrial Sensor Networks and Underwater Sensor Networks. Moreover, the two categories belong to entirely different domains with respect to numerous factors such as communication cost, latency and error likelihood of all the nodes. UWSNs incur high cost, latency and error rate in comparison to TWSNs. Though Underwater sensor networks, along with being a buzz word, also is a promising and an emergent domain in the technology world yet is not explored completely. UWSNs comprise numerous sensor nodes in the network (Erol et al., 2007). Sensor nodes can effectively sense the information stored and hence, are capable enough to proceed forward in the direction of upper surface sink node. Based upon the application domain, packets forwarded towards the sink nodes are manipulated accordingly. Error ratio hugely decides an accuracy of information packets. Limitation lies in the fact that an ocean comprises various noise factors as well as dynamic and time-dependent link properties (Sozer et al., 2000), which eventually descends the performance and accuracy of data transmission. Moreover, water has an absorbent properties, specifically for radio waves henceforth, UWSNs utilizes acoustic signals, which further leads to an optimum performance (Maqsood et al., 2015; Wahid and Kim, 2012).

Figure 1 depicts the model for underwater sensor networks. The network comprises m sensor nodes lying underwater and one underwater sink node. Each node has been assigned unique ID [1,2,..m]. Moreover, an underwater sink node has an ID value of 0 and the parent node of the sink node is allocated the value of -1. These IDs will help in sending the notifications to underwater sink node at the time of deployment.Wireless communication is used for building connections between sensor nodes and underwater sink on land. Also, serial ports are used to make connection of sensor nodes with a server system (SS) on land. Henceforth, SS manages the data reception from underwater sensor nodes on land and commands are sent via sensor nodes on land along with the data display on the monitor/screen. Afterwards, the collected information is uploaded on an Internet server which can be utilized by researchers in their problem domains.Unlike underwater sensor nodes, underwater sink nodes consists of other concerns as well such as channel distribution, route maintenance and generation, thus, congestion control is required for sink nodes in order to provide reliable network. Occurrence of node failure is comparatively high in underwater sensor networks in comparison to ground-based networks due to harsh conditions and environment in underwater scenario.

Surroundings of the sensor nodes comprise salt water and movement of currents and waves affect an overall network. There are numerous differences in terrestrial sensor networks comprising traditional wires and underwater communications comprising sensor nodes. First and foremost is the variation in energy consumption for the two networks as some of the applications demands huge data transmission (Heidemann et al., 2006).Instead of an independent user representation, such networks rely on a common task.The main focus here lies in throughput maximization instead of targeting fairness among all the nodes. One more difference lies in that these networks have prominent relationship between hop count, link distance as well as reliability. In order to reduce energy consumption over network, packets are sent over multiple short hops instead of utilizing long links.Moreover, transmitting packets over multiple hops leads to performance degradation and end-to-end reliability function, specifically for the frail underwater environment. Last one is based on the deployment constraints. Deployment of networks is done by a single organization while considering economical hardware; thus far, there is no requirement of rigid

Figure 1. Underwater sensor network models

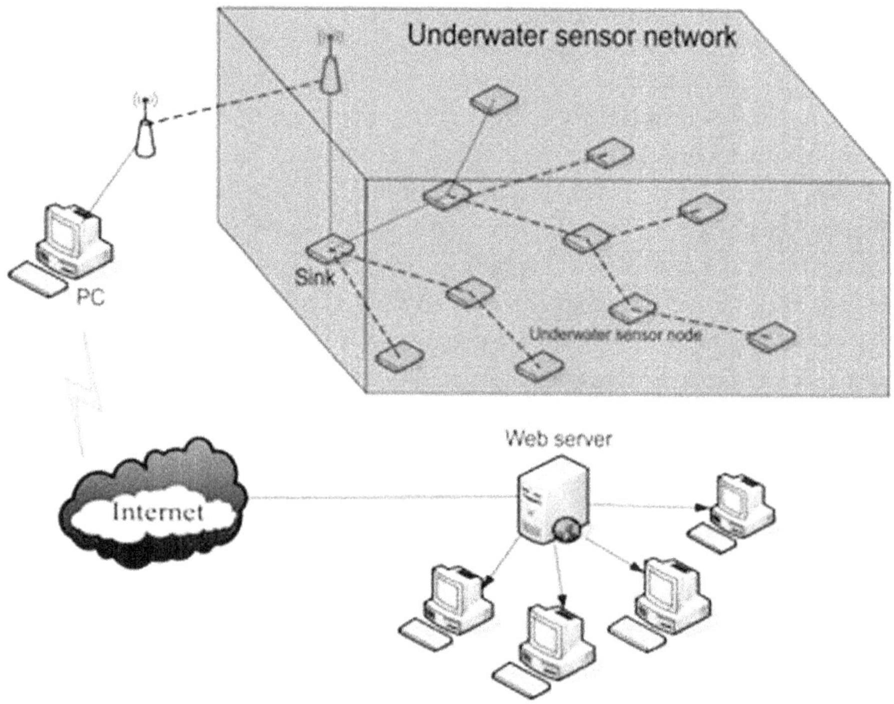

inter-operability amongst existing hardware. For such reasons, UWSNs assimilates the traditional communication protocols' existing structure and review it for bridging the gaps being captured.

Numerous researchers put forward efforts in an underwater environment in order to enhance transmission throughput. Only few protocols works on the basis of traditional acknowledgment (ACK) mechanism, still the difficulty with them is the degraded throughput at large extent.

Numerous researchers targeted the problem of collision avoidance by designing suitable methods and protocols yet none of the approaches is reliable. Thus, the major problem domain becomes network reliability. In order to perform data extraction from the received packets, reliability provides assurance of minimizing adverse effects on the delivered data packets.Reliable communication becomes a challenging task in underwater scenario due to numerous factors related to underwater environment such as water salinity, multi-path fading, signal attenuation, noise along with the presence of marine animals. Communication is being done with the help of acoustic waves instead of radio waves because of harsh underwater environment and unlike acoustic waves, radio waves are largely affected by attenuation and absorption in underwater scenario (Lanbo et al., 2008).Latency of UWSNS is comparatively higher in comparison to ground-based networks, reason being the speed of an acoustic signal is approximately five times slower than the speed of radio signals in water (Davis and Chang, 2012).Detecting the correct position of nodes in underwater communication is a difficult task because of dynamic nature of water currents.

As, it is clear from the above facts that reliability of sensors nodes in underwater environment is a major issue of concern and requires special attention. Thus, the study in this chapter has focused on the comparative analysis of various reliability models or algorithms available for the UWSN. Further, the next section has reviewed the findings and proposed approaches from various researchers followed by the section discussing various reliability concerns for underwater communication. The forth section has covered the detailed study of existing reliability models and their comparison based on various parameters. The last section concluded the findings and their basis.

RELATED WORK

Park and Shin developed the Selective Multiple Acknowledgement (SMA) method, which further relied on Multiple Acknowledgement (MA), targeted to significantly minimize the amount of data being transmitted in an underwater environment. The techniques for transmission collaboratively transmitted Acknowledgement (Ack) or Negative Acknowledgement (Nack) in order to avail updates about the presence (or absence) of transmission errors. The concept of Ack and Nack were inculcated in the Beacon frame. The Cluster Head supervised regarding collection of data and scheduling information. It collected data via Uplink Transmission inside a cluster of unit size. The approach significantly reduced the frame length as well as number of transmissions (which was further directly proportional to the count of networking nodes comprising Ack). The proposed method was compared with many existing approaches namely, Block Acknowledgement (BA), conventional Automatic Repeat-reQuest (ARQ), MA methods and block response; when analyzed results obtained were improvised. The SMA method considered smallest size frame and minimized number of transmissions with varying number of variables. The performance was significant when transmission targets were increased, eventually leading to enhanced efficiency of the underwater sensor network, forming large cluster (Shin and Park, 2011).

Lior and Malumbres did research on acoustic propagation models and how they impacted the performance and efficiency of high level communication protocols. At initial stage, they studied the underwater acoustic prediction models evaluation. Afterwards, they evaluated various MAC protocols based upon their performances, and evaluated the impact of environmental changes on network performance, such as delay time, collision ratio and throughput. Further, they analyzed numerous high layer network protocols such as routing protocols and MAC precisely such that the effect of environmental factors could be measured on the results so obtained. It was observed that high level protocols are more sensitive to different physical layer based parameters based upon network scenario such as floor sediment composition, ocean surface activity and scenario location along with an acoustic propagation model in relation to complexity and accuracy level. Proposed approach also dealt out with the capturing influence of time-varying environmental scenario on high layer protocols. They described the significance of defining propagation model accurately and with low complexity for network simulation (Lior and Malumbres,2012).

Gao et al. did research on network connection by selecting range of transmission and energy efficiency trade off. Authors developed an analytical framework in order to measure the relation between average energy consumption, range of transmission as well as connectivity lying in Underwater Wireless Sensor Network (UWSN). To reach a specific network connectivity level would result in additional cost for deployment of node otherwise it would lead to raised energy consumption because of functionality getting deviated from an optimum range of transmission. The analytical framework proposed in the paper entailed guidelines for designing functionality for energy-efficient packet transmission constrained under specific network connection requirements in order to deploy Underwater Wireless Sensor Network (Gao et al., 2012).

Recent Sensor Networks, that is, localization problem which lead to innovation of numerous ideas and solutions. Han et al. surveyed different localization algorithms for Underwater Wireless Sensor Networks. Based upon mobility of sensor nodes, they categorized localization algorithms into three types namely, mobile localization algorithms, stationary localization algorithms and hybrid localization algorithms. Comparison between different localization algorithms was done. It was concluded that such algorithms would work effective in the domain of large scale Underwater Wireless Sensor Networks as well as for the realistic models relying on the mobility of sensor node in accordance with the respective underwater information such as water salinity, depth etc. (Han et al., 2012)

Moradi et al. investigated a novel scheme for 3D centralized localization for mobile UWSN, which was named Reverse Localization Scheme (RLS).The scheme was an event-driven localization method which was sparked by sensor detectors for initiation of localization process which eventually lead to improved average response time in UWSN. It was proven to be fast and efficient approach for transferring the estimates of location from sensors towards the base station. The sensor nodes passed the event in the direction of surface anchors immediately after detection. Reception of location information from anchors was not considered significant. The proposed algorithm considerably affected energy consumption as well as computational complexity in positive ways. Designing of mechanism for message exchange was being presented (Moradi et al., 2012). The scheme yielded reduced count of localization messages resulting in reduced average response time and improved accuracy while considering the mobility parameters of water currents. The proposed approach was proven suitable in the domain of surveillance applications which demands quick responses to events. The results so obtained concluded the minimized response time for localization, reduced energy consumption along with an increased accuracy level as far as mobility features of water currents are concerned. Such components might play significant role in real life applications such as disaster prevention.

Mohamed et al.reviewed and analyzed various network architectures on the basis of their characteristics, enhancement strategies, and challenges in reliability advantages as well as disadvantages. Evaluation of reliability was done in context with security constraints of physical network, power availability for the network and network connectivity. An analytical model was designed for the architecture comparison in terms of reliability and underwater pipelines monitoring. Hierarchical sensor network framework was also being developed and evaluated for monitoring underwater pipelines. The architecture comprised the node batteries utilization if the fault would be detected in the wires while providing power to the different nodes, thus far resulting in an increased probability of power availability in an integrated wireless or wired sensor networks in comparison to the wireless sensor networks. An integrated wireless or wired sensor networks also yielded improved reliability in terms of sensing and communication. Moreover, limitations of physical security constraints of underwater sensor networks got eliminated in this network, reason being the physical attacks would not be able to hamper the network connectivity in such networks on a large scale (Mohamed et al., 2011).

Climent et al. described and designed a novel architecture and routing protocol, named Energy-efficient aDaptive hiErarchical and robusT Architecture (EDETA) and EDETA-e protocol respectively, for underwater sensor networks. An architecture was modified in such a way that scheduling techniques for delay-aware and non-delay aware could be applied by combining numerous re-transmission techniques along with them to measure their performance level. In order to cluster the nodes and reduce the power consumption when node is not awake, EDETA-e was used which was a power-aware routing protocol. The protocol also included mechanisms for fault-tolerance and properties such as time-constraints. To accommodate changes or variations in topology and node failure, dynamic path adaptation was used making it energy-efficient exclusive of precise information of location. Results obtained showed firm and well-organized cluster behavior and tree structure. EDETA-e protocol was examined in the sub-aquatic medium. When simulated, it was observed that energy saving was improved while reducing the delays. The packet delivery ratio was also proven to be consistent when acquainted with packet errors. The comparison with the existing approaches such as Depth Based routing (DBR) protocol, yielded that EDETA-e was more energy savvy as the nodes were preserved in low-power state when data transmission was not taking place (Climentet al., 2012).

Due to critical and harsh environment in underwater sensor networks, node failure ratio remains significantly high. H. Min et al. developed check pointing scheme to enhance the network reliability as this scheme considered the head nodes from the clustering routing protocols such that the head nodes could be easily recovered from any failure. The concept of back up nodes was used where the routing information was saved and refreshed periodically as well as data collection was also done here which was sent by head nodes. If in case, the head node stopped working due to any failure occurrence, one of the backup nodes was used to replace the head node and would start acting as a new head node. This way an overall network remained firm and stable without losing the collected information by re-electing a new head node. No matter the fault was permanent or transient, backup node could easily be used to replace the affected head node on immediate basis. Simulation results proved the reliability and efficiency of the network by making it energy efficient. Recovery latency was also improved when the check pointing scheme was applied (Min et al., 2012).

Yoon et al. discovered a novel routing scheme for underwater networks named Autonomous Underwater Vehicles aided Underwater Routing Protocol (AURP). In order to minimize the power consumption and to increase the ratio of delivered data packets in underwater sensor nodes, this scheme utilized controlled mobility concept of multiple Autonomous Underwater Vehicles (AUVs) and heterogeneous acoustic communication channels. The architecture comprised numerous units such as gateway nodes (GWs), mothership or surface unit, Autonomous Underwater Vehicles (AUVs) and U-sensor nodes. AURP aimed to reduce the data transmission rate. By utilizing midrange underwater acoustic channel, the data was sensed by underwater sensors and was sent to Gateway Nodes either in direct manner or in a multi-hop approach. Gateway Nodes forwarded an assimilated data to an Autonomous Underwater Vehicles (AUVs). AUVs passed on the data to the sink using a short-range high rate acoustic channel which was further forwarded to the mothership via surface unit (Yoon et al., 2012). The proposed strategy was compared with Directed Diffusion based routing protocol and it was concluded that delivery ratio of the proposed approach was high along with reduced energy consumption, which, in turn was used in the domains of mass data transmissions namely, multimedia applications.

Caiti et al. proposed and presented cooperative algorithm as well as security suite; cooperative algorithm was designed considering response of mobile underwater sensors, which were installed on AUV, to the localized rules defined on the basis of existing information and security suite targeted minimizing communication overhead which was incurred by size and number of messages being exchanged between autonomous mobile underwater sensors. Cooperative algorithm relied on the information that if communication loss occurred, then the overall coverage performance would not get lost, but degraded up to some extent whereas cooperative algorithm imbibed that that the information from other sensors would be reliable and trustworthy, though not fully and completely. Both the algorithms defined and designed aptly targeted the reliability of network's communication links in the underwater scenario (Caiti et al., 2012).

Wu et al. suggested time-slot based balancing network coding (TSBNC) routing algorithm. In this paper, time-slot routing algorithm was considered in which lot of improvements were imposed to obtain complete routing protocol known as TSBNC for underwater scenario. In order to target energy efficiency and lifeline of network, probability-balanced mechanism was implemented on to time-slot based routing algorithm which eventually resulted in reduction of node confliction ratio, routing path construction shortening as well as reduced power consumption (Wu et al., 2012).

Balandrón et al. put forward architecture for data retrieval prediction known as Artificial Neural Network (ANN)-based architecture for underwater sensors. The proposed architecture utilized Artificial Intelligence methods to enhance the data quality and hence diminished data incompleteness. Numerous factors namely density water salinity and temperature were considered as these values are in close proximity in terms of values for locations and depth which made it easy to predict the missing values from the available information. The approach was apt for scenario where there is an issue of missing data and long time failures as it guaranteed data quality at negligible cost, minimized maintenance and operational cost (Baladrón et al., 2012).

Javaid et al. proposed two new techniques aiming to reduce energy consumption and occurrence of void zone. The concept of mobile sink nodes was utilized in the first technique to reduce energy consumption of the nodes. Routing with cooperation was embedded with the first technique to define and implement second technique. Multi-path routing used in cooperative scheme increased the reliability of the network. First scheme raised the nodes lifeline and was energy savvy. Moreover, second scheme reduced the packet drop count as well as transmission delay rate (Javaid et al., 2017).

Sajid et al. defined a cooperative scheme in order to enhance the energy efficiency and reliability. DSN and RLN were made to co-operate with each other. The criteria which were considered for DSN and RLN selection were quality of channel, highest value of residual energy and minimum depth value. Mobile sink node was used in the scheme for reducing power consumption. The sink nodes captured the information received from other sensor nodes and forwarded them to their final reach. BER check was done at the receiving end. If value reached the optimum criteria, REQ was sent to the RLN. Packet was eventually sent to the desired destination with ACK. Limitations of the proposed scheme lied in raised energy consumption and few nodes failed to stay alive for longer duration; although ratio of alive node was maintained to high value and was proven to be capable enough to send maximum amount of information (Sajid et al., 2016).

Co-operative scheme was also proposed by Nasir et al. to obtain reliable network. The criteria considered for data forwarding were residual energy and depth (Nasir et al., 2014). Along with destination nodes, two relay nodes were considered, where the criteria for selection of such nodes was lowest depth. The packets being sent were received by these nodes. BER was checked by DSN, in case high BER was registered, negative acknowledgment (NACK) was sent back. Destination request was respond back by the relay nodes and eventually, the packet was forwarded along with the acknowledgement (ACK) towards the destination node. Merging of the received packets was done on the basis of diversity technique and all the captured information was eventually forwarded to the sink node. The concept of multi-hop was utilized in case communication could not be made with the sink node. Although the proposed approach resulted in maximized transmitted information, minimized information loss and reliable network yet lifeline of few nodes was not prolonged and power consumption was also high due to co-operation scheme utilized in it.

Chen et al. developed REBAR algorithm, short name for Reliable and Energy Balanced Routing algorithm, which embedded the factor of node movement on continuous basis. The algorithm captured the varying transmission range of nodes which was calculated on the basis of distance of the nodes from the sink node. The transmission range was inversely proportional to the distance of the node from the sink. Although energy consumption was reduced yet dynamic nature of nodes was considerably high which further hampered an overall performance of the proposed algorithm (Chen and Wu, 2008).

Hwang et al. structured and proposed one more algorithm named Directional Flooding-based Routing (DFR) algorithm, which was developed considering the nodes' geographic information imbibing the nodes mobility into it. Quality was an important criterion which was taken care of in order to enhance the network reliability. Packets were flooded in the flood region thus far, adapting network topology as and when required. In this manner, network was energized and made more reliable (Hwang and Kim, 2008).

Mohamad et al. investigated greedy routing protocols for maximizing the delivery rate of packets while minimizing an overall cost of the network. As the system being studied was an underwater network which incurred mobile nature of nodes, thus yielding hop-by-hop routing mechanism based on greedy approach as the most reliable one. As greedy approach makes the decision which is locally optimal, it didn't work well in an underwater scenario as there might come a point when positive progression towards the sink node becomes infeasible after some specific node. There could be many reasons for such situation; one of them could be shadow zones. This problem or scenario was named as communication void problem (Mohamad and Kheirabadi, 2013).

While considering the benefits as well as limitations of the previous work, the protocol which was designed in the proposed approach tried to overcome such problems and situations as the already existing work was not embedding any information related to depth or location. Reliable energy-efficient cross-layer routing protocol called RECRP, was developed which was a location-free single-copy protocol. In order to improve the delivery rate of two-hop packet and to balance energy among the networks, an optimal max–min method was presented which managed to control channel frequency as well as transmission power dynamically. Since the Doppler scale shift can be measured from a single message exchange between two nodes, information from the physical layer such as Doppler scale shift measurement, Received Signal Strength Indicator (RSSI), etc. (Lu et al., 2010; Li et al., 2008) to select next hop without the use of any additional hardware (e.g., depth probes, inertial sensors) was adapted for the protocol. Thus far, RECRP reduced the equipment cost significantly.

Though an extensive research has been done in the field of underwater reliability concerns yet the field demands more exhaustive work in order to attain satisfactory parametric values. Few such reliability models have been explored in the chapter discussing which model would work best in considered underwater scenario while assuring system performance in terms of network delays, power consumption, packet loss as well as network lifespan.

RELIABILITY CONCERNS IN UNDERWATER COMMUNICATIONS

As far as wireless networks are concerned (specifically multi-hop wireless networks), protocols such as Transmission Control Protocol (TCP) and various mechanisms for congestion control does not provide any effective solution for a reliable transmission of data. TCP, a connection-oriented protocol, relies heavily on establishing three-way handshake mechanism between sender and receiver for sending and receiving the data. In underwater environment, the task of three-way handshake mechanism would be a burdensome task, precisely if data to be transmitted is merely of few bytes. TCP demands acknowledgement policies along with re-transmission mechanism, in case data is not received or got corrupted while sending towards the destination. In UWSNs such mechanisms would yield transmission delays as well as reduced throughput. Moreover, TCP considers congestion as a sole factor which results in loss of packet, thus, it works mainly on controlling the network congestion by reducing the transmission speed at sender side. As far as UWSN is concerned, congestion is not the only parameter, resulting in packet loss, rather numerous other factors also hampers the data transmission, namely, node failure, error occurrence in acoustic channel etc. If packets are getting lost because of problematic channel, in such cases, transmission rate is not desired to be reduced in order to support throughput consistency.

On the contrary, User Datagram protocol (UDP), being connection-less protocol, neither support any policies for congestion control nor provides flow control. When congestion occurs, packets are dropped in case of UDP that too without providing any recovery mechanism for packet loss. Also, the concept of acknowledgement is not used in UDP, and hence, it has dependency on its neighboring layers for recovery mechanism of lost packets; thus, proving it as an infeasible approach for UWSN.

Following are the domains where challenges lie in underwater networks:

- **Bandwidth Constraint:** Due to harsh underwater environment, only specific range of frequencies is admissible in an acoustic spectrum in order to carry out the flow of information.
- **Speed limitation of Acoustic Waves:** Numerous parameters affect the speed of acoustic waves in underwater network such as water depth, temperature and salinity of the water. All such factors hamper the speed of acoustic waves, hence forth affecting the data transmission speed and data delivery time at the time of routing.
- **Reduced Network Lifespan:** As underwater communication relies on the battery power, henceforth resulting in shortened network lifetime.
- **Signal Attenuation:** Spreading loss and Absorption loss results in channel attenuation in underwater environment. As a consequence, signal strength gets decreased. Moreover, data extraction also becomes difficult at the receiving end.
- **Underwater Noise:** Noise gets added in the delivered information in underwater environment which degrades the quality and efficiency of signal. In order to overpower its effects, routing protocols makes decision of finding the path which is least affected by the noise.

Although reliable communication is an important concern for any network whatsoever, yet it demands special focus as far as UWSN is concerned. The two approaches used for the purpose are: hop-by-hop and end-to-end. In a wired network scenario, an end-to end delivery mechanism works efficiently as nodes stays in static state most of the times; whereas in underwater communications, there is a dynamic behavior of nodes along with an increased propagation delays, an end-to-end delivery mechanism seems inappropriate and leads to the resource wastage. On the contrary, hop-by-hop data transfer mechanism proves efficient for reliable communications in the absence of internet in UWSN environment and accommodating topological variations on frequent basis in underwater situations.

RELIABILITY MODELS

The reliability of an UWSN depends upon the reliability of its sensor nodes. In this section, the reliability algorithms used in underwater sensor networks have been discussed:

RSAR

The Reliable and Stability-Aware Routing (RSAR) Algorithm comprised total of 5 energy grades formation on the basis of information related to depth. Grade was inversely affected by the depth of the nodes, that is, nodes with lowest value of depth constituted the highest value of grade as well as comprised of the most powerful batteries at initial stage. The focus behind such energy grade formation and providing high values of energy at initial level for the nodes was based on the fact that nodes with minimum depth values had the maximum data traffic on the paths and hence, died quickly. Afterwards, the basis for filtering the forwarded candidate relied on the function F_n as described below:

$$V(s,\pi) = \sum_{t=0}^{\infty} E^{\pi}(r(s_t, a_t) \mid \pi, s_0 = s) \tag{1}$$

where $F_n(i)$ was used to describe Function value for node i, remaining energy was notated by Ri, Di represented depth value and Gi was used to tell the grade value of node i. These were the parameters used for calculating the upcoming forwarder candidate function. The node which was selected as the next forwarded candidate had the highest function value. If the nodes were considered only on the basis of minimum depth value, it resulted in over burdening nodes with lowest depth value, which eventually would hamper the network stability. Thus far, such nodes died instantly because of data over-burden. Reliability issues persisted in RSAR algorithm as no verification was done for reliable delivery of sent data. It happened because only single layer was considered for data transmission (Ali et al., 2019).

Data Forwarding in RSAR

Sensor nodes were used for parameter conversions into the desired data. Data was further transmitted to the base stations or data centers via sink for the purpose of analysis and manipulations. As the range of transmission was insufficient for direct data transmission towards the sink; hence, the concept of multi-hopping was used for data delivery at the surface of water. Eventually, data from the sink was sent to the data center. If the sink falls in the sender's range, then direct data transfer could be done. If this was not the case, then next forwarder was considered in order to send the packet; where next forwarder considered amongst the neighboring list was the one with the maximum value of function Fn. Two IDs were added by the sender in the packet header, where first one was the ID of sender itself and second was an ID of the next node which would be forwarding the data packet further. The neighboring nodes would be receiving the data packets. Once the packets were received by all the neighbors, then they checked for their IDs on the received packet. If their IDs were not there, in such case, packets were ignored by those respective neighboring nodes. If this is not the case, and the neighboring node found its ID on the received packet, then the packet was accepted and eventually forwarded to the upcoming forwarded node on a similar pattern. Figure 2 illustrate the RSAR concept of data forwarding.

Co-RSAR Algorithm

The RSAR algorithm suffered from an issue of reliable data delivery as it utilized only a single link for data delivery. In order to overcome this limitation of RSAR algorithm, Co-RSAR, Co-operative Reliable and Stability-Aware Routing was proposed. However, initial parameters of RSAR were preserved

Figure 2. Data forwarding flowchart of RSAR (Ali et al.,2019)

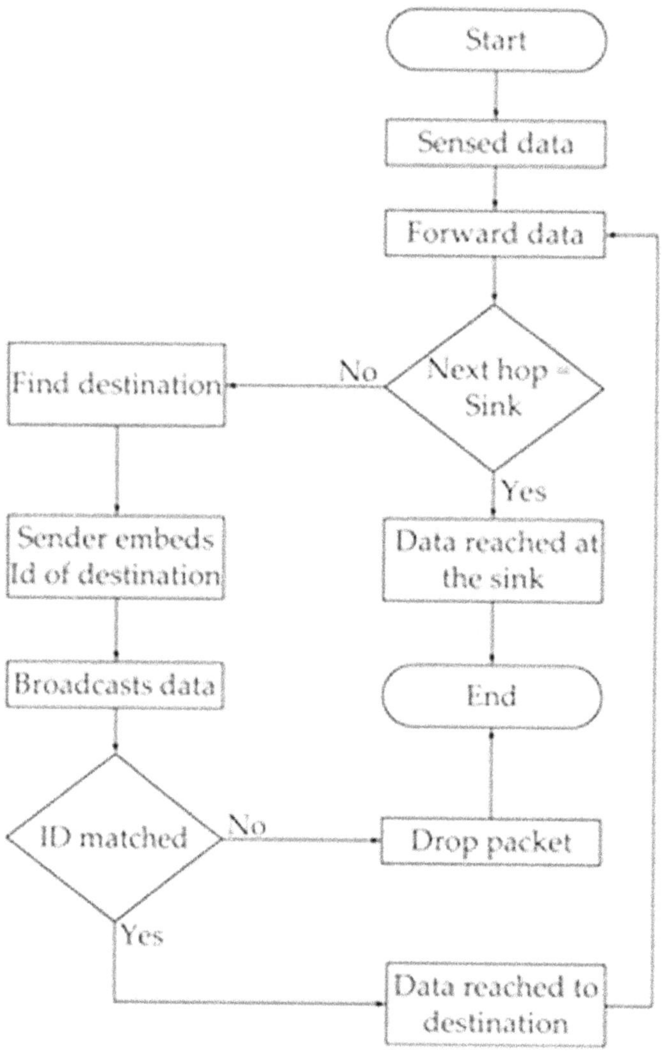

as it was in Co-RSAR algorithm as well, that is, parameter initializations and an overall architecture of network were same for both the approaches.

Finding Relay

While sender sent the data, it considered the node from the neighboring list which had the highest value of function along with the two relays which were at the closest distance to the destination. The packet was broadcasted by the sender and all its neighboring nodes received the same at their respective ends. Neighboring nodes matched their IDs on the received packet in order to make a decision of whether to consider or discard the packet. If the two IDs matched together, the packet was further forwarded to

Figure 3. Data FORWARDING FLOWCHart of CoRSAR(Ali et al., 2019)

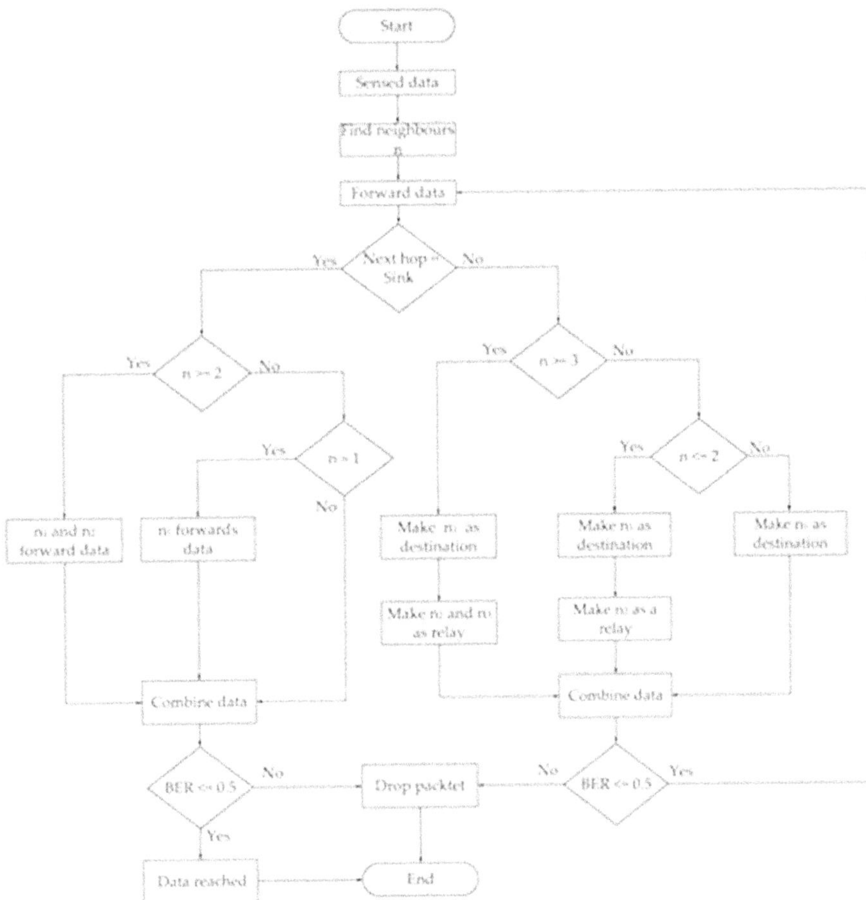

the destination node via relay nodes. Figure 3 depicts the data forwarded mechanism used in Co-RSAR algorithm.

Directional Flooding-Based Routing Protocol

One of the geographical routing protocols were DFR protocol, which worked under an assumption that every respective node has complete information about the location of its own as well as its neighbors. Thus far, the nodes for data forwarding towards the sink were based on geographical advances. The probability of packet loss stays high in underwater environment. For such reasons, if the current forwarder found weak connections to their neighboring nodes with geographical advances, DFR utilized more nodes to work as a forwarder node so that packet loss probability should get minimized and thus, data packet reached the sink in a reliable manner. This mechanism of forwarding the packet was done per hop.

Packet was broadcasted to its neighboring nodes. Moreover, the nodes belonging to the flooding zone were considered for the packet transmission. In this manner, flooding of packet in the whole network was overpowered. That is, DFR utilized flooding technique instead of creating and maintaining an es-

Figure 4. Packet transmissions in DFR (Hwang & Kim, 2008)

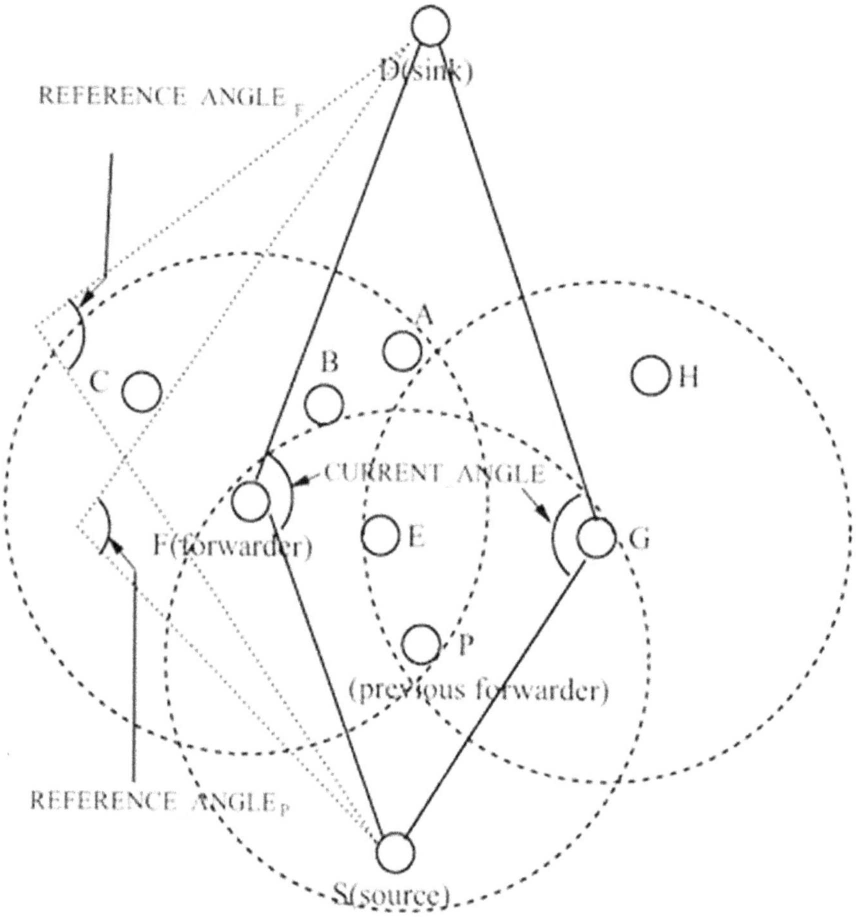

tablished connection. The concept of Reference angle was used in order to enhance the flood zone area; which further resulted in numerous benefits. First benefit lied in the minimized routing overhead. No controlling mechanism was required for maintaining path as there did not exist established connection for data transmission. Next was reduced overhead of transmission. Eventually, the main concern, that is, reliability was obtained. The participating nodes were the nodes which belonged to flooding zone. The size of flood zone was increased in case if the connection with the existing node did not work properly. As a consequence, the reliability factor was enhanced to a great extent (Hwang & Kim, 2008).

Hop by Hop (2H-ACK) Reliability Protocol

An important concern for every network remains transferring information reliably. As far as UWSN is concerned, it demands special attention. Two methods which are used for such purpose are: hop-by-hop and end-to-end. For static environment such as wired networks, an end-to-end approach was found suitable. Whereas for an environment where nodes position was not fixed and there were more intense propagation delays such as underwater scenario; conventional end-to-end approach was not found to be reliable and resulted in a resource wastage. On the contrary, hop-by-hop approach worked better in frequently varying topology and missing internet facility environment, for reliable data transmission.

Thus the major goal was to target hop-by-hop data transfer mechanism to attain reliable solution for such underwater scenario (Ayaz et al., 2010). Hop-by-hop ACK scheme utilized two nodes; where the node at the receiver end was expected to send an acknowledgement if the packet was received successfully without any loss or error detection. As soon as the sender received acknowledgement, it started processing next data packet while discarding the current data packet. This mechanism worked efficiently in terrestrial-based networks (Ayaz et al., 2010). In underwater environment, it became difficult to follow this traditional acknowledgement mechanism because in such scenario, node might get lost or die quickly. In UWSN, it became difficult for the receiving node to transmit the packet further to the next hop because of dynamic nature of nodes and sparseness, meanwhile the node might die due to insufficient power or battery and moreover, failure might also take place due to corrosion, fouling or any such similar issues. As the packet was only held by the receiving node, in such cases, all the held packets by the current node would get lost on permanent basis, reason being there was no back up for the sent nodes. For such reason, 2H-ACK model was proposed. The method was based on the concept of acknowledgement again but with little modifications. The node which received the packet would hold the packet until it found the next node for forwarding the data packet. Moreover, it did not send the acknowledgement to the sender until it found the next node in the routing path. Therefore, there were two nodes carrying the same data packet resulting in reliable overall network. Figure 5 elaborated the data forwarding mechanism. The network comprised various nodes where N9, with HopID 56, held a data packet to be forwarded to the sink node. It would query its neighboring nodes for their respective HopIDs. Nodes N7 and N8 responded back as they were falling in the range of Node N9. On the basis of HopID comparison, node was further transmitted to N7 as its HopID value was lesser in comparison to N8 HopID. The same process was repeated by N7 and the next hop found was N4 instead of N6. The acknowledgment was not sent to N9 until N7 found its next forwarder, which was N4. The process was repeated until the packet reached its final destination.

Enhanced Network Coding-Based Energy Efficient and Reliable Multi-path Routing protocol (EERCM)

EERCM was the communication model, which incorporated the benefits of Enhanced Network Coding based Energy Efficient and Reliable Multi-path routing to enhance an energy efficiency level as well as network reliability in UWSN. EERCM was developed along with braided model where every node forming a path had one node as a backup. Node disjoint path was longer on frequent basis because of multi-hop concept and its influence on UWSN and hence, resulted in raised energy or battery consumption. In the concept of disjoint multi-path, same node could be used for n number of times for the routing path, which yielded more power consumption. Number of paths were also got raised due to braided multi-path. An effective ways could be to equally utilize every node belonging to main routing path, thus far, would result in optimized energy consumption and balanced an overall load of the network. This protocol embedded clustering hierarchy technology with distributed routing technology. EERCM was designed by combining MAC protocol with distributed underwater routing protocols(Gnanasingh & Leavline, 2013).

Figure 5. selecting next hop for data packet forwarding

Probability-based Multi-Path Optimization in EERCM

EERCM communication resulted in reliable data transmission for UWSN along with raised battery life, network with balanced load and re-distribution of packets mechanism by making use of optimization methods. The agenda was optimizing power consumption for reception as well as transmission of data packets towards the sink. It made use of numerous sub-optimum paths to enhance the network lifespan. The selection criterion for path establishment was probability function, which was further dependent on the power consumption of individual routing path. An approach called Network Survivability, dealt out with the path with minimized energy level all the time, leading to the node energy destruction on the respective routing path (Gnanasingh & Leavline, 2013). The lifespan of an entire network was enhanced by utilizing multiple paths with a specific probability value. The protocol assumed addressability of all the nodes with a network coding-based addressing comprising the location (latitude/ longitude) and the nodes of basic types.

COMPARATIVE STUDY OF VARIOUS RELIABILITY MODELS

The Table 1 shows the detailed comparison amongst the various above discussed reliability algorithms applicable in UWSN on the basis of various parameters like their underlying concepts, focus, limitations, scopes and performance.

RESULTS AND DISCUSSION

MATLAB has been used in order to perform simulation and obtain the results for the comparison of above discussed five approaches i.e. RSAR, Co- RSAR, EERCM, Hop by Hop, DFR Model.

Table 1. Comparison chart

Sr. No	Comparison Parameters	Reliability Models				
		RSAR (Ali et al., 2019)	Co-RSAR (Ali et al., 2019)	EERCM (Gnanasingh & Leavline, 2013)	Hop by Hop (2H-ACK) Model (Ayaz et al., 2010)	DFR Model (Hwang & Kim, 2008)
1	Underlying concept	Concept of energy grades were used which were inversely proportional to the depth of the nodes.	Broadcast mechanism was used to forward the packet towards the sink. Moreover, IDs were allocated to match the respective receiver.	Backup node was utilized while forming the routing path as well as sub-optimum paths were considered to increase the life of the nodes.	Current node sent back an acknowledgement to the sender but only after finding the next node in the routing path towards the destination.	Nodes forwarded data to the node while considering the geographical locations as well as on the basis of flooding, precisely maintained the flood zone area.
2	Prime focus	Load balancing of various nodes on the routing path.	Reliable data delivery at the receiver end.	Reduced energy consumption and network fault tolerance.	Congestion avoidance and minimized packet loss.	Reduced communication overhead as well as latency time reduction.
3	Assumptions made	Energy grades were allocated to every respective node based on the node depth value.	Along with the concept of energy grades, co-operation amongst nodes was inculcated.	Network-coding based addressing mechanism was utilized for addressing all the nodes which comprised the basic type of nodes as well as location in the terms of longitude and latitude.	Two nodes were considered to hold the same packet in order to overcome an issue of packet loss.	• Geographic information: Every node held the information about its own location, its one-hop neighbors' location and the sink location. • Link Quality: Measurement of link quality for all the nodes with their respective nodes.
4	Limitations	Reliable data delivery was not guaranteed.	More energy consumption by nodes.	Redundancy level got increased.	Acknowledgements and re-transmissions led to the over-burdening of the nodes.	Being a geographical routing protocol, if the node did not exist in the close proximity of the sink, it resulted in void problem.
5	Extensibility scope	Creation of direct link with sink nodes in order to reduce the delays incurred.	More focus on channel quality and raising the nodes' transmission range.	Deployment of sensor nodes in sea-bed experimental setup so that processed information could be uploaded and distributed in cloud environment.	Ways to overpower energy death of nodes.	In-depth analysis of pre-defined set of values in order to enhance performance.
6	Performance Metrics	• Network Dimension. • Modulation Type. • Energy Grades • Data Packet Size • Transmission Range.	• Transmission Range. • Data Packet Size • Network Dimension. • Energy Grades • Modulation Type.	• Location error. • Prediction Parameters. • Path setup. • Energy Grades.	• Packet delivery ratio with varying number of nodes. • Packet delivery ratio at specified time intervals. • Packet loss and redundancy impacts at different intervals of time.	• Communication overhead. • Average end-to-end delay. • Packet Delivery Ratio.

Simulation Parameters

Numerous simulation parameters with their specified ranges are depicted in the Table 2 below:

Basis of Comparison

The comparison is performed based upon packet delivery ratio, network delay and energy efficiency as discussed below:

Packet Delivery Ratio

The packet delivery ratio has been proved as an important metric for estimation of the contribution of the solution for improving the network reliability (Goyal et al., 2019b). It can be defined as the quotient between the packet counts at the receiver side and the packet count at the sender side at the time of simulation. That is,

Delivery ratio = Received packet count/ Sent packet count

The packet delivery ratio of various protocols for a network size of 100 has been plotted in Figure 6. As a consequence, every node would encounter more substitutes for re-transmission of the packets, which would further minimize the packet loss probability. When the models were compared, Co-RSAR(Ali et al., 2019) was proven to achieve best packet delivery ratio. This was a predictable outcome as delay minimization helps incur reduced packet loss along with an enhanced transmission power, improved link quality which further leads to lesser transmission count for packet delivery. However, reduced delays may result in packet redundancy. In fact, if the wait time between various re-transmissions would be minimized, it would engender redundancy in terms of same packet transmission.

Delay Occurrence in Network

Occurrence of network delays is one of the prominent concerns and performance characteristic of underwater sensor network. The time taken by data bits to span across the network from the sender end to the receiver end is termed as the network delay, measured in seconds. Moreover, delay incurred may vary from one location to another for the respective pairs of communicating nodes.

Point-to-point networks for multiple hops are created by the communication link formation only amongst the neighboring nodes. Data transfer takes place from sender to receiver via packet hopping from one node to another node. Intelligent algorithms (Goyal et al., 2019a).are used for data routing in order to tackle the varying conditions and underlying scenario. In this way, huge areas can be covered with the help of multi-hop networks as network range is now dependent on the node count instead of modem range. There also exists the drawback of increased node count in terms of packet delays. There is a demand of special attention for the applications where delays are critical. Increased delays between the message transfer leads to significant reduction in system throughput, if not handled properly. When compared to RSAR model, Co-RSAR incurs large delays due to relay nodes along with the end-points communications. Also, EERCM and RSAR suffer from reduced delays in comparison to 2H-ACK and DFR schemes since there exist a single path between the communication points. The comparison chart of the packet delay in various models i.e. RSAR, Co-RSAR, EERCM,2H-ACK, DFR are shown in Figure 7.

Table 2. Simulation parameters

Parameter	Value
Max. number of nodes in the network	1000
Simulation time	50 s
Antenna type	Omni antenna
Channel capacity	2 Mbps
Packet size	50 to 250 bytes
Initial energy	1000 J
Sensor Nodes Deployment	Random Deployment
Sensing Range (R)	25m
Transmission Range	10-20 m

Energy Efficiency

The major concern in proliferation of UWSN lies in energy efficiency as battery concerns can be easily resolved as far as terrestrial sensor networks are concerned; whereas battery replacement is nonetheless, a difficult task to perform; thus far, demanding maintenance of battery life for longer duration (Goyal et al., 2019c). In order to provide fair and equal chance to all the acoustic nodes for network participation, battery life of every sensor node should be maintained in accordance with other sensor nodes. Figure 8

Figure 6. Packet delivery ratio

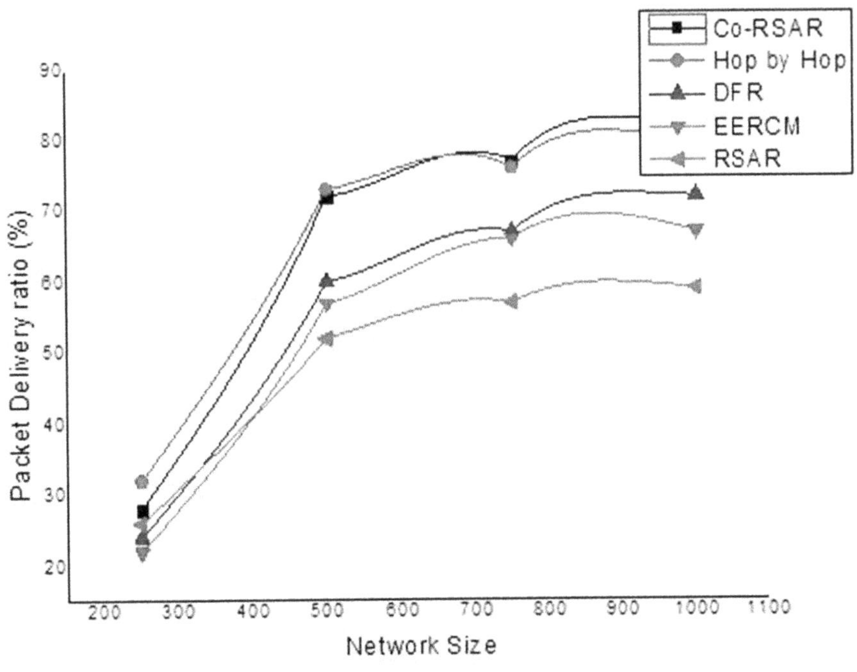

Figure 7. Packet delay in network

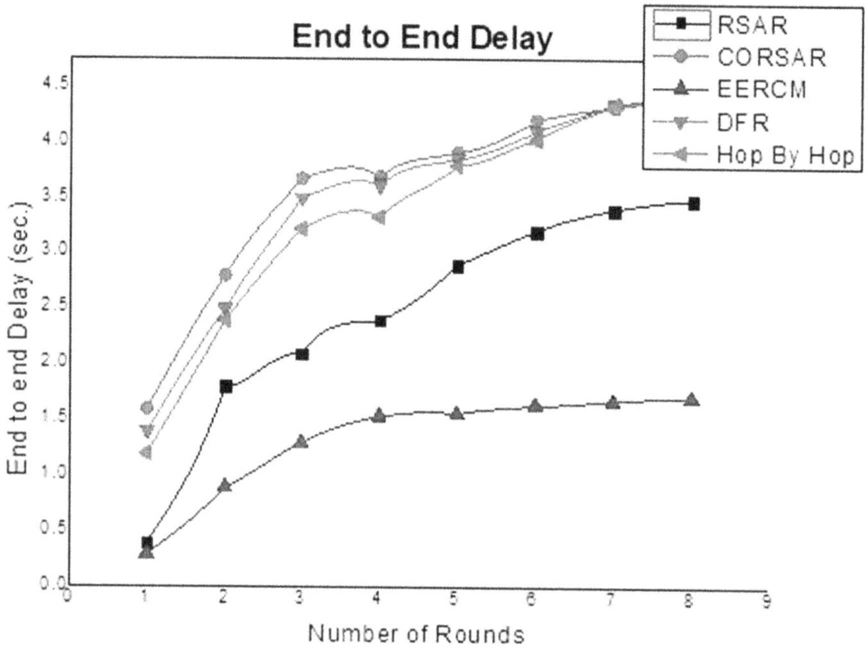

represents the energy consumption ratio of different algorithms. Initially RSAR was found to be giving optimized results in terms of energy efficiency. RSAR model was utilizing less energy in comparison to other models and 2H-ACK Model consumed highest energy amongst all in initial phase. After certain rounds, the energy consumption became comparable as all the algorithms reached to a certain specific value and remained consistent over there.

CONCLUSION

In order to obtain an efficient network, reliability plays a prominent role. It becomes more crucial as far as underwater sensor networks are concerned as conditions underwater are more adverse and harsher in comparison to terrestrial network. For such reasons, it becomes mandatory to consider the reliability factors which impact an underwater overall environment and the reliability of the acoustic nodes. There are numerous reliability models which have been proposed by the researchers in the past targeting path delays, total energy consumption and an overall reliability of the network. Such reliability models, namely, RSAR (Reliable and Stability-Aware Routing), Co-RSAR(Co-operative Reliable and Stability-Aware Routing), 2H-ACK (2 Hop-Acknowledgement), DFR (Directional Flooding-Based Routing Protocol) and EERCM (Enhanced Network Coding Based Energy Efficient and Reliable Multi-path routing) have been discussed and compared on the basis of various parameters and metrics. There is a trade-off amongst all the approaches when analyzed on different parameters. Co-RSAR performed efficiently well when packet delivery ratio was considered. The packet loss ratio was less in this model in comparison to other models. But the same approach suffered drastically from network delays. As far as network delay

Figure 8. Energy consumption

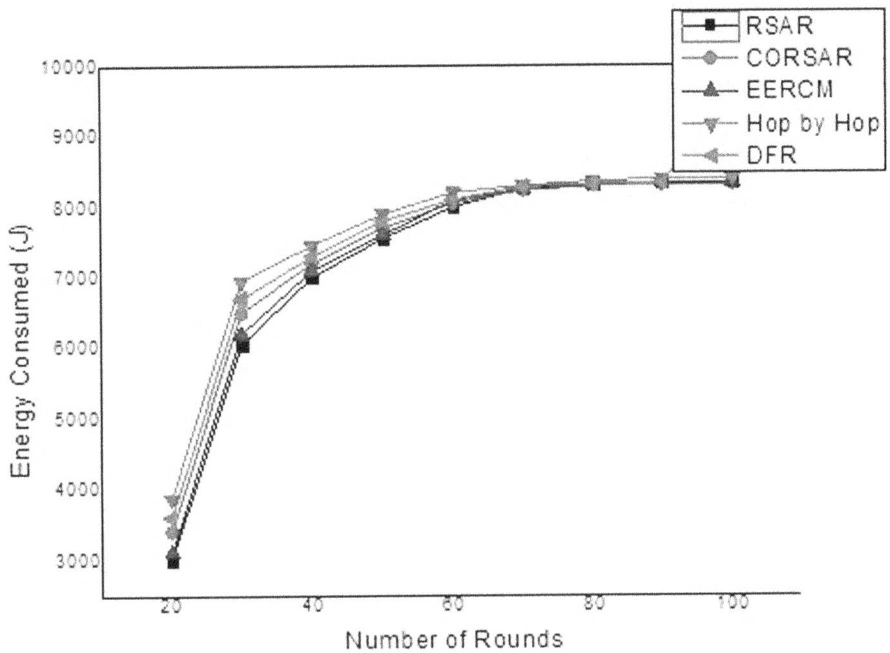

is concerned, RSAR and EERCM outperformed other approaches. If efficiency was analyzed in terms of energy consumption, all the algorithms resulted in approximately equal value after certain rounds; whereas in initial phase, RSAR consumed less energy and 2H-ACK consumed the highest energy. Thus, overall performance could be a trade-off between the various performance metrics for the different models.

REFERENCES

Al-Dharrab, S., Uysal, M., & Duman, T. (2013). Cooperative underwater acoustic communications. *IEEE Communications Magazine*, *51*(7), 146–153. doi:10.1109/MCOM.2013.6553691

Alghamdi, R., Saeed, N., Dahrouj, H., Alouini, M. S., & Al-Naffouri, T. Y. (2019, September). Towards Ultra-Reliable Low-Latency Underwater Optical Wireless Communications. In *2019 IEEE 90th Vehicular Technology Conference (VTC2019-Fall)* (pp. 1-6). IEEE. 10.1109/VTCFall.2019.8891506

Ali, M., Khan, A., Mahmood, H., & Bhatti, N. (2019). Cooperative, reliable, and stability-aware routing for underwater wireless sensor networks. *International Journal of Distributed Sensor Networks*, *15*(6), 1550147719854249. doi:10.1177/1550147719854249

Ayaz, M., Abdullah, A., & Faye, I. (2010, November). Hop-by-hop reliable data deliveries for underwater wireless sensor networks. In *2010 International Conference on Broadband, Wireless Computing, Communication and Applications* (pp. 363-368). IEEE. 10.1109/BWCCA.2010.97

Baladrón, C., Aguiar, J. M., Calavia, L., Carro, B., Sánchez-Esguevillas, A., & Hernández, L. (2012). Performance study of the application of artificial neural networks to the completion and prediction of data retrieved by underwater sensors. *Sensors (Basel), 12*(2), 1468–1481. doi:10.3390120201468 PMID:22438720

Caiti, A., Calabro, V., Dini, G., Lo Duca, A., & Munafo, A. (2012). Secure cooperation of autonomous mobile sensors using an underwater acoustic network. *Sensors (Basel), 12*(2), 1967–1989. doi:10.3390120201967 PMID:22438748

Chen, J., Wu, X., & Chen, G. (2008, October). REBAR: a reliable and energy balanced routing algorithm for UWSNs. In *2008 Seventh international conference on grid and cooperative computing* (pp. 349-355). IEEE. 10.1109/GCC.2008.12

Climent, S., Capella, J. V., Meratnia, N., & Serrano, J. J. (2012). Underwater sensor networks: A new energy efficient and robust architecture. *Sensors (Basel), 12*(1), 704–731. doi:10.3390120100704 PMID:22368492

Davis, A., & Chang, H. (2012, October). *Underwater wireless sensor networks. In 2012 Oceans*. IEEE.

Erol, M., Vieira, L. F., & Gerla, M. (2007, September). Localization with Dive'N'Rise (DNR) beacons for underwater acoustic sensor networks. In *Proceedings of the second workshop on Underwater networks* (pp. 97-100). ACM. 10.1145/1287812.1287833

Gao, M., Foh, C. H., & Cai, J. (2012). On the selection of transmission range in underwater acoustic sensor networks. *Sensors (Basel), 12*(4), 4715–4729. doi:10.3390120404715 PMID:22666054

Goyal, N., Dave, M., & Verma, A. K. (2019a). Protocol stack of underwater wireless sensor network: Classical approaches and new trends. *Wireless Personal Communications, 104*(3), 995–1022. doi:10.100711277-018-6064-z

Goyal, N., Dave, M., & Verma, A. K. (2019c). Data aggregation in underwater wireless sensor network: Recent approaches and issues. *Journal of King Saud University-Computer and Information Sciences, 31*(3), 275–286. doi:10.1016/j.jksuci.2017.04.007

Goyal, N., Sandhu, J. K., & Verma, L. (2019b). Machine learning based data agglomeration in underwater wireless sensor networks. International Journal of Management. *Technology And Engineering, 9*(6), 240–245.

Han, G., Jiang, J., Shu, L., Xu, Y., & Wang, F. (2012). Localization algorithms of underwater wireless sensor networks: A survey. *Sensors (Basel), 12*(2), 2026–2061. doi:10.3390120202026 PMID:22438752

Heidemann, J., Ye, W., Wills, J., Syed, A., & Li, Y. (2006, April). Research challenges and applications for underwater sensor networking. In *IEEE Wireless Communications and Networking Conference, 2006. WCNC 2006* (Vol. 1, pp. 228-235). IEEE. 10.1109/WCNC.2006.1683469

Hwang, D., & Kim, D. (2008, September). DFR: Directional flooding-based routing protocol for underwater sensor networks. *Oceans, 2008*, 1–7.

Hwang, D., & Kim, D. (2008, September). DFR: Directional flooding-based routing protocol for underwater sensor networks. *Oceans, 2008*, 1–7.

Javaid, N., Ejaz, M., Abdul, W., Alamri, A., Almogren, A., Niaz, I., & Guizani, N. (2017). Cooperative position aware mobility pattern of AUVs for avoiding void zones in underwater WSNs. *Sensors (Basel), 17*(3), 580. doi:10.339017030580 PMID:28335377

Kheirabadi, M. T., & Mohamad, M. M. (2013). Greedy routing in underwater acoustic sensor networks: A survey. *International Journal of Distributed Sensor Networks, 9*(7), 701834. doi:10.1155/2013/701834

Lanbo, L., Shengli, Z., & Jun-Hong, C. (2008). Prospects and problems of wireless communication for underwater sensor networks. *Wireless Communications and Mobile Computing, 8*(8), 977–994. doi:10.1002/wcm.654

Li, B., Zhou, S., Stojanovic, M., Freitag, L., & Willett, P. (2008). Multicarrier communication over underwater acoustic channels with nonuniform Doppler shifts. *IEEE Journal of Oceanic Engineering, 33*(2), 198–209. doi:10.1109/JOE.2008.920471

Llor, J., & Malumbres, M. P. (2012). Underwater wireless sensor networks: How do acoustic propagation models impact the performance of higher-level protocols? *Sensors (Basel), 12*(2), 1312–1335. doi:10.3390120201312 PMID:22438712

Lu, F., Mirza, D., & Schurgers, C. (2010, September). D-sync: Doppler-based time synchronization for mobile underwater sensor networks. In *Proceedings of the Fifth ACM International Workshop on UnderWater Networks* (p. 3). ACM. 10.1145/1868812.1868815

Maqsood, H., Javaid, N., Zain-ul-Abidin, M., Ejaz, M., & Shakeel, U. (2015, November). Energy balanced interference aware energy efficient depth base routing protocol for UWSNs. In *2015 10th international conference on broadband and wireless computing, communication and applications (BWCCA)* (pp. 13-19). IEEE. 10.1109/BWCCA.2015.85

Min, H., Cho, Y., & Heo, J. (2012). Enhancing the reliability of head nodes in underwater sensor networks. *Sensors (Basel), 12*(2), 1194–1210. doi:10.3390120201194 PMID:22438707

Mohamed, N., Jawhar, I., Al-Jaroodi, J., & Zhang, L. (2011). Sensor network architectures for monitoring underwater pipelines. *Sensors (Basel), 11*(11), 10738–10764. doi:10.3390111110738 PMID:22346669

Moradi, M., Rezazadeh, J., & Ismail, A. S. (2012). A reverse localization scheme for underwater acoustic sensor networks. *Sensors (Basel), 12*(4), 4352–4380. doi:10.3390120404352 PMID:22666034

Nasir, H., Javaid, N., Ashraf, H., Manzoor, S., Khan, Z. A., Qasim, U., & Sher, M. (2014, November). CoDBR: cooperative depth based routing for underwater wireless sensor networks. In *2014 Ninth International Conference on Broadband and Wireless Computing, Communication and Applications* (pp. 52-57). IEEE. 10.1109/BWCCA.2014.45

Sajid, M., Wahid, A., Pervaiz, K., Khizar, M., Khan, Z. A., Qasim, U., & Javaid, N. (2016, July). SMIC: Sink mobility with incremental cooperative routing protocol for underwater wireless sensor networks. In *2016 10th International Conference on Complex, Intelligent, and Software Intensive Systems (CISIS)* (pp. 256-263). IEEE.

Shin, S. Y., & Park, S. H. (2011). A Cost Effective Block Framing Scheme for Underwater Communication. *Sensors (Basel), 11*(12), 11717–11735. doi:10.3390111211717 PMID:22247689

Singh, D. A. A. G., & Leavline, E. J. (2013). Eercm: energy efficient and reliable communication model for achieving qos in underwater sensor networks. *Communications, 4*(5).

Sozer, E. M., Stojanovic, M., & Proakis, J. G. (2000). Underwater acoustic networks. *IEEE Journal of Oceanic Engineering*, *25*(1), 72–83. doi:10.1109/48.820738

Wahid, A., & Kim, D. (2012). An energy efficient localization-free routing protocol for underwater wireless sensor networks. *International Journal of Distributed Sensor Networks*, *8*(4), 307246. doi:10.1155/2012/307246

Wu, H., Chen, M., & Guan, X. (2012). A network coding based routing protocol for underwater sensor networks. *Sensors (Basel)*, *12*(4), 4559–4577. doi:10.3390120404559 PMID:22666045

Yoon, S., Azad, A. K., Oh, H., & Kim, S. (2012). AURP: An AUV-aided underwater routing protocol for underwater acoustic sensor networks. *Sensors (Basel)*, *12*(2), 1827–1845. doi:10.3390120201827 PMID:22438740

Chapter 16
Real-Time Data Transfer in Marine Environment Monitoring Applications

Renu Popli
Chitkara University Institute of Engineering and Technology, Chitkara University, Punjab, India

Isha Kansal
Chitkara University Institute of Engineering and Technology, Chitkara University, Punjab, India

Kanwal Garg
Department of Computer Science and Applications, Kurukshetra University, India

ABSTRACT

Due to immense application scenarios emerged by underwater communication, the researchers have shown their ardent interest in this field. Underwater wireless sensor network (UWSN) provides a promising approach for realism of such type of applications such as pollution monitoring, data agglomeration, natural calamities prediction, etc. In UWSN, sensor nodes are deployed at a number of different points of depths based upon the application requirements. Nodes at different locations collect the information and forward it to the sink node at surface. As a node cannot directly communicate to the sink node at surface, the multi hop communication takes place which poses great challenges underwater such as limited resource, bandwidth unavailability, end-to-end delays, temperature, water pressure, etc. However, congestion control is considered to be fundamental issue to be solved for effective data transfer underwater. This chapter presents the importance of real-time data in marine environment for plenty of applications and discuss the major issue in efficient data communication.

DOI: 10.4018/978-1-7998-3640-7.ch016

INTRODUCTION

In UWSN, the sensor nodes are deployed in particular acoustic regions for the core purpose of data collection as well as collaborative monitoring. The nodes in this network are equipped in such a way that they can communicate with the base station located on shore. This communication is performed through messages by effectively utilizing the sound propagation mechanism in sea water environment, which is called acoustic communication (Akyildiz et al., 2006). Acoustic communication is a type of physical layer mechanisms used in underwater networks. It is considered that radio wave covers a large space at very small frequency of about 30 Hz, which creates a need for high transmission power and big antennas. One of the other alternatives is through optical waves, which are not affected by high attenuation but these waves suffer from scattering problem. Therefore, the communication in underwater is effectively performed by acoustic mechanisms (Kumar et al., 2014).

As compared to traditional under water networks, the nodes in the UWSN are of dynamic nature. The nodes can freely move from one location to another based on the requirements of the activity or application scenario. However, in traditional networks, sensors were deployed at static locations for purpose of monitoring and data collection. The nodes are then recovered after the completion of the mission. There are a couple of disadvantages with this approach, as pointed below:

1. **Lack of Interaction:** The connectivity between underwater deployed sensors and monitoring equipments is not maintained on real time basis. However in some applications such as environmental monitoring applications, real time monitoring is paramount.
2. **Unavailability of Data During The Mission:** The sensed data can be accessed only after the recovery of the nodes which may take long time.
3. **Difficulty to Detect Failures:** It is not possible to detect failures or mis-configuration before the recovery of the instruments which may sometimes lead to the loss of sensed data or failure of complete mission.

These shortcomings demand for deployment of underwater networks which can support real time monitoring, configuration as well as interaction with onshore receiving station. This can be achieved by connecting wireless sensor nodes with base stations at water surface through acoustic communication as shown in Figure 1.

A network of sensor nodes can be set up for monitoring and recording temperature, pressure as well as pollutants from underwater eco-system which results into a plethora of applications (Srinivas et al., August 2019). Some of the major applications are disaster prevention, military surveillance, assisted navigation, aquatic data monitoring, sampling and many more as shown in Figure 2.

The major objective in underwater wireless sensor network is to accomplish efficient data collection and communication (Yan et al., 2008). The congestion in the network hinders the performance of the network. In UWSN, congestion control is a significant and difficult task as it also helps in improving QoS in the network.

To achieve this, there are lots of issues or challenges to be faced by researcher fraternity as compared to terrestrial wireless sensor networks (Akyildiz et al., 2005). For incidence, the congestion control mechanisms specifically for terrestrial networks do not suit for UWSNs. This is due to the poor performance of TCP in UWSN as it finds difficult to differentiate loss of packets because of congestion or channel conflicts (Goyal et al., 2020). Thus the design of congestion control mechanism for UWSN is

Figure 1. A network deployment of sensor nodes in UWSN

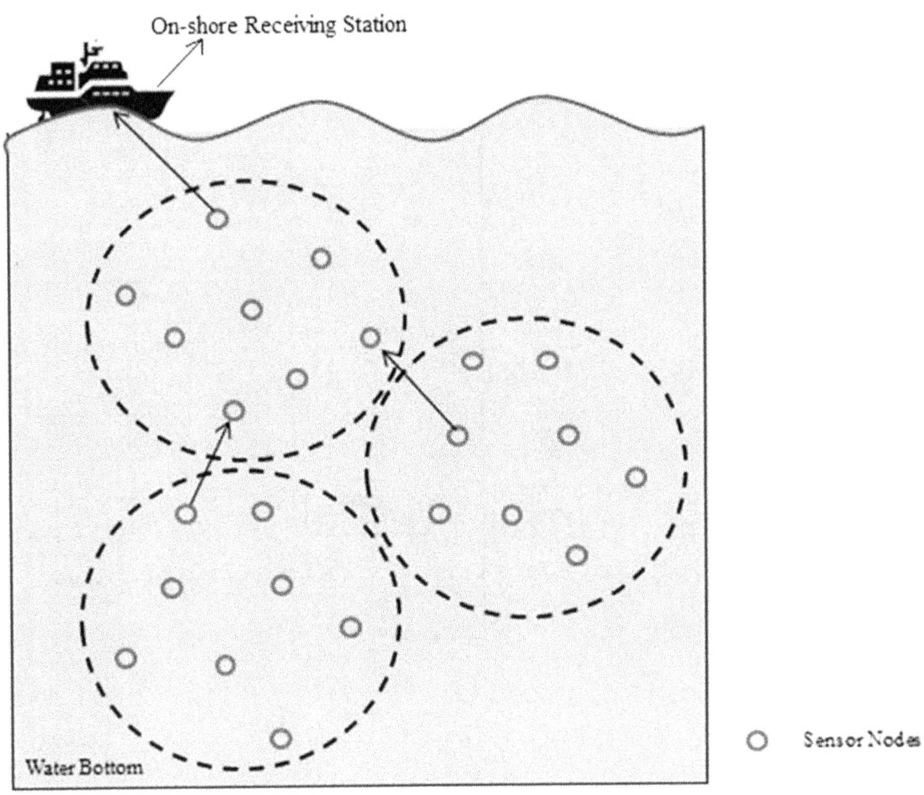

restricted because of the specific properties of the acoustic communication media, some of which are: variant error rates, propagation delays, high attenuation and multi path etc. Although a huge amount of research work has been done in the field of improving the performance of WSN, but the peculiar nature of underwater communication channel restricts their usage in UWSN.

A comparison of distinctive properties of WSN and UWSN is provided in Table 1 below:

The majority of applications of UWSN in marine environment involve the use of instant data related to the event occurring under sea. The paper addresses the issue of data transfer in real time applications of UWSN in marine environment. The major hurdle in real time applications is to avoid the network from being choked. So to maintain efficient data communication, it is mandatory to control congestion in the network. The following sections introduce basics of acoustic communication in Section 2, further section 3 describe the issues related to various layers of communication beneath the sea. Section 4 illustrates the congestion control approaches by focusing on their key parameters and finally section 5 presents the significance of real data in monitoring applications of UWSN.

FUNDAMENTAL OF ACOUSTIC COMMUNICATION

The underwater communication through acoustic channels is characterized by some factors including Doppler Effect, variable propagation speed, loss of path, dynamic paths, noise etc. Due to the above

Figure 2. Application areas of UWSN

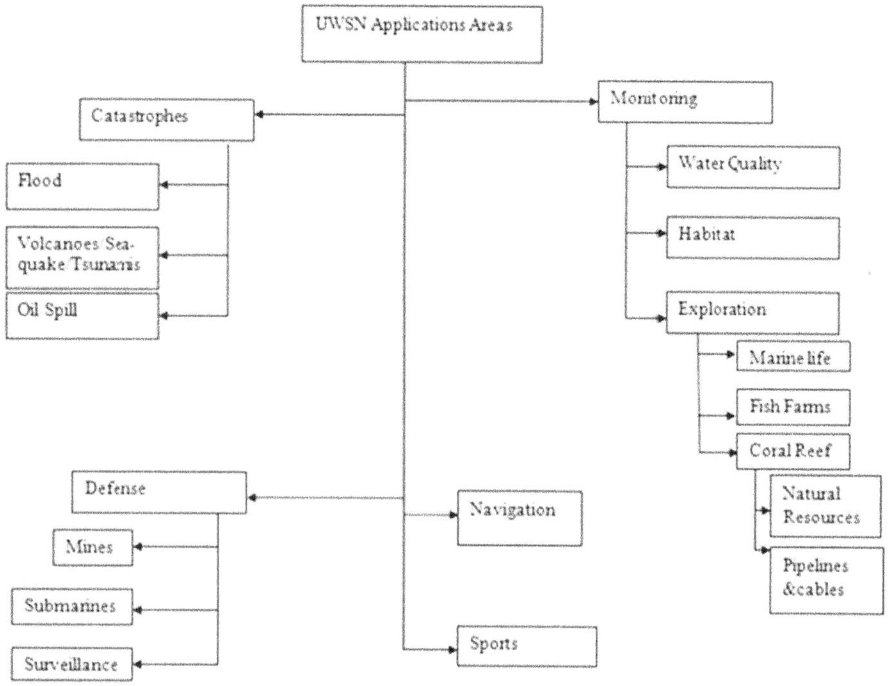

Table 1. Comparison between WSN and UWSN

WSN	UWSN
1. The network consists of dense deployment of nodes	1. Because of costly instruments, network chooses sparse deployment of nodes.
2. Different network topologies can be applied as nodes are considered to be static.	2. The topology is quite dynamic, as nodes continues to move with water currents at approximate speed of 1-3m/s.
3.Link stability is not a rich issue in stable networks.	3. In a network with dynamic nodes, link stability becomes a challenging task.
4. More reliable networks.	4. Reliability is a big concern due to frequent losses and high bit error rate.
5. Follows deterministic deployment pattern.	5. Commonly, non-uniform and random deployment is followed.
6. Better efficiency as nodes are homogeneous.	6. A variety of technical issues are there due to heterogeneous nature of nodes.
7. Low propagation delays as communication medium is radio waves with speed of light.	7. High propagation delays as communication is done through acoustic waves with speed of sound.
8. Less energy consumption	8. Due to complex signal processing high energy consumption.
9. High data rate is feasible usually in order of MHz.	9. Supports low data rate in the order of KHz. However with frequency and transmission range attenuation of acoustic signal increases.
10. Nodes can work for longer time without any failure or errors.	10. Nodes can die as they are more prone to failures which demands for self recovery and more reliable algorithms for routing purposes.

properties, these channels supports low bandwidth and it depends on the frequency and availability of range. For incidence, systems that work in Long-range specifically in the range of tens of kilometers have a low bandwidth of order of a few kHz, on the other hand, systems with short range may support higher bandwidth of order of hundred kHz (Akyildiz, et al., 2004). Furthermore, acoustic communication channels supports very less communication range in comparison to radio channel terrestrial communication systems.

The acoustic channels communication medium can be categorized based on their range as very low, slow, medium, high and very high. Based on their direction of motion, acoustic channels are divided into horizontal and vertical. The propagation features of this type of medium vary in correspondence to delay variance, multi-path spreading etc.

The implementation of acoustic signals under water poses a great challenge as there are many factors which need to be considered specifically. These are explained below:

1. **Path loss:** Due to propagation of acoustic waves under sea environment, there occurs a loss of energy which can be in any of the following forms:
   a. **Attenuation:** It is described as the conversion of wave energy into different kind of energy. For incidence, in terms of acoustic communication, conversion of sound waves into heat is termed as attenuation. This property varies proportionally with respect to frequency and distance.
   b. **Geometric Spreading Loss:** This refers to the technique that when acoustic signals are generated, they expand in form of wave fronts. It is directly proportional to the distance; however, frequency has no effect over it. Mainly they are of two types which are spherical, that is suited for deep ocean communication and other is cylindrical that is suited for shallow water communication.
   c. **Scattering Loss:** This is a physical property which involves the deviation from line of sight of signals. In acoustic communication, this is related to roughness of surface caused by high wind. It dramatically affects power as well as delays.
2. **Noise:** This is an unwanted characteristic in any communication medium which diminish the quality of the system. This can be divided into the following types:
   a. **Ambient Noise:** This is caused by movement of water and it is quite complex to identify the source of noise. This is sometimes defined as background noise which can be categorized into four types:
      i. Wind noise
      ii. Noise due to large ships
      iii. Turbulence e.g. tides
      iv. Thermal noise (self noise)
   b. **Noise Caused by Human Beings:** There are various human activities which disturbs the sea environment. This specifically comprise of noise due to fishing activities, shipping, defense activities or machinery utilization. Sometimes to and fro of data under sea also interferes the communication medium.
3. **Multi-path:** This creates a hurdle in effective transmission of data, as it reduces the strength of the signals due to inter symbol disturbances.
   a. Acoustic communication in deep water is more affected by multi-path as compared to shallow water.

   b. Due to the multiple paths, data communication is affected by propagation delays.
4. **Doppler Spread:** Due to relative motion of transmission and receiving, there is a mean frequency shift called as Doppler shift and Doppler spread is caused by oscillation of frequency in the area. Doppler spread results into the following consequences in acoustic signals:
   a. Compression and stretching of pulse width
   b. Frequency offset

In the beginning of acoustic underwater networks, submarine communication system gained the popularity which was introduced in USA during the Second World War. The technology behind it was based on analog modulation which was single sideband amplitude modulation. With the advancement of digital system, the modern communication systems were equipped with digital modulation techniques. Nowadays, a variety of acoustic modems offers a great speed for data transfer. There are a number of examples of deployment of underwater sensor networks such as Sea-web, which was specifically deployed for defense applications. It is an acoustic communication and navigation network.

MULTI-LAYER ISSUES IN MARINE ENVIRONMENT COMMUNICATION

The underwater communication between source node and sensor nodes takes place through multiple layers of protocol stack. The protocol stack for marine environment mainly comprise of five layers which are physical layer, data-link layer, network layer, transport layer and application layer as shown in Figure 3.

These layers are explained as follows:

1. **Physical Layer:** This layer comprise of the underlying communication channel e. g. in marine monitoring applications, acoustic communication channel performs much better as compared to radio waves as they are capable of navigating long enough in such type of environment. The major goal at this layer is to improve performance due to dispersed channels.
2. **Data Link Layer:** The main function at this layer is to provide error control functions for effective transmission of data. The characteristics of acoustic channel such as path loss and multipath fading lead to large bit errors which can be calculated with the help of ARQ(Automatic Repeat Request). This layer deals with various issues related to multiple path, load balancing etc. which greatly affect the performance of communication medium. Various MAC protocols are also implemented at this layer. These protocols tend to perform with large propagation delay also. For small range acoustic networks, there is no requirement for MAC protocols as sensor nodes are dispersed sparsely.
3. **Network Layer:** The main concern of this layer is data routing. In harsh marine environment, where bandwidth of the network is quite low, efficient routing of data is quite tedious task. In order to reduce propagation delay and possibility of errors, MAC protocols should be sensitive to battery power and time management.
4. **Transport Layer:** This layer is concerned with reliable connectivity with end to end communication, flow control and congestion control under sea. Existing network layer protocols mainly focus on energy efficiency by providing connectivity with small connection delays. In underwater networks, this is a tedious task as TCP performs very poor under sea.

Figure 3. Multi-layer stack for marine communication

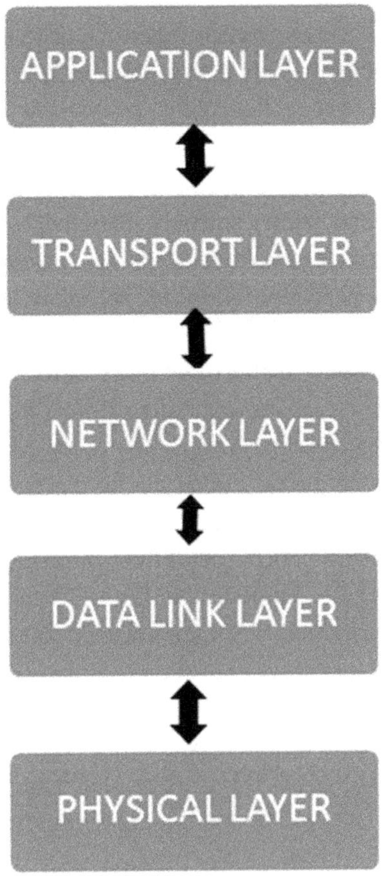

5. **Application Layer:** This layer deals with various network management protocols which utilize hardware and software information at lower layers to manage applications. Various functions of this layer includes finding out resource availability, synchronization, event advertisement etc.

Recently, many researchers are engaged in effective utilization of underwater wireless sensor networks by fixing the issues of the network deployment. Congestion control is one of the mechanisms which are required for providing efficient data communication. Congestion control mechanisms thwart the network from being choked with voluminous data as compared to the capacity of the network. The major objective of congestion control in sensor networks is to ensure successful delivery of data to the destination without compromising network performance. However, it is also equally crucial to assure and focus on reliability factor with achievement of energy efficiency.

STATE OF THE ART APPROACHES FOR CONGESTION CONTROL

Congestion control and flow control are two major functions of transport layer which helps in avoiding overloading of data transmitted through the network by constantly keeping an eye on the network devices. It is considered that Flow control operation relies mainly on windows based methods which itself depends upon the precise evaluation of Round Trip Time (RTT), but, in particular case of aquatic communication, it is quite challenging to set a timeout for windows based mechanism due to varying of RTT. In literature various methods are proposed for congestion control by focusing on different parameters. The standard TCP protocols available for congestion control cannot perform well in UWSN due to its unique aquatic environment. In UWSN, network congestion and high bit error rate are two major causes of packet drop (Wahid et al., 2010). Due to congestion, more packets cannot be injected into the network which ultimately reduces the transmission rate. Furthermore bad channel quality can also adversely affect the transmission rate. But TCP protocols consider that congestion is the only reason for loss of packets and reduce transmission rate whenever packet loss occurs. However, several TCP solutions are proposed in literature which distinguishes between packet losses due to congestion or channel link errors. Despite of their improved performance as compared to conventional TCP protocols (Chiang et al., 2005), they are not suitable for aquatic communication because of their throughput reduction, energy overheads and scalability issue with long distance between nodes. In the following paragraphs, it is defined that how congestion control algorithms in Terrestrial WSN are not able to be used in UWSN.

1. **Congestion Control in Terrestrial WSN**: There are a plenty of congestion control techniques specifically designed for terrestrial sensor networks. An Event-to-sink Reliable Transport (ESRT) protocol (Sankarasubramaniam et al., 2003) is defined in which mainly focuses on controlling congestion in the network. The results prove that the algorithm is quite reliable and energy efficient.

The performance evaluation of an algorithm CODA is presented in (Wan et al., 2003). It is an appropriate algorithm for congestion identification and control. In this, congestion is identified by using two parameters which are length of queue and channel utilization factor in-between nodes. Once congestion is identified at any intermediate node, backpressure warning messages are forwarded to the neighboring nodes to instruct to reduce the sending rate. Also in case of crossing the boundary of threshold, the source node maintains a feedback of acknowledgement to cope with data rate. If congestion still persists, the source node itself cut down its rate. However, in particular case of UWSN, the congestion identification based upon acknowledgement is not consistent as variable delays can result in poor system response. Also in multisource system, if congestion starts in the network, Acknowledgements are lost and then the source is required to decrease its rate. Rather, sink node can also take decision of not forwarding ACKs according to conditions of the network. However, in UWSN, due to high and random delays, the uniform delivery of Acknowledgements between the source and the sink act badly which ultimately increases the system respond time considerably.

Another resource oriented protocol TARA (Kang et al., 2007) provides an efficient approach to control congestion. During the interval of congestion, to increase capacity of the network, more nodes are activated using resource control. Hence new paths are created to redirect the traffic of packets. However, these deviations are quite time-taking, energy-consuming and even sometimes non feasible. That is why this approach does not suit well in UWSNs.

Furthermore, in congestion control approaches presented in (Pant et al., 2017; Shams et al., 2018), congestion occurrence due to channel interference is not considered which is a critical issue in underwater acoustic network due to high rate of packet losses.

Another congestion control approach based on availability of queue is proposed in (Zawodniok et al., 2007). In this congestion is reduced by diminishing the flow of the traffic according to the availability of queue and channel capacity. In this approach, a desired data rate is maintained by varying the backward rate according to the network statistics e.g. different number of neighbors and channel loss etc. However, this algorithm does not suit to UWSNs because of large propagation delays.

For wireless home automation networks, an adaptive congestion control algorithm is described in (Lee et al., 2010). In this, on the detection of congestion, the receiving speed of destination is improved as well as the transmission speed of the source node is reduced. Thus the duty-cycle is maintained to get rid of congestion. To utilize energy efficiently, each sensor node occasionally shifts from a dormant state to an active state and vice versa. Such type of methods can be adopted in UWSNs for long-lasting time-independent underwater observational applications but it performs poor in temporary time-dependent underwater research applications e.g. hurricane disaster recovery, underwater natural resource discovery etc (Srinivas et al., June 2019).

Furthermore, the congestion control approaches proposed in; (Xie et al., 2007; Ayaz et al., 2010; Shams et al., 2018) do not take into consideration the occurrence of congestion caused by shadow underwater regions, which results in a large number of packet losses in UWSN.

2. **Congestion control techniques in Underwater WSN:** Due to harsh underwater environment, the congestion control techniques described above cannot be adopted in UWSN as the communication medium is acoustic channel. In UWSN, a number of nodes relay data towards multiple sink nodes. In these types of networks, data is collected based on the events, which results in frequent increment of the load that the nodes can handle (Goyal et al., 2019). Thus a congestion scenario is created in the network at the sink or in the vicinity of the sink. Due to congestion, the quality of service of the network is also affected. Hence, it is identified that in UWSN, the following are the standard causes of congestion:

    a. Lack of synchronization
    b. Buffer overflow
    c. High bit error rate
    d. Multiple sources to a common sink problem

During last decade, researchers are finding out flexible and sophisticated solutions to cope up with the above challenges for controlling congestion in UWSNs. A detailed overview of the research is presented below in Table 2.

The effective utilization of congestion control methods results in better performance of the UWSN, where bandwidth is very low. These methods mainly focus on reducing propagation delay and energy consumption as described above.

Table 2. Congestion control techniques in UWSN

Paper References	Methodology used	Type	Key Parameters	Implementation Results
(Jin et al., 2019)	Reinforcement learning	Congestion avoidance routing protocol	Energy usage and end-to-end delay	Outperforms in terms of balancing of load and reduces end to end delay
(Ali et al., 2018)	Utilizing relay sensors to incorporate opportunistic routing to thwart network from congestion.	Congestion control routing protocol	Network stability, Node existence, PDR(Packet Delivery Ratio), PRR(Packet Received Ratio)	Energy proficiency, high throughput and network life time
(Dong et al., 2016)	A mixture of various solutions including center manifold theorem and normal form theory	Congestion control	Time delay	Network stability is achieved
(Goyal et al., 2016)	An efficient congestion detection protocol from intra and inter cluster in UWSN	Congestion Detection	Load balancing	Improved network performance
(Cai et al., 2013)	A reliable transport protocol based on diverse paths and network coding techniques	Congestion Avoidance	Data reliability and energy constrained	Higher throughput and data delivery rate
(Domingo et al., 2013)	A biologically motivated algorithm for congestion control in UWSN	Congestion Detection as well as control method	Throughput and packet losses	Achieves a stable state by differentiating loss of packets due to congestion or channel link faults
(Ayaz et al., 2012)	Effective utilization of 2-hop Acknowledgement model considering resource scarcity	Congestion Avoidance	Bit error rate, Reduced packet size, Data reliability	Increased data transfer rate, improved throughput
(Shin et al., 2009)	forward error correction (FEC) mechanism, so as to provide reliable and fast delivery of packets	Congestion avoidance	link quality toward neighbors, High bit error rate	guarantee the reliable and fast message delivery
(Liu et al., 2008)	A hybrid approach of forward error correction and binary BCH code to generate an energy efficient protocol	Congestion avoidance	High bit error rate, low propagation delay	Provides a better probability and energy tradeoff for both single- and multi-hop communications
(Xie et al., 2007)	Random forward error correction methodology to generate reliability of the nodes for successful packet reception	Congestion avoidance	Transfer latency	minimizing the number of transmissions
(Chiang et al., 2005)	A hybrid approach of reduction of congestion with power control technique	Congestion control	Non-linearly constrained utility maximization	Achieves high throughput and energy efficiency in the network

SIGNIFICANCE OF REAL TIME DATA COLLECTION IN MARINE APPLICATIONS

1. **Guided Navigation:** In oceans, various types of detectable or hidden obstacles occur. A guided navigation helps in identifying these types of obstructions such as sea shoals and moorage etc. An assistance of navigation can act as life saving opportunity if the real-time data can be communicated well in advance. For incidence, history witnessed that the tragic destruction of Titanic could be avoided by the prior identification of location of iceberg.
2. **Catastrophe Avoidance:** In general, the natural catastrophes are irrevocable. The impact of underwater disasters is quite horrible as compared to other natural ones. The tectonic movements under sea can cause calamities under water such as tsunami, volcanic eruptions, underwater earthquakes called as seaquakes etc. The detailed survey of underwater natural disasters monitoring applications comprising volcano and sea-quakes is presented in (Lloret et al., 2013). The real-time data regarding their movements would be helpful in avoiding the evil circumstances of these predicaments. The potential of UWSN for congestion free communication of real time data makes them suitable for such kind of life saving applications. A robust architecture for identifying tsunamis is proposed by (Casey et al., 2008).
3. **Defence Scenarios:** The submarines take a major role in strike warfare under water. UWSN helps in prevention of ally submarines from entering the unauthorized region. The real time data related to ally submarines can be used by defence force to plan a strategy to take defensive measures and to route their own submarines on safe track.

Figure 4. Sense-ational Submarine

For incidence, the submarine shown in figure 4 is equipped with fish inspired sensors. These sensors have the capability of taking the 3-D view of their surrounding objects and the real time data related to the environment can be shared with the defence personnel to choose the out of harm's way.

4. **Surveillance:** The use of UWSNs for military intelligence system helps in recognizing the unauthorized access of any entity in the network. The importance of real time data related to intrusion activity is manifold. An efficient mechanism for underwater surveillance is suggested in (Cayirci et al., 2006). The system comprises of many layers of sensors at different depths starting from surface to different regions where coverage is maximum.
5. **Oil Leakage Monitoring:** The leakage of oil in sea has an adverse effect on the health of marine species. The UWSN has a significant role in monitoring the thickness of oil spills, which helps in the initiation of quick cleaning process. The real time data collection helps in avoiding such kind of risk to marine life.

CONCLUSION

In UWSN, due to acoustic channel communication, the deployment of nodes is quite complex. Moreover the dynamic topology of the network creates a big challenge in managing the network. In these types of networks, congestion in the network creates a blunder of problems in achieving the mission i.e. generating the sensed data from the nodes at sea shore. Hence this issue has become a center of attraction for many researcher's community. However, the effective implementation of UWSN provides a bundle of applications in the harsh environment of sea. Hence this paper describes the significance of real time data for monitoring applications in marine environment, by focusing on the congestion related concerns.

REFERENCES

Akyildiz, I. F., Pompili, D., & Melodia, T. (2004). Challenges for efficient communication in underwater acoustic sensor networks. *ACM Sigbed Review*, *1*(2), 3–8. doi:10.1145/1121776.1121779

Akyildiz, I. F., Pompili, D., & Melodia, T. (2005). Underwater acoustic sensor networks: Research challenges. *Ad Hoc Networks*, *3*(3), 257–279. doi:10.1016/j.adhoc.2005.01.004

Akyildiz, I. F., Pompili, D., & Melodia, T. (2006, September). State-of-the-art in protocol research for underwater acoustic sensor networks. In *Proceedings of the 1st ACM international workshop on Underwater networks* (pp. 7-16). ACM. 10.1145/1161039.1161043

Ali, M. T., Rahim, S. S., Jan, M. A., Ishtiaq, A., Ahmed, S., Ahmad, M., ... Khan, M. A. (2018). Dist-Coop: Distributed cooperative transmission in UWSNs using optimization congestion control and opportunistic routing. *Int. J. Adv. Comput. Sci. Appl*, *9*(6), 356–368. doi:10.14569/IJACSA.2018.090649

Ayaz, M., Abdullah, A., & Faye, I. (2010, November). Hop-by-hop reliable data deliveries for underwater wireless sensor networks. In *2010 International Conference on Broadband, Wireless Computing, Communication and Applications* (pp. 363-368). IEEE. 10.1109/BWCCA.2010.97

Ayaz, M., Jung, L. T., Abdullah, A., & Ahmad, I. (2012). Reliable data deliveries using packet optimization in multi-hop underwater sensor networks. *Journal of King Saud University-Computer and Information Sciences*, *24*(1), 41–48. doi:10.1016/j.jksuci.2011.09.001

Cai, S., Gao, Z., Yang, D., & Yao, N. (2013). A network coding based protocol for reliable data transfer in underwater acoustic sensor. *Ad Hoc Networks*, *11*(5), 1603–1609. doi:10.1016/j.adhoc.2013.02.001

Casey, K., Lim, A., & Dozier, G. (2008). A sensor network architecture for tsunami detection and response. *International Journal of Distributed Sensor Networks*, *4*(1), 27–42. doi:10.1080/15501320701774675

Cayirci, E., Tezcan, H., Dogan, Y., & Coskun, V. (2006). Wireless sensor networks for underwater survelliance systems. *Ad Hoc Networks*, *4*(4), 431–446. doi:10.1016/j.adhoc.2004.10.008

Chiang, M. (2005). Balancing transport and physical layers in wireless multihop networks: Jointly optimal congestion control and power control. *IEEE Journal on Selected Areas in Communications*, *23*(1), 104–116. doi:10.1109/JSAC.2004.837347

Domingo, M. C. (2013). Marine communities based congestion control in underwater wireless sensor networks. *Information Sciences*, *228*, 203–221. doi:10.1016/j.ins.2012.11.011

Dong, T., Hu, W., & Liao, X. (2016). Dynamics of the congestion control model in underwater wireless sensor networks with time delay. *Chaos, Solitons, and Fractals*, *92*, 130–136. doi:10.1016/j.chaos.2016.09.019

Goyal, N., Dave, M., & Verma, A. K. (2016, December). Congestion control and load balancing for cluster based underwater wireless sensor networks. In *2016 Fourth International Conference on Parallel, Distributed and Grid Computing (PDGC)* (pp. 462-467). IEEE. 10.1109/PDGC.2016.7913240

Goyal, N., Dave, M., & Verma, A. K. (2019). Protocol Stack of Underwater Wireless Sensor Network: Classical Approaches and New Trends. *Wireless Personal Communications*, *104*(3), 995–1022. doi:10.100711277-018-6064-z

Goyal, N., Dave, M., & Verma, A. K. (2020). SAPDA: Secure Authentication with Protected Data Aggregation Scheme for Improving QoS in Scalable and Survivable UWSNs. *Wireless Personal Communications*, *113*(1), 1–15. Advance online publication. doi:10.100711277-020-07175-8

Jin, Z., Zhao, Q., & Su, Y. (2019). RCAR: A reinforcement-learning-based routing protocol for congestion-avoided underwater acoustic sensor networks. *IEEE Sensors Journal*, *19*(22), 10881–10891. doi:10.1109/JSEN.2019.2932126

Kang, J., Zhang, Y., & Nath, B. (2007). TARA: Topology-aware resource adaptation to alleviate congestion in sensor networks. *IEEE Transactions on Parallel and Distributed Systems*, *18*(7), 919–931. doi:10.1109/TPDS.2007.1030

Kumar, M., & Goyal, N. (2014). Reviewing underwater acoustic wireless sensing networks. *International Journal of Computer Science and Technology*, *5*(2), 95–98.

Lee, D., & Chung, K. (2010). Adaptive duty-cycle based congestion control for home automation networks. *IEEE Transactions on Consumer Electronics*, *56*(1), 42–47. doi:10.1109/TCE.2010.5439124

Liu, B., Garcin, F., Ren, F., & Lin, C. (2008, June). A study of forward error correction schemes for reliable transport in underwater sensor networks. In *2008 5th Annual IEEE Communications Society Conference on Sensor, Mesh and Ad Hoc Communications and Networks* (pp. 197-205). IEEE. 10.1109/SAHCN.2008.33

Lloret, J. (2013). *Underwater sensor nodes and networks*. Academic Press.

Pant, N. (2017). A comparative study of congestion control in wireless sensor networks using efficient resource management. In *Proc. the 2nd Int. Conf. Advancement in Engineering, Applied Science and Management (ICAEASM)* (pp. 320-325). Academic Press.

Sankarasubramaniam, Y., Akan, Ö. B., & Akyildiz, I. F. (2003, June). ESRT: event-to-sink reliable transport in wireless sensor networks. In *Proceedings of the 4th ACM international symposium on Mobile ad hoc networking & computing* (pp. 177-188). ACM. 10.1145/778415.778437

Shams Shamsabad Farahani, S. (2018). Congestion Control Approaches Applied to Wireless Sensor Networks: A Survey. *Journal of Electrical and Computer Engineering Innovations*, *6*(2), 125–144.

Shams Shamsabad Farahani, S. (2018). Congestion Control Approaches Applied to Wireless Sensor Networks: A Survey. *Journal of Electrical and Computer Engineering Innovations*, *6*(2), 125–144.

Shin, D., & Kim, D. (2009, December). FRT: Fast and reliable transport protocol for underwater wireless sensor networks. In *2009 IEEE Asia-Pacific Services Computing Conference (APSCC)* (pp. 402-407). IEEE. 10.1109/APSCC.2009.5394096

Srinivas, K., & Dua, M. (2019, June). Early Discovery of Disaster Events from Sensor Data Using Fog Computing. In *International Conference on Intelligent Computing, Information and Control Systems* (pp. 114-121). Springer.

Srinivas, K., & Dua, M. (2019, August). Fog Computing and Deep CNN Based Efficient Approach to Early Forest Fire Detection with Unmanned Aerial Vehicles. In *International Conference on Inventive Computation Technologies* (pp. 646-652). Springer.

Wahid, A., & Dongkyun, K. (2010). Analyzing routing protocols for underwater wireless sensor networks. *International Journal of Communication Networks and Information Security*, *2*(3), 253.

Wan, C. Y., Eisenman, S. B., & Campbell, A. T. (2003, November). CODA: congestion detection and avoidance in sensor networks. In *Proceedings of the 1st international conference on Embedded networked sensor systems* (pp. 266-279). ACM. Retrieved from https://www.defencetalk.com/ntus-sense-ational-invention-helps-underwater-vessels-navigate-with-ease-45958/

Xie, P., & Cui, J. H. (2007, August). An FEC-based reliable data transport protocol for underwater sensor networks. In *2007 16th International Conference on Computer Communications and Networks* (pp. 747-753). IEEE.

Xie, P., Zhou, Z., Peng, Z., Cui, J. H., & Shi, Z. (2010). SDRT: A reliable data transport protocol for underwater sensor networks. *Ad Hoc Networks*, *8*(7), 708–722. doi:10.1016/j.adhoc.2010.02.003

Yan, H., Shi, Z. J., & Cui, J. H. (2008, May). DBR: depth-based routing for underwater sensor networks. In *International conference on research in networking* (pp. 72-86). Springer. 10.1007/978-3-540-79549-0_7

Zawodniok, M., & Jagannathan, S. (2007). Predictive congestion control protocol for wireless sensor networks. *IEEE Transactions on Wireless Communications*, *6*(11), 3955–3963. doi:10.1109/TWC.2007.051035

Chapter 17
Demur and Routing Protocols With application in Underwater Wireless Sensor Networks for Smart City

Arun Kumar
Maharishi Markandeshwar University, Mullana, India

Sharad Sharma
https://orcid.org/0000-0001-6667-8170
Maharishi Markandeshwar University, Mullana, India

ABSTRACT

As everyone knows the internet of underwater things (IoUT) is an innovative discussion of internet of things (IoT). To help the idea of IoUT, underwater wireless sensor networks (UWSNs) have risen as a promising framework. IoUT is characterized as the system of brilliant consistent submerged items. IoUT is required to empower different functional applications, for example, natural observing, submerged investigation, and fiasco counteractive action. IoUT is viewed as one of the latentinventions toward creating savvy urban communities. The internet of underwater things (IoUT) is characterized as an overall system of savvy interconnected submerged items that empower to screen tremendous unexplored water zones. The motivation behind this chapter is to investigate how to profit by the IoUT to gain from adventure and safeguard the normal submerged assets, and the authors bring up the contrasts among UWSNs and customary territorial wireless sensor networks (TWSNs), and these distinctions are the primary difficulties for IoUT with diverse protocol.

DOI: 10.4018/978-1-7998-3640-7.ch017

INTRODUCTION

Underwater Wireless Sensor Network (UWSN) indicates an incredible potential for the future yet, in addition, presents new difficulties for IoUT. Furthermore, the steering convention will assume a noteworthy job in dealing with how to advance information among the generous number of "things". In this section, an inside and out perspective on the IoUT is given. Approximately an idea as 71% of the Earth's surface is secured by sea, a persistent waterway that is generally partitioned into a few head seas and littler oceans. Sea temperatures decide atmosphere and wind designs that influence life ashore (Goyal et al., 2016; 2017a; 2017b; 2017c). Freshwater in lakes and waterways covers less than 1%. It's sullying genuinely harm biological systems. The IoUT empowers to screen tremendous unexplored water territories. The motivation behind this chapter is to break down how to profit by the IoUT to gain from, adventure and safeguard the common submerged assets. Logical, modern, military and home security applications are required Be that as it may; the Internet of Underwater Things stays in its early stages (Kumar et al., 2014; Goyal et al., 2018a; 2018b). Undertakings like Sunrise and Nu Monet have made some amazing progress yet there are still some huge difficulties to survive. Furthermore, it will require additional time and considerably increasingly diligent work to completely build up this innovation. Educator Basagni clarified, "The fundamental specialized difficulties for a far-reaching dissemination/sending of advancements for the Internet of Underwater Things stem essentially by the very idea of the submerged condition. This is especially unfriendly to the standard thing, earthbound types of imparting. Particularly in the remote space" For example, radio correspondence can't go for in excess of a couple of meters submerged. Furthermore, dim waters make optical correspondence – normally used to get high information rates – inconceivable. Another NSF-subsidized venture called SEANet was as of late begun at North-eastern University. The primary center is to fabricate submerged acoustic modems. They will ideally differentiate the difficulties of moderate information rates average of current innovation. Educator Basagni stated, "For this reason, we are investigating new, unexplored ways for acoustic correspondence as shown in Fig. 1. We are depending on the properties of certain materials, and so forth."

Figure 1. The acoustic modems will pass on information and data (Nasir et al., 2019)

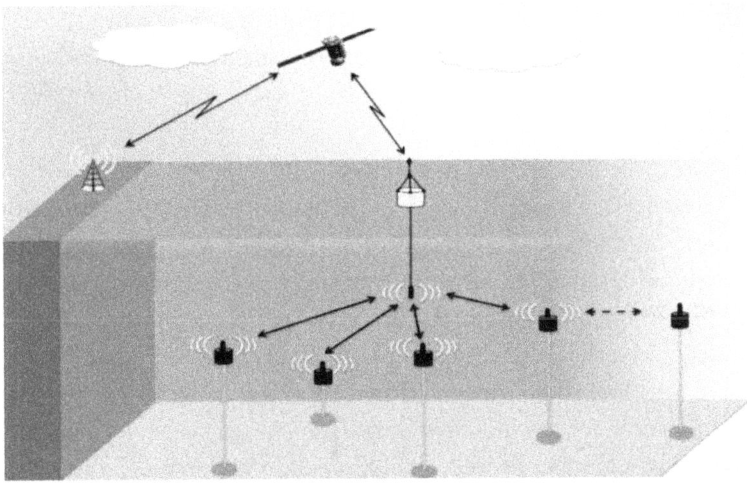

"Answers for the Internet of Underwater Things as real frameworks will bear some significance with an unmistakable fragment of the purchaser advertise from the start, he accepts. This portion will incorporate Defence Departments. For example, beachfront security, harbour observation. Also, the angling business for application identified with fisheries and aquaculture."

BACKGROUND

In this segment, we present three best in the class Internet of Underwater Things -related works. These works can give additional data of Internet of Underwater Things, and bid the latent research bearings in the close to highlight in (Xu et al., 2016; Goyal et al., 2019a; 2019b), the examination exertion takes concentrated on the advance MAC-layer bundle booking for the Internet of Underwater Things. Bearing in mind the plan difficulties of MAC conventions, the creators gave better Energy-Aware booking spatial-Temporal reuse convention for the Internet of Underwater Things. The creators referenced the plan difficulties of the Internet of Underwater Things MAC conventions, together with (i) tiny sensor organize era; (ii) smart acoustic channel usage; and (iii) parcel misfortune issues. As we know EAST uses the data of information cradle and traffic flow burden for imperativeness modes for submerged sensor centres and appropriately realizes both high essentialness viability and extraordinary channel use. As discussed In (Zhou et al., 2016), the makers focus on the framework layer controlling show for IoT Pondering their challenges in information communication. E-CARP abuses the exciting technique to pass on groups ricochet by-choice. Precisely, E-CARP improves exchange centre point decision methodology and lessens control packs, and as such diminishes the correspondence cost and imperativeness usage. Consequently, this protocol can spare essentialness for the Internet of Underwater Things In (Berlian et al., 2016), the examination focused on the arrangement and use of the sharp city in Underwater. The makers planned the super Smart Environment Monitoring and Analytics in Real-time framework (SEMAR) and it is a blend of water excellence checking structure, coral reef watching structure, remote work masterminds system, and Big Data structure. SEMAR accumulates statistics by using the submerged sensors, remotely worked vehicles and submerged cameras. The gathered information would be conveyed to the observing focus. For information handling (in the observing focus), the creators call attention to that utilizing for the Hadoop Distributed File Systems (HDFS) can perform superior to utilizing SQL frameworks, particularly in the instances of preparing enormous/colossal measures of information.

IoT and UWSN

The awareness of the Internet of Things" is never again the matter of sci-fi yet a basic part of the reality of our steady actualities. Currently, there are in excess of 13 billion consistent computerized and electronic gadgets in activity all around, the likeness almost 2 gadgets for each human on earth. Relations completed with IoT-empowered tools inspire the quick and effective discussion of data expected to help a wide scope of training and tasks (kumar et al., 2019). Thusly, the use of IoT innovations can prompt critical operational upgrades, for example, expanded productivity, better execution, and improved security. Many think about IoT as a solitary worldwide system (Rana et al., 2018; kumar et al., 2018). However, the IoT really includes a huge number of free yet corresponding systems in different help segments, for example, those supporting effective vitality conveyance and use (alleged shrewd matrices), home mechanization frameworks, vehicular traffic the board, and social insurance administrations as shown in Fig. 2. Surely, innovation research and counselling firm Beecham Plainly, utilizing the growth capability of the IoT is probably going to be a basic section of the methodology of novelty organizations in the years to come. The possibility of submerged correspondence is anything but another one (Arora et al., 2014).

Figure 2. Basic flow of information in UWSN (Nasir et al., 2019)

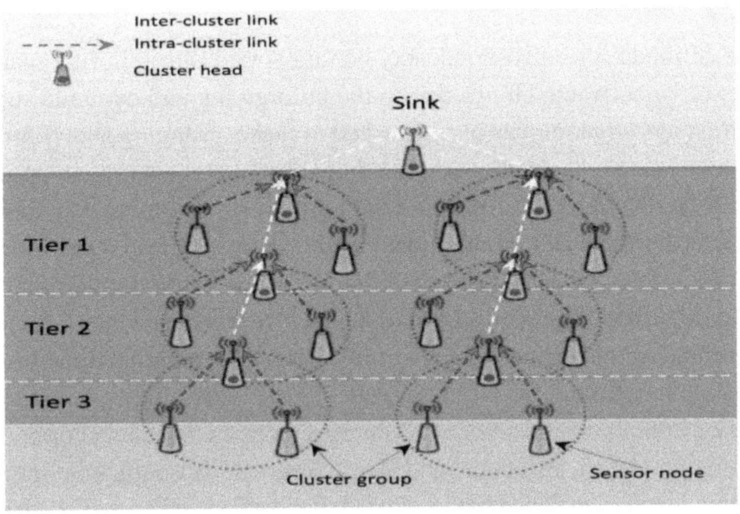

. In World War II, a submerged phone system was utilized by the US military to speak with submarines. Parts of submerged remote sensor systems incorporate sensors, vehicle and sink hubs that are set up to team up and convey for performing different errands. In UWSN the correspondence is done Utilizing acoustic waves or a mix of acoustic and radio waves. The acoustic correspondence has a few constraints, for example, low data transfer capacity and high mistake rates. While the focal points incorporate longer correspondence range and it is less influenced by the commotion. It is considered as a dependable innovation for sending submerged remote sensor systems. UWSN frameworks are envisioned as self-designed applications as shown in figure 3. For instance, cabled sea observatories are based on submarine links to convey a broad fiber-optic system of sensors, for example, cameras, wave sensors and seismometers covering miles of the seafloor. As of late sensors, innovation has advanced enough to empower the arrangement of UWSN effectively.

CHALLENGES FOR IoUT

The constrained step of natural investigation on IoUT Things is deferred because issues ascending out of the uniqueness of Underwater Wireless Sensor Networks. Standard troubles for the Internet of Underwater Things are the distinctions among Underwater Wireless Sensor Networks and TWSNs. In any case, Underwater Wireless Sensor Networks, for the most part, depend upon acoustic exchanges rather than radio correspondences. This is considering the way that the radio sign would be eaten up by water rapidly. In that capacity, how to ensure the limited the whole distance postponement would be an inconvenient issue. Third, the transmission degree of Underwater Wireless Sensor Networks could be on various events lengthier than TWSNs. This source more impedance and impacts during information transmission. The mechanical troubles for the Internet of Underwater Things can be widely assembled into correspondence, frameworks organization, and restriction.

Networking and Localization Challenges

Other than the sort of mode for correspondence, getting sorted out is in like manner a troublesome task for the Internet of Underwater Things due to the unforgiving and dynamic submerged condition. In spite of the way that investigation has been finished to make arranging shows for the acoustic-based Internet of Underwater Things, few works exist for frameworks organization in optical correspondence based IoUT, for instance, (Celik et al., 2018) and the positions in that moreover, payable to the absence of GPS banner underneath the water, confinement of the sharp articles on the Internet of Underwater Things is a huge need use of IoUT. For the case, in one of the continuous works, we have evaluated the zone of sensors with limited accessibility for optical correspondence based submerged sensor masterminds with multidimensional scaling (Saeed et al., 2018). An additional instance of restriction for optical correspondence based Internet of Underwater Things fuses the examination of abnormalities and stays geometry on containment exactness. Submerged optical channel shortcomings, for instance, absorption, scattering, unsettling influence, and air pockets can cause gigantic range estimation goofs (special cases). In this manner, in (Saeed et al., 2019), we have taken care of the issue of special cases to improve the general accuracy of the limitation. In like manner, each territory finding methodology depends upon the geometry of the stays, and if the geometry of the catches isn't upgraded, it can incite a gigantic limitation botch. Along these lines, we have improved the catch's geometry to all the almost certain measures the region of submerged sensors.

Communication Challenges

Internet of Underwater Things claims involve a mediocre to pass into submerged conditions plus from this condition to the outdoor domain. From this time forward, correspondence for the Internet of Underwater Things is of major importance, regardless, it speaks to a movement of troubles. For example, radio waves experience the evil impacts of high ingestion in view of their high frequencies and can go for two or three meters outwardly of the water. On the other hand, acoustic waves give a long transmission expand, yet as a result of the little quickness of sound waves, the information rate is small. As such, optical waves have starting late been familiar with giving a high information rate to submerged trades. Regardless, optical waves in submerged experience the evil impacts of the low broadcast run and require exact indicating and obtainment frameworks. (Lloret et al., 2019)

Internet of Underground Things (IoUGT)

Consistently expanding interest to normal assets for fresh advancements to improve the dissident investigation and to deliver more harvests. In this for the subsurface condition with horticultural terrains give different regular assets, for example, earth minerals, petroleum products, metal metals, groundwater, and nourishment as shown in Fig. 3. To screen and improve the making of these advantages

Figure 3. MI-based IoUGT for oil and gas monitoring (Nasir et al., 2019)

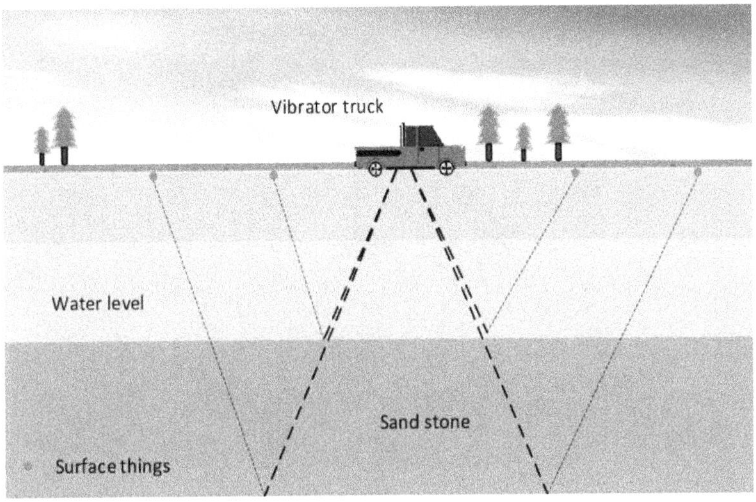

IoUGT is an engaging advancement that can give sharp gas and oil fields, insightful cultivating fields, and splendid seismic quality control (Saeed et al., 2019). In any case, the use of Internet of Underground Things is a troublesome task on account of the brutal underground multiplication condition, which needs very low power and minimal size underground little sensors, since a long time ago run correspondence development, successful frameworks organization game plans, and exact constraint procedures.

Internet of Space Things (IoST)

NASA needs to twitch the primary humanoid area on Mars by 2025.for overall system past earth to the individuals on Mars, IoST is an empowering development (Saeed et al., 2019). Thusly, IoST will power gives interplanetary trades. Likewise, uses, for instance, PC produced authenticity used at earth to walk around MARS with IoST. Thus, for creating an eagerness to grow the IoST by the space corporate that is for the most part in the improvement organized. Also, IoST moreover gives extended consideration and improve the identifying limits of present satellite frameworks. For example, Iridium correspondences have put 66 satellites (Sensor Pods) to offer accessibility to the end customers similarly as identifying the space and earth. So likewise, Kepler correspondences have starting late impelled KIPP satellite broadcasting in Ku-band for planetary missions.

NASA needs to twitch the principle human region on In the latest decade, researchers have shown different possible Internet of Underwater Things applications (Tan et al., 2015)Function divide into five sorts: (1) ecological observing (2) submerged examination, (3) catastrophe balancing activity, (4) armed, and others, as showed up in Fig. 4. A. Normal monitor largely generally used the Internet of Underwater Things use forms is eco-friendly checking (Vuran et al., 2018)counting water superiority watching, substance and natural tainting watching, warm pollution watching, weight and temperature checking, and fishes following. Furthermore, oil and gas pipelines watching can be practiced by using UWSNs. B. Web of Underwater Things is required to recognize flood and seismic tremor (submerged) and to offer applications to the officials and repugnance of such cataclysms (Singer et al., 2017)D. Military The

Figure 4. Architecture of IoST (Nasir et al., 2019)

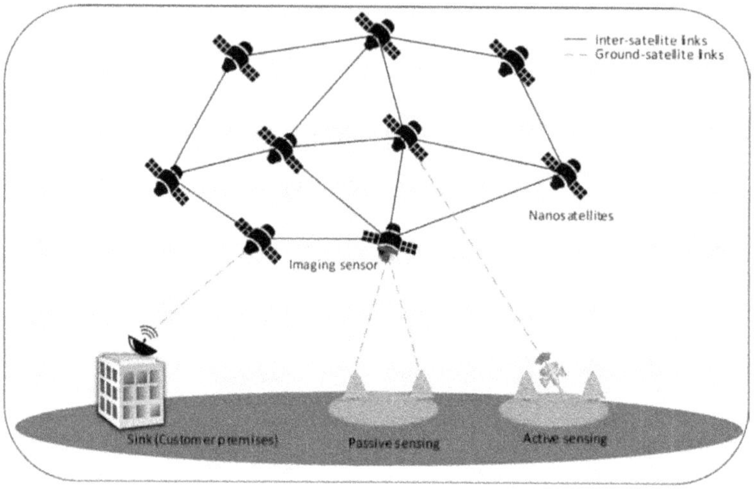

military as often as possible mirrors the limit of a country to secure itself against an ambush, including the kind of submerged attack. Web of Underwater Things is required for assurance determinations and can be useful to submarine revelation and submerged mine acknowledgment (Abrudanet al., 2016)E. Others with the movements in UWSNs, progressively more Internet of Underwater Things uses are getting the opportunity to charm, for instance, the games (Saeed et al., 2019) and course uses. See, for sample, submerged sensors used as region allusion centers, and along these lines can give swimmers, jumpers, transports, and submerged vehicles with significant information.

APPLICATIONS OF INTERNET OF UNDERWATER THINGS

Investigators have obtained able many real-world and potential Internet of Underwater Things applications as shown in Fig. 5. Five types are: (1) environmental monitoring; (2) Submerged Exploration; (3) Fiasco Prevention; (4) Armed improvement; and (5) others

Environmental Monitoring

for the best part generally utilized the Internet of Underwater Things use types is ecological observing with water superiority checking (Menon et al., 2012), substance and natural contamination observing (Khan et al., 2008), warm contamination observing, pressure checking, and temperature checking (Alippi et al., 2017). What's more, oil pipelines checking (Saeed et al., 2014) can likewise be accomplished by utilizing UWSNs. Natural checking applications/frameworks have turned out to be progressively well known and are currently in incredible interest for worldwide savvy urban areas. The creators of (Menon et al., 2012)give a pragmatic case of the water quality observing framework in India. In view of remote sensor systems, they propose a reasonable system design for a stream water excellence observing framework. In addition, this chapter displays the specialized plan of a tiny sensor core that can be

Figure 5. Applications of IoUT

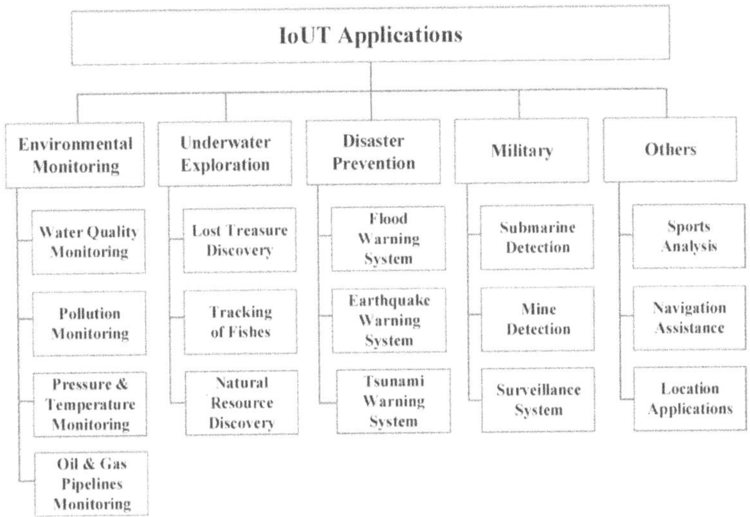

utilized for checking the pH of water. Note that the pH of water is one of the principal parameters that demonstrate the nature of water. The detected pH worth will be remotely communicated to the base position by utilizing Zigbee interchanges. Creators likewise referenced for the following period of the venture will incorporate water-conductivity sensors, broke up oxygen sensors, and temperature sensors (notwithstanding the pH sensors). In (Khan et al., 2008), concentrating on sea contamination aversion, the creators announced a specially appointed remote tiny sensor system to screen sea contamination. Contrasted and the customary long-run submerged correspondence, this work focuses on the short-extend multi-jump correspondence. Short-run correspondence infers that the sensor system can reuse a portion of the acoustic data transmission and furthermore maintain a strategic distance from a considerable lot of the difficulties of long-extend correspondence. All the more significantly, the short-go auditory modems are more affordable than long-ago modems. What's more, the creators propose a synchronization convention that can expand the QoS and viably upgrade the lifespan of the submerged sensor arrange. In (Alippi et al., 2017), the shot exhibits a checking framework conveyed in Queensland, Australia, for observing the submerged temperature and glow (splendor), i.e., the data important to infer the wellbeing status of the coralline hindrance. In view of remote sensor organizes, the creators present a certifiable down to earth structure. They planned and actualized all parts of the ecological observing framework, including the detecting movement, information stockpiling, nearby transmission, remote broadcast, what's more, representation. For the nearby broadcast between the sensors and sinks, the creators propose a power-mindful TDMA convention that can ensure the heartiness and versatility when the system topology changes. In (Saeed et al., 2014), the oil-and-gas related applications are featured. The creators first call attention to that the fast headway in remote correspondence advancements has made an opportunity to get better in every aspect of mechanical practices identified with oil and gas preparing. Starting here of view, the creators explore and propose an oil with gas pipeline observing arrangement. By utilizing remote sensor organizes, the proposed arrangement can give dependable checking of oil and gas pipelines. In particular, the proposed framework is equipped for announcing pipeline wellbeing related insights over huge regions. In the creators present various submerged observing applications. The creators characterize submerged checking as a system of sensors sent submerged to screen the submerged situations, qualities, possessions, or any objects of intrigue. The creator's group the submerged observing uses into three distinct classes and talk about every one of them. The three classes are (I) water quality observing; (ii) living space checking; and (iii) observing submerged investigations. Peruses inspired by these sorts of uses are alluded to for a progressive point by point depiction.

Submerged Exploration

The possibility of the IoUT can be applied to the submerged covered fortune revelation. For example, the disclosure of the Titanic in 1985 benefitted by the use of self-administering submerged vehicles (AUVs). By using UWSN headways, IoUT can moreover is applied to the accompanying of fishes. Also, submerged standard resource disclosure, for instance, raw materials (Srinivas et al., 2012), metals, corals, and coral reefs (Bainbridge et al., 2012), would all have the option to benefit by the structure of UWSNs moreover. For the next decade, Davis and Chang present a survey of UWSNs (Lloret et al., 2013). They give a huge outline of various UWSN models and exhibit how the plans support unmanned submerged examination. As referenced in this chapter, different productive covered fortune disclosures completed with the help of UWSNs. The impartial of (Srinivas et al., 2012) is to make impelled correspondence frameworks for capable constant assessment of gigantic abandoned oceans. To research the sea condition, the makers suggest a gigantic gauge structure that usages UWSNs.To explore the submerged assets, the planned sensor sections are presented with the video getting contraptions. Similarly, the planned structure can be used to determine and reveal the mineral properties submerged. The re-enactment results are engaging as the proposed system is valuable in tremendous scale perception and examination.

Fiasco Prevention

Adversity shirking uses the most vital IoUT applications. This is in light of the fact that these uses protect our lives. Water-based calamitous occasions are possibly unsafe. The Japanese nuclear setback, Fukushima Daiichi nuclear cataclysm, was begun essentially by the tsunami following the Tohoku seismic tremor (For example). To give uses for the neutralizing activity of such cataclysms, the Internet of Underwater Things is extensively expected to perceive flood (Preez et al., 2012), seismic tremor (Kumar et al., 2012), and wave (Casey et al., 2008) submerged, and to offer early notice organizations. In (Preez et al., 2012), Perez et al. present the arrangement of a useful consistent estimation structure for flood watch. They sent a genuine checking framework with a couple of nonstop estimation policy in Spain. The testing grades be evidence for that the planned checking structure is imperativeness beneficial and has an incredible correspondence capacity for whole deal amassing of water-level data in various regions. The makers in (Kumar et al., 2012) underline that the progression of the early notice age structure dependent on UWSNs would beyond question add to saving people. The counsel age structure protects when a destructive occasion, e.g., a tremor, occurs. One of the rule responsibilities of this chapter is to raise the physical layer challenges in working up a UWSN system for notice age. The makers assume that: (1) trustworthy correspondence; (2) low-control plan; and (3) capable source the board will remain the huge troubles for the UWSN-based reprobation age frameworks. In (Manjula et al., 2013), the makers talk about different approaches for wave area and propose a beneficial plan uses. The proposed designing uses seismic weight sensors to envision waves submerged and to propel/move the distinguished in sequence by means of the organized scattering coordinating the show. The makers moreover talk about and delineate the assessment and answer frameworks in this effort.

Armed Improvement

Armed every now and again reflects the limit of a country to shield itself against an ambush, including a submerged attack. Web of Underwater Things is necessary for defensive point and practical to submarine distinguishing proof (Manjula et al., 2013), submerged mine acknowledgment (Khaledi et al., 2013), and submerged observation structures (Cayirci et al., 2006). These uses have the mind-boggling possible for prospect sea powers. For the submarine area, the makers in (Manjula et al., 2013) look at the structure of the sensor association and propose an arrangement to improve the watching incorporation. By taking the water significance, communication range, and tightening into an explanation, the proposed arrangement uses Particle Swarm Optimization (PSO) to choose the spots of the sensor center points. The makers give both the numerical assessment and NS3-based amusements to affirm the display of the proposed arrangement with respect to the watching consideration and number of sensor center points necessary. In (Khaledi et al., 2013) analyze the arrangement of submerged extract area systems. Especially, they believe two sonar decisions and five towing-vehicle substitutes (i.e., one submerged vehicle elective, two surface vehicle decisions, and two airborne vehicle choices). PC model to re-enact the route toward recognizing submerged excavation and test every one of the structure alternatives. The results show that the submerged vehicle alternative uses a negligible proportion of essentialness. They in like manner take the security, speed, mileage, and prospect of the area into the record, and give the relating assessment to exhibit the best other option. In (Cayirci et al., 2006), the makers propose a novel UWSN structure for submerged perception scheme. In the planned plan, when tiny sensor center points are sent, the sensor cent er points are first set in quite a while; by then, the sensor centers are brought down to different profundities. The profundities are settled by the proposed arrangement for boosting the perception incorporation. Furthermore, every center is outfitted with different of littler scale sensors. Examining the data assembled from the littler scale sensors, the makers further present a data burrowing based arrangement for gathering the submarines, mines, and jumpers.

Others

with the developments in UWSNs, a broadening numeral of Internet of Underwater Things applications are getting the chance to progress, for example, sports (Magalhaes et al., 2015), course (Waldmeyer et al., 2011), and repression (Guo et al., 2017) uses. Note that the imperative uses are especially trying yet extremely noteworthy. In particular, in reality, seeing the zone submerged is attempting to consider the way that the GPScannot work in submerged conditions. Envision, for instance, submerged sensors can be utilized as an area reference focuses, and in this manner can give swimmers, jumpers, conveys, and submerged vehicles with noteworthy zone data. The creators in (Magalhaes et al., 2015) give a beneficial audit, concentrating on the biomechanical assessment of swimming with utilizing the inertial sensors submerged. As referenced in this paper, beginning late, the swimming contenders, mentors, and sports researchers have been endeavouring to accomplish a "restricted amount of a second" improvement. The utilization of sensors acknowledges a basic movement in supporting execution updates from a biomechanical perspective. This paper presents distinctive existing procedures those utilization wearable inertial tiny sensors to evaluate the biomechanics of swimming execution. The outcomes display that the submerged inertial sensors are reasonable and solid devices for swimming biomechanical evaluations As featured in (Waldmeyer et al., 2011), for a gathering of utilizations, the recognized information must be made an understanding of definitively when referenced to the district of the sensors; for instance, the course is such an application. The producers propose a multi-sort out AUV-helped hindrance conspire for UWSNs. They drove re-institutions to assess the showcase of the proposed game plan to the degree the obstacle thought, accuracy, and correspondence costs. The outcomes show that fantastic execution can be developed by appropriately picking the correspondence to grow. In (Guo et al., 2017), the producers recommend an Anchor-Free Localization Algorithm, called AFLA, for submerged conditions (Tayal et al., 2017). A large portion of the standard submerged control figuring envisions that momentous submerged focuses (with exceptional mechanical assembly) or AUVs can be utilized as "get focus focuses" to help see the region submerged. The planned AFLA needn't sit around with data from the extraordinary snare focus focuses. Or on the other hand, possibly, AFLA utilizes the adjoining relationship of customary sensor focus focuses to see the region submerged (Partap et al., 2017) the re-enactment results show that AFLA can be utilized with the static and dynamic UWSNs.

ROUTING PROTOCOLS

A steering convention chooses the ways for information bundles to pursue (from the sensor hubs to the sink hubs in UWSNs) and assume a fundamental job in accomplishing alluring Quality of Service (QoS) for the Internet of Underwater Things. In this way, it is important to inspect different sorts of directing conventions as there have been increasingly steering conventions that can possibly be utilized on the Internet of Underwater Things. We order 21 forthcoming steering conventions for the Internet of Underwater Things into five unmistakable sorts: (1) table-driven, (2) On-Demand Based, (3) Geographic Based, (4) Artful Based, and (5) multicast-tree-based as appeared in Fig.2. Around 33% of the potential steering conventions are outstanding in IoT or in UWSNs.

Table-Driven

To limit the constant inactivity, the table-driven directing convention keeps up records of goals and their courses by occasionally appropriating steering data/tables entirely through the correspondence organize. Along these lines, forward-thinking steering data from each hub to each other hub can be kept up. When there is information prepared to convey, the sensor hubs can look into the directing tables to advance information through the proper courses.

On-Demand Based

On-request directing convention finds a course on request. In particular, a course revelation procedure is started by a sensor hub just when it needs to transmit information to a goal hub. In the first place, the course revelation procedure would flood the system with a course demand parcel. At that point, if the goal hub gets the solicitation bundle, the goal hub will send back the coordinating course answer bundle. Likewise, a sensor hub can discover a course to its goal and afterward transmit information bundles through this course.

Geographic Based

Geographic steering has all the earmarks of being a more sensible answer for the Internet of Underwater Things than table-driven and on-request directing since geographic steering conventions can significantly lessen the telecom overhead by abusing the geographic area data of the sensor hubs, together with the data on source hubs, neighboring hubs, and goal hubs. A hub can choose its next bounce as per the area of its neighboring node(s) and of the goal node(s) and forward information bundles morally justified (or a reasonable) bearing.

Artful Based

Artful directing adventures the telecom idea of remote correspondences. At the point when a sensor hub conveys information parcels, all its neighboring hubs get the bundles; at that point, every one of the recipients chooses whether to advance the parcels as indicated by a likelihood model or a foreordained condition. For instance, the condition could be the separation between the recipient and the sink. In the event that the condition is satisfied (e.g., the collector is nearer to the sink than the sender), the beneficiary would choose to re-communicate the bundles. Along these lines, on the grounds that the recipients will decide if to advance parcels without anyone else when a sender has bundles to transmit, it can convey the parcels legitimately.

Multicast-Tree-Based

Multicast-tree-based directing is a cutting edge elective strategy for steering in submerged situations. Unmistakable from past directing conventions, this class of steering convention would manufacture a tree topology for course advancement. To accomplish an exact objective (e.g., defer minimization or vitality minimization), these conventions develop the relating explicit tree (e.g., the briefest way tree or least traversing tree). Note that, rather than flooding bundles all through the system, multicast tree-based directing conventions are increasingly similar to multicasting parcels on the particular tree topology to accomplish the particular objective. With the developments in UWSNs, an expanding number of Internet of Underwater

A BATTERY-FREE SENSOR FOR UNDERWATER OUT EXPLORATION

A sans battery sensor for submerged examination looks at the inconceivably uncharted oceans cover nearly all our planet, authorities intend to make a submerged arrangement of organized sensors that drive information to the surface — a submerged "web of things." other than how to supply the steady ability to achieve of sensors proposed to keep on for extended ranges in the ocean's significant? MIT investigators reply: a sensor battery submerged correspondence structure that brings into play near-zero ability to broadcast sensor information. The structure may be used to screen sea hotness to think about ecological modify and pathway marine life over broad stretches — and still model waters on faraway planets. The structure uses two input marvels. One, call the "piezoelectric effect," happens at what time sensations in explicit resources make an electrical indict. The other is "backscatter," a correspondence methodology routinely worn for RFID names, that broadcast information by reflecting balanced remote banner off a tag and back to a peruse. In the examiners' structure, a spreader transmits acoustic impressions through the water toward a piezoelectric sensor that has taken care of data. Precisely when the wave hits the sensor, the material vibrates and stores the ensuing electrical charge. By at that point, the sensor utilizes the set away centrality to mirror a wave back to a beneficiary or it doesn't reflect one in any capacity whatsoever. Moving back and forth between appearances in this manner relates to the bits in the transmitted information: For a reflected wave, the gatherer disentangles a 1; for no reflected wave, the beneficiary translates a 0. "Right when you have an approach to managing transmitting 1s and 0s, you can send any data," says co-producer Fadel Adib, an accomplice instructor in the MIT Media Lab and the Department of Electrical Engineering and Computer Science and working up to a leader of the Signal Kinetics Research Group. "Essentially, we can converse with submerged sensors subject to the advancing toward sound flag whose essentialness we are gathering." The specialists indicated their Piezo-Acoustic Backscatter System in an MIT pool, utilizing it to collect water temperature and weight estimations. The framework had the decision to transmit 3 kilobits for each second of accurate information from two sensors at the same time a separation away of 10 meters among sensor and gatherer.

CONCLUSION

This chapter examines the Internet of Underwater Things, a new class of IoT. This chapter provides practical information regarding the Internet of Underwater Things (1) challenges, (2) routing protocols, as well as (3) the association connecting the challenges and routing protocols. Firstly, discuss the Internet of Underwater Things challenges into distinct perspectives and classify the prospective Internet of Underwater Things routing protocols into different categories. Then, we illustrate the common advantages and disadvantages shared by the routing protocols of the same categories and present the link between the Internet of Underwater Things challenges and routing protocols and they remain wide open for future investigation.

REFERENCES

Abrudan, T., Kypris, O., Trigoni, N., & Markham, A. (2016). Impact of rocks and minerals on underground magneto-inductive communication and localization. *IEEE Access: Practical Innovations, Open Solutions*, *4*, 3999–4010. doi:10.1109/ACCESS.2016.2597641

Alippi, C., Camplani, R., Galperti, C., & Roveri, M. (2011). A robust adaptive solar-powered WSN framework for aquatic environmental monitoring. *IEEE Sensors Journal, 11*(1), 45–55. doi:10.1109/JSEN.2010.2051539

Arora, A., Gupta, S., & Kumar, A. (2018). Survey of Machine Learning Methods, What's More, Their Sensor and IoT Applications for Future World. *AJCS Journal, 21*(2).

Bainbridge, S., Eggeling, D., & Page, G. (2016). Lessons from the field—Two years of deploying operational wireless sensor networks on the great barrier reef. *Sensors (Basel)*, 6842–6855. PMID:22163988

Berlian, M. H., & Sahputra, T. E. R. (2016). Implementation of smart environment monitoring and analytics in real-time system framework based on the internet of underwater things and big data. *Proceedings of the IEEE International Electronics Symposium (IES)*. 10.1109/ELECSYM.2016.7861040

Casey, K., Lim, A., & Dozier, G. (2008). Sensor network architecture for tsunami detection and response. *International Journal of Distributed Sensor Networks, 4*(1), 27–42. doi:10.1080/15501320701774675

Cayirci, E., Tezcan, H., Dogan, Y., & Coskun, V. (2006). Wireless sensor networks for underwater surveillance systems. *Ad Hoc Networks, 4*(4), 431–446. doi:10.1016/j.adhoc.2004.10.008

Celik, N., & Saeed, T. Y. (2018). Modeling and performance analysis of multihop underwater optical wireless sensor networks. *IEEE Wireless Communications and Networking Conference (WCNC)*, 1-6. 10.1109/WCNC.2018.8377388

Goyal, N., Dave, M., & Verma, A. K. (2016, December). Congestion control and load balancing for cluster-based underwater wireless sensor networks. In *2016 Fourth International Conference on Parallel, Distributed and Grid Computing (PDGC)* (pp. 462-467). IEEE. 10.1109/PDGC.2016.7913240

Goyal, N., Dave, M., & Verma, A. K. (2017a). Improved data aggregation for cluster-based underwater wireless sensor networks. *Proceedings of the National Academy of Sciences, India Section A: Physical Sciences, 87*(2), 235-245. 10.100740010-017-0344-y

Goyal, N., Dave, M., & Verma, A. K. (2017b). Data aggregation in an underwater wireless sensor network: Recent approaches and issues. *Journal of King Saud University-Computer and Information Sciences*.

Goyal, N., Dave, M., & Verma, A. K. (2017c). The trust model for cluster head validation in underwater wireless sensor networks. *Underwater Technology, 34*(3), 106–113. doi:10.3723/ut.34.107

Goyal, N., Dave, M., & Verma, A. K. (2018a). A novel fault detection and recovery technique for cluster-based underwater wireless sensor networks. *International Journal of Communication Systems, 31*(4), e3485. doi:10.1002/dac.3485

Goyal, N., Dave, M., & Verma, A. K. (2018b). Adaptive Error Control Technique for Cluster-Based Underwater Wireless Sensor Networks. In *International Conference on Wireless Intelligent and Distributed Environment for Communication* (pp. 269-280). Springer. 10.1007/978-3-319-75626-4_20

Goyal, N., Dave, M., & Verma, A. K. (2019a). Protocol Stack of Underwater Wireless Sensor Network: Classical Approaches and New Trends. *Wireless Personal Communications, 104*(3), 995–1022. doi:10.100711277-018-6064-z

Goyal, N., Sandhu, J. K., & Verma, L. (2019b). *Machine Learning based Data Agglomeration in Underwater Wireless Sensor Networks.* Academic Press.

Guo, Y., & Liu, Y. (2017). Localization for anchor-free underwater sensor networks. *Computers & Electrical Engineering, 39*(6), 1812–1821. doi:10.1016/j.compeleceng.2013.02.001

Khaledi, S., Mann, H., Perkovich, J., & Zayed, S. (2013). Design of an underwater mine detection system. *Proceedings of the IEEE Systems and Information Engineering Design Symposium (SIEDS).*

Khan, A., & Jenkins, L. (2008). Undersea wireless sensor network for ocean pollution prevention. *Proceedings of the IEEE International Conference on Communication Systems Software and Middleware (COMSWARE).* 10.1109/COMSWA.2008.4554369

Kumar, A., & Sharma, S. (2019). Enhanced Energy-efficient heterogeneous Routing Protocols in WSNs for IoT Application. *IJEAT, 9*(1).

Kumar, K., & Kumar, A. (2018). Wireless Sensor Networks: A review on "Challenges and Opportunities for the Future world-LTE". *AJCS, 1*(2).

Kumar, M., & Goyal, N. (2014). Reviewing underwater acoustic wireless sensing networks. *International Journal of Computer Science and Technology, 5*(2), 95–98.

Kumar, P., & Kumar, P., & Priyadarshini, S. P. (2012). Underwater acoustic sensor network for early warning generation. *Proceedings of the IEEE OCEANS.* 10.1109/OCEANS.2012.6405009

Lloret, J. (2013). Underwater sensor nodes and networks. *Sensors (Basel), 13*(9), 11782–11796. doi:10.3390130911782 PMID:24013489

Magalhaes, F. A., Vannozzi, G., Gatta, G., & Fantozzi, S. (2015). Wearable inertial sensors in swimming motion analysis:A systematic review. *Journal of Sports Sciences, 33*(7), 732–745. doi:10.1080/02640414.2014.962574 PMID:25356682

Manjula, R. B., & Manvi, S. S. (2013). Coverage optimization based sensor deployment by using PSO for anti-submarine detection in UWASNs. *Proceedings of the International Symposium on Ocean Electronics.* 10.1109/SYMPOL.2013.6701906

Menon, K. A. U., Divya, P., & Ramesh, M. V. (2012). Wireless sensor network for river water quality monitoring in India. *Proceedings of the IEEE International Conference on Computing, Communication and Networking Technologies (ICCCNT).* 10.1109/ICCCNT.2012.6512437

Perez, R. M., Pintado, J. G., & Gómez, A. S. (2012). A real-time measurement system for long-life flood monitoring and warning applications. *Sensors (Basel), 12*(4), 4213–4236. doi:10.3390120404213 PMID:22666028

Pratap, N., & Arya, S. (2015). A Review of Devnagari Character Recognition from Past to Future. *International Journal of Computer Science and Telecommunications*, *3*(6), 77–82.

Rana, A., & Salau, A. (2018). Recent Trends in IoT, its Requisition with IoT Built Engineering: A Review. Springer.

Saeed, H., Ali, S., Rashid, S., Qaisar, S., & Felemban, E. (2014). Reliable monitoring of oil and gas pipelines using wireless sensor network (WSN): REMONG. *Proceedings of the IEEE International Conference on System of Systems Engineering (SOSE)*. 10.1109/SYSOSE.2014.6892493

Saeed, N., & Alouini, M. (2019). Outlier detection and optimal anchor placement for 3-D underwater optical wireless sensor network localization. *IEEE Transactions on Communications*, *67*(1), 611–622. doi:10.1109/TCOMM.2018.2875083

Saeed, N., & Alouini, M. (2019). Performance analysis of connectivity and localization in multi-Hop underwater optical wireless sensor networks. *IEEE Transactions on Mobile Computing*, *18*(11), 2604–2615. doi:10.1109/TMC.2018.2878672

Saeed, N., & Alouini, M. (2019). On achievable accuracy of localization in magnetic induction-based internet of underground things for oil and gas reservoirs. CoRR, vol. abs/1803.02442

Saeed, N., & Alouini, M. (2019). Towards the internet of underground things: A systematic survey. CoRR, vol. abs/1902.03844

Saeed, N., & Alouini, M. (2019). *Towards the Internet of X-things: New Possibilities for Underwater.* Underground, and Outer Space Exploration.

Saeed, N., & Celik, A. (2018). Underwater optical wireless communications, networking, and localization: A survey. CoRR, vol. abs/1803.02442

Singer, A., Yang, S., & Oelze, M. (2017). Acoustic communications: Through soils, sands, water, and tissue. *The Journal of the Acoustical Society of America*, *141*(5), 3986–3987. doi:10.1121/1.4989117

Srinivas, S., Ranjitha, P., Ramya, R., & Narendra, K. G. (2012). Investigation of oceanic environment using large-scale UWSN and UANETs. *Proceedings of the IEEE International Conference onWireless Communications, Networking and Mobile Computing (WiCOM)*. 10.1109/WiCOM.2012.6478552

Tan, X., Sun, Z., & Akyildiz, I. F. (2015). Wireless underground sensor networks: MI-based communication systems for underground applications. *IEEE Antennas & Propagation Magazine*, *57*(4), 74–87. doi:10.1109/MAP.2015.2453917

Tayal, S., & Nandi, A. (2017). Effect of FIBL in-conjunction with channel parameters on analog and RF FOM of FinFET. *Superlattices and Microstructures*, *105*, 152–162. doi:10.1016/j.spmi.2017.03.018

Vuran, M., Salam, A., Wong, R., & Irmak, S. (2018). Internet of underground things in precision agriculture: Architecture and technology aspects. *Ad Hoc Networks*, *81*, 160–173. doi:10.1016/j.adhoc.2018.07.017

Waldmeyer, M., Tan, H. P., & Seah, W. K. G. (2011). Multi-stage AUV-aided localization for underwater wireless sensornetworks. *Proceedings of the IEEE Workshops of Advanced Information Networking and Applications (WAINA)*.

Xu. M., Liu, L. (2016). Sender-receiver role-based energy-aware scheduling for Internet of Underwater Things. *IEEE Trans*.

Zhou, Z., Yao, B., Xing, R., Shu, L., & Bu, S. (2016). E-CARP: An energy efficient routing protocol for UWSNs in the internet of underwater things. *IEEE Sensors Journal*, *16*(11), 4072–4082. doi:10.1109/JSEN.2015.2437904

Chapter 18
Issues in Underwater Cables Deploymentby Risk Analysis

Jitender Grover
École de technologie supérieure (ÉTS), Canada

ABSTRACT

Underwater cable deployment facilitates the coverage and speed of internet all over the world for various applications like international trade, various communication transfers, and other daily end user requirements. Also, critical network infrastructure below the water remains largely unexplored to end users because they are not directly related with its development. However, the risk of damage or destruction by unintentional underwater accident or intentional malicious threats leads to costly and communications disruptions. This chapter solves two primary goals. The first is to highlight the issues in underwater cable deployment that may be seen or unseen leads to various potential risks that could interrupt cable-supported services. The second goal is to explore the various possibilities for securing the organization to ensure business continuity.

INTRODUCTION

In around the world, it is normal that interest for "Under Unmanned Vehicles" units is to develop by 49 percent up to 2020 from 2016 (of which 73 percent are relied upon to be utilized for the military purposes, yet there is quick development in the gas and oil segment). Some portability and drive structures are taken from the normal world, and several look progressively like torpedoes. Energy units have expanded UUV go and in any event, going at a relaxed 2 bunches, most types can meander aimlessly a long way from their shoreline, effectively arriving at the 12-mile nautical point of confinement and past it (Goyal et al., 2018; Ellis and Mohan, 2019).

While trading firms and research teams have been successful in enhancing power and manoeuvrability, the rules and regulations of physics still constrain communications: high-speed wireless "underwater internet" of the kind which is now penetrating in the biosphere is still just theory. The data just doesn't spread simply through water without cables. Be that as it may, the innovation is growing quickly. Investigations in the North Sea and Baltic in 2006 demonstrated that submerged unmanned vehicle (UUV)

DOI: 10.4018/978-1-7998-3640-7.ch018

control sign will transmit at 100 bits for every second by means of audible channels. A study distributed by RAND in 2009 suggested seven military missions for UUVs and assessed the empowered frameworks required for them. Fast specialized advances in installed sensors and correspondences fill the plans of a universal yearly gathering and the item inventories of business organizations both little and huge around the world. Acoustic interchanges systems are the favoured channel and function admirably over short separations; still they are perilous to danger and can undoubtedly be stuck. The NATO STO 2016 Underwater Communications and ebb and flow condition of-craftsmanship and Networking gathering talked about the close term difficulties, similar to task by a group of Italian who have created secure submerged remote interchanges and marketed it with the goal that it can achieve 10Mb/s at a scope of 32 feet, in harbour water conditions (Reiber, 2018; Vancouver, 2018; McGillivary, 2018).

This chapter consists of 4 sections mainly with section II to classify the risks involved in hacking during underwater cable deployment. Further in section III the ways to cyber secure the organizations are detailed. Furthermore the chapter leads to the conclusion of these parameters.

ISSUES IN UNDERWATER CABLE DEPLOYMENT

The application backbone of underwater communication network is the submarine cables, these cables are laid almost around the world. It helps to communicate and transfer data between different countries. But researchers investigated that these submarine cables are becoming risky day by day because the attackers nowadays becoming more vulnerable as they can access the data transfer within the cables (Goyal et al., 2017c). Similarly in UWSN the security of these networks is shrinking day by day, as each sensor node contains important data that cannot be shared, although this data is available in decrypted form (Burnett, 2018; Qiu, 2011; US Report, 2018). In today's world, the attackers are also using prominent technologies and methods to get into it or to access the link which connects the adjoining sensor nodes with other nodes. The link between the nodes should be deployed in a safe manner as it is only path to transfer data from main substation to a receiver point (Bimbaum, 2017; Goyal et al., 2019; Young, 2018).

Due to some insignificant activities under the deep sea like cyclones, tsunami which affects adversely these connected nodes like through the rate of data transfer (Liam, 2018). Some attacks are also there, which are generally carried out by most of the hackers to make the user unable to get services of a particular system. The biggest fear for UWSN is DOS (denial of service attack). DOS attack puts a great effect on the UWSN, as it makes the sensor nodes affected to that level that they are unable to process and send data to the receiving end. This attack results in lack of communication services, important data breaching and huge cost for again deployment of the sensor nodes. Once the service is impaired it also affects all other sensor nodes as they are unable to find path for sending data to a particular end (James, 2018; James, 2017). As the interference of individual cables is commonplace, there are two kinds of disruptions that can result in major impacts:

1. **Simultaneous Cable Breaks:** Natural calamities like earthquakes, tsunami puts an adverse effect on the submarines cables, e.g. imagine the earthquake held in 2006 which abruptly damaged around 8 submarines and the cables associated results in 90 percent of loss of traffic between Europe, United States and China (Mauldin, 2017). This results in a great amount of loss to the communication service provider. UWSN can also be effected by this simultaneous cable breaks. As in UWSN every sensor node is connected to each other, but due to some natural disturbances in water sometimes

these sensor nodes displace from their initial position where it placed or get cut off due to huge pressure under the deep sea water which results in the loss of data and communication services (Telegraphy, 2018; Jonathan, 2015).

2. **Regional Points of Failure:** Areas having restricted connectivity usually have high exposure for a single point of failure. Similarly in UWSN there are certain connected nodes that contain huge amount of data which provide data services to many different countries (The Economist, 2008). Sometimes, due to some unusual circumstances under the deep water the main node which contains almost 90 percent of the internet data may damage which results in changing the path of data transmission node, due to change in path some factors like amount of time to transfer the data, data losses during transmission will get changed (Matsakis, 2018).
3. **Foresight: The Attack Surface is Growing:** In this growing world every big network partners or organization wants a secret space to deploy or install their network services. Big companies are installing long data cables for proper communication across different countries. So, companies are installing cables under deep sea water which become very easy for the attackers to find and access various communication lines below water contains very sensitive data that can be used inappropriately by the attackers. In UWSN more number of sensor nodes is placed deep under the sea water more will be the risk of attacks on the sensor nodes (Roach, 2018a; Roach, 2018b).
4. **How can you Reduce the Ris:** UWSN mainly composed of sensor nodes that are placed under the deep sea water, to make sure that these nodes are free of risk there will be a proper maintenance check up every week to ensure that every node is working properly? There should be a proper communication among every neighboring nodes, whereas in some cases if there is symptoms of some data loss or some type of connection problems then maintenance team first check the point of problem in their user interface and also check the percentage of loss occurred during the start of the problem and take initiatives steps to solve a problem as soon as possible so as to start the network services without any big loss.

SECURE YOUR ORGANIZATION

Organizations now days don't have the liberty to choose, that to be equipped with cybersecurity tools, devices or policies (Goyal et al., 2017a). Till now it's assumed only to be mandatory because of any cyber-attack that can target anyone anyplace anyhow. (Kumar et al., 2014; Goyal et al., 2017b) Whereas it's not possible to become 100% cyber secure, but still there exists some schemes those an organization can embed to attain optimum cybersecurity (Goyal et al., 2016).

1. **Create Cyber Awareness:** Since the increase in technology usage and various internet or web services, victims usually are unaware of the system and the security policy used. That's why they are on the major threat from the cyberattacks. Today employee is so indulge in the work that they even don't bother to know whether the system is safe to do a particular work/task or not. For this cyber awareness practices should be implemented in the organizations which makes or helps people to be aware enough to tackle a particular attack. Cyber awareness generally include various practices like passwords basis e.g. password restoring, password management, password cracking time and also the password length. Password is considered to be the major key point or a factor by which we can keep our network secure. Each employee working in an organization is being

warned to keep the password strong enough so that it will take much time and efforts to crack, also password should be changed after every 2 to 3 weeks, and there should be proper check on password restoration databases to determine the stability of the password. Organization can also develop a secure environment by awareness within employees about various attacks like Phishing which generally involves sending a fake mail and accessing data from the host system. So there should be knowledge to each and every employee of a company that how to recognize these fake mails and take action according to that. All these types of cyber awareness practices make a huge role in an overall security of the system (Simola et al., 2018).

2. **Protect Against Data Leaks:** In this technological world the main requirement is some type of specific data, whether it is your personal details or some type of username and passwords. Similarly an organization have huge amount of data sets in their databases which generally includes a person name, address, age, or any identity proofs. Security of data is the utmost step to prevent it from data leaks. Data is considered to be a very sensitive part of a network system which when get misuse, creates a big problem. A data leak generally occurs when we have full access to specific databases. So to prevent the data leaks an organization must limit the public access to data. There should be a proper policy of not using the public data on the various blogging or the chatting sites. Only those people can have access to the company data which are on the higher responsible side like manager or database administrator and also there should be proper authentication criteria to access that database. Only by denying access to data cannot solve these problems of data leaking because there are still some employees in the company which aims to destroy a company at a huge level by data breaching. For that, they send data to other companies, add some type of malwares to data so that some of the important data can be corrupted or sometimes it may be inaccessible to the employee of the organization. By implementing access control measures these problems can be resolved (Lal et al., 2016).
3. **Protect Against Ransomware Attacks:** These attacks fall under the top category and are very dangerous for the data security in underwater communication. A Ransom attack generally encrypts the data present in a system and asks for the ransom in the form of money to decrypt that data. These attacks generally perform on the server sites which includes huge amount of usage of data related services like banks and companies working on databases. To prevent these types of attacks organization must have to initiate some procedures like backup of the data in the form of hard drives. In today's world data being as important as it can be needed anywhere anytime so for cloud backup technology is being implemented where we can store data in a virtual cloud the best thing is it can be accessed anywhere from the world. Backup perform an important task so as to continue the day to day operations of a particular company. Another procedures or functions that an organization can perform to protect data or a system from ransom attacks is use of advance firewalls and antiviruses. Firewall helps to filter data that are coming into our system, the best firewall works when it has complex and strict rules of package filtering.it generally remove data from the space where firewall find something inappropriate for the system as well as software. The major role of the antiviruses is to remove or to stop ransom in the form of viruses or malwares to enter in the system (Qiu et al., 2019).
4. **Prevent Phishing and Social Engineering Attacks:** Under water sensor networks always required best and advance technology which helps to protect the sensor nodes to get attacked. Sensor nodes transfer a huge amount of communication data from one node to another and that sensitive data reaches to the base station where the data converts into the decrypted form so that it must be

easily understandable. The base station should be secure and authorized enough so that only the administrator people can enter that station. The major threat to these systems is the attacks like phishing and social engineering attacks. Phishing involves the superscripted emails or some type of animation which looks as a real one. These mails generally include the malware or the viruses attached to them. An example when an employee working in a UWSN data base station receives an email from a department regarding some issues that have to be corrected as soon as possible when employee click on that mail which is basically a phishing mail with malware attached to it then it automatically redirected to a new web page results in the installation of the malware to that system and that system is effected enough such that it creates discrepancy in data. Now day's attackers are using advance tools or software which generally works on artificial intelligence (AI). AI helps the attacker to make fake mails same as the original one so it is really difficult for a employee to find the original or a fake mail. So to prevent these types of attacks organization must suggest employee not to post their emails on an online platforms. There should be proper secured emails conversation platform in a company so as to destroy entering of fake mails and therefore the system remain secure and functions properly (Xiao et al., 2014).

5. **Adopt Policies for Securing Emerging Technologies:** Advancement in technology puts an adverse effect in the systems we are using. Advance technology is sometimes incompatible to the hardware we are using and makes it easy for the attackers to access. It also effects or decreases the vulnerability level of a system. UWSN comprises of set of sensor networks that includes nodes and special type of sensors. But when a new technology is implemented on these types of sensor networks the main motive is to remember the compatibility of technology that whether the technology can work efficient on the network devices or not. So, there will be proper policies that governed the technology we are using. Every device that should be made for future working firstly governs a particular technology policy that makes attackers often hard to access those devices. At last every employee working in an organization must work under the technology policies govern by the company (Thayer, 2012).

CONCLUSION

This chapter deals with underwater cable deployment to facilitate the coverage and maximized speed of internet all over the world. This chapter also described about various possible attacks and their effects on various networks or computer system for underwater signal transfer. Network infrastructure below the water is explored to the risk of damage or destruction by unintentional underwater accident or intentional malicious threats leads to costly and communications disruptions. It will help to understand how to manage and secure the sensor nodes from various attacks performed by the attacker like phishing, advanced engineering tools and how we can get rid of these attacks. This chapter highlights the issues in underwater cable deployment and explores the various possibilities for securing the organization. The overall dependency is upon the public community that how to secure their networks and organization.

REFERENCES

Birnbaum, M. (2017). Russian submarines are prowling around vital undersea cables. It's making NATO nervous. *The Washington Post*, 22.

Burnett, D. R. (2018). OSPAR and Coastal State Encroachment on High Seas Submarine Cable Freedoms. *Sustainable Ocean Resource Governance: Deep Sea Mining, Marine Energy and Submarine Cables*, 234.

Ellis, R., & Mohan, V. (Eds.). (2019). *Rewired: Cybersecurity Governance*. John Wiley & Sons.

Goyal, N., Dave, M., & Verma, A. K. (2016, December). Congestion control and load balancing for cluster based underwater wireless sensor networks. In *2016 Fourth International Conference on Parallel, Distributed and Grid Computing (PDGC)* (pp. 462-467). IEEE. 10.1109/PDGC.2016.7913240

Goyal, N., Dave, M., & Verma, A. K. (2017a). Improved data aggregation for cluster based underwater wireless sensor networks. *Proceedings of the National Academy of Sciences, India Section A: Physical Sciences, 87*(2), 235-245. 10.100740010-017-0344-y

Goyal, N., Dave, M., & Verma, A. K. (2017b). Data aggregation in underwater wireless sensor network: Recent approaches and issues. *Journal of King Saud University-Computer and Information Sciences.*

Goyal, N., Dave, M., & Verma, A. K. (2017c). Trust model for cluster head validation in underwater wireless sensor networks. *Underwater Technology, 34*(3), 106–113. doi:10.3723/ut.34.107

Goyal, N., Dave, M., & Verma, A. K. (2018). Adaptive Error Control Technique for Cluster-Based Underwater Wireless Sensor Networks. In *International Conference on Wireless Intelligent and Distributed Environment for Communication* (pp. 269-280). Springer. 10.1007/978-3-319-75626-4_20

Goyal, N., Sandhu, J. K., & Verma, L. (2019). *Machine Learning based Data Agglomeration in Underwater Wireless Sensor Networks*. Vol. 9, Issue 6, pp. 240-245, IJMTE. Academic Press.

James, E. P. (2017). *Russia a 'risk' to undersea cables, defense chief warns, BBC*. Retrieved from https://www.bbc.com/news/uk-42362500

James, E. P. (2018). *Australia to cement internet deal with Solomon as leaders meets in Canberra, SBS News*. Retrieved from https://www.sbs.com.au/news/australia-to-cement-internet-deal-with-solomons-as-leaders-meet-in-canberra

Jonathan, H. (2015). *Are Russian Submarines Attacking the Internet? Telegeography*. Retrieved from https://blog.telegeography.com/can-russian-submarines-attack-submarine-cables

Kumar, M., & Goyal, N. (2014). Reviewing underwater acoustic wireless sensing networks. *International Journal of Computer Science and Technology*, *5*(2), 95–98.

Lal, C., Petroccia, R., Conti, M., & Alves, J. (2016, August). Secure underwater acoustic networks: Current and future research directions. In *2016 IEEE third underwater communications and networking conference (UComms)* (pp. 1-5). IEEE.

Liam, F. (2018). *Australia, Solomon Islands, PNG sign undersea cable deal amid criticism from China, ABC.* Retrieved from https://www.abc.net.au/news/2018-07-12/australia-solomon-islands-png-sign-undersea-cable-deal/9983102

Matsakis, L. (2018). *What would really happen if Russia Attacked Undersea Internet Cables, Wired.* Retrieved from https://www.wired.com/story/russia-undersea-internet-cables/

Mauldin, A. (2017). *Cable Breakage: When and How Cables Go Down, Telegeography.* Retrieved from https://blog.telegeography.com/what-happens-when-submarine-cables-break

McGillivary, P. (2018). Why Maritime Cybersecurity Is an Ocean Policy Priority and How It Can Be Addressed. *Marine Technology Society Journal, 52*(5), 44–57. doi:10.4031/MTSJ.52.5.11

Qiu, J., Xing, Z., Zhu, C., Lu, K., He, J., Sun, Y., & Yin, L. (2019). Centralized fusion based on interacting multiple model and adaptive Kalman filter for target tracking in underwater acoustic sensor networks. *IEEE Access: Practical Innovations, Open Solutions, 7*, 25948–25958. doi:10.1109/ACCESS.2019.2899012

Qiu, W. (2011). *Submarine cables cut after Taiwan earthquake in Dec 2006.* Submarine Cable Networks Website.

Reiber, J. (2018). The Fastest Way Across the Seas: Cyberspace Operations and Cybersecurity in the Indo-Pacific. In Eurasia's Maritime Rise and Global Security (pp. 83-94). Palgrave Macmillan.

Roach, J. (2018a). *Under the sea, Microsoft tests a datacenter that's quick to deploy, could provide internet connectivity for years, Microsoft.* Retrieved from https://news.microsoft.com/features/under-the-sea-microsoft-tests-a-datacenter-thats-quick-to-deploy-could-provide-internet-connectivity-for-years/

Roach, J. (2018b). *Project Natick, Microsoft.* Retrieved from https://natick.research.microsoft.com/

Simola, J., & Rajamäki, J. (2018). *Improving Cyber Situational Awareness in Maritime Surveillance.* Academic Conferences and Publishing International Limited.

Telegeography. (2018). *Submarine Cable Map.* Retrieved from https://www.submarinecablemap.com/#/

Thayer, C. A. (2012, February). The rise of China and maritime security in southeast Asia. In *Presentation to 11th IDE Forum*, (p. 11). Academic Press.

The Economist. (2008). *Of cables and conspiracies.* Retrieved from https://www.economist.com/international/2008/02/07/of-cables-and-conspiracies

U.S. (2018). *Treasury Sanctions Russian Federal Security Service Enablers, U.S. Department of the Treasure.* Retrieved from https://home.treasury.gov/news/press-releases/sm0410

Vancouver, C. A. S. I. S. (2018). Russian Deep-Sea Operations and Canadian Cybersecurity Issues. *The Journal of Intelligence, Conflict, and Warfare, 1*(2), 6–6. doi:10.21810/jicw.v1i2.641

Xiao, S., Gong, W., & Towsley, D. (2014). Communication Security and Key Safety. In *Dynamic Secrets in Communication Security* (pp. 5–11). Springer. doi:10.1007/978-1-4614-7831-7_2

Young, M. (2018). *How an Australian spy stopped China from growing internet influence South Pacific, News Limited*. Retrieved from https://www.news.com.au/technology/online/security/how-an-australian-spy-stopped-china-from-growing-internet-influence-south-pacific/news-story/4eb83151f54c66a5091 7a95096015666

Chapter 19
Smart City in Underwater Wireless Sensor Networks

Saira Banu Atham
https://orcid.org/0000-0002-4469-5153
HKBK College of Engineering, Bangalore, India

Kalpna Guleria
Chitkara University Institute of Engineering and Technology, Chitkara University, Punjab, India

ABSTRACT

The earth is covered 71% by water and the human utilizes the remaining 29% for their shelter and living. The idea of living under the water is possible only in fiction movies for our ancestors. But in this century, the idea of living in an underwater city has become a reality with the development of the existing technology. The exploration of the undersea is booming in the science community, which shows the path for underwater cities, underwater museum, and underwater hotels. This chapter contributes the information related to underwater smart cities in three folds: (1) discusses the major challenges in developing the underwater infrastructure, (2) introduces the internet of underwater things components involved in interconnecting the devices for underwater acoustic communication, (3) list the examples of the existing masterpiece architecture constructed underwater.

INTRODUCTION

The global warming is a major threat to the future generation .Many living areas in the coastal cities are gradually sinking into the sea as a result of the change in climate and the sea level is raised as an outcome of global warming (Vince et al., 2018). For example, the capital of Maldives, is sinking at an average of 17cm (6.7 inches) every year. The coastal cities called Sunny Isles Beach located on the north of Miami are started sinking (bbb.com). Number of cities located near the coastal areas are sinking gradually into the water due to the global warming which caused the rise in sea level. As a result, the architects with innovative architectural ideas, tried to address the ongoing global warming problem inclusive of its damaging concerns through original and alternative architecture design. One among the

DOI: 10.4018/978-1-7998-3640-7.ch019

contemporary innovative concepts suggested by research architects and designers are "Underwater Cities". This is a creative and imaginative answer to the upcoming environmental problems (Vince et al., 2018). It is a creative idea that aims at using the space in the ocean/sea. This is a method that can cause the human residents settlement in the oceans (Kieth et al., 2015), since the land became more and more limited in some countries.

CHALLENGES IN DEVELOPING THE UNDERWATER INFRASTRUCTURE

This section, discusses the major challenges (liou et al., 2018) for telecommunication in underwater based on the following aspects:

1. Transmission media
2. Propagation speed
3. Transmission range
4. Transmission rate
5. Difficulty to recharge
6. Mobility
7. Reliability.

Transmission Media: The communication under the sea is based on acoustic communications instead of radio communications. The acoustic communication is the method of sending and receiving the information below the water. The properties of the waves below the sea are different from the radio waves. So different protocols are required for establishing the communication system under the sea water. Additionally establishing the transmission media is another major challenge. Various researches related to the transmission channel for multimedia communication (Saira Banu et al.,2019).using the VoIP Protocol is in progress.

Propagation Speed: The propagation speed of objects in underwater networks is approximately 200,000 times slower than in land network. For example the propagation speed of radio channels is 300,000,000 m/s in land whereas in acoustic channels it is only around 1500 m/s. So addressing the delay in packet transmission will be a major issue.

Transmission Range: The transmission range of underwater sensor network will be approximately ten times longer than in the normal land network. The signals will be absorbed by the water in the acoustic environment. So the signals have to be transmitted in low frequency. Lower frequency indicates longer transmission range; which in turn results in high risks for interferences and collisions during the transmission of data. Hence, avoidance of interferences and collisions is considered as another challenge in underwater communication.

Transmission Rate: The acoustic environment goes for narrow bandwidth, which lowers the transmission rate (approximately 10 kbps). Hence, efficient utilization of bandwidth is very important challenge to be addressed in underwater communication.

Difficulty to Recharge: The batteries are difficult to recharge under sea environment. So the cost of recharging the batteries in the underwater has an impact on the energy efficiency of the device.

Reliability: The link reliability in underwater environment is low and unstable. Thus the delivery ratio of the packets is rigorously affected by the transmission loss. This is because the data signals are absorbed by water in acoustic environments.

COMPONENTS INVOLVED IN IOUT COMMUNICATION

The trend in recent years is, smart cities in the terrestial areas, and this has initiaited the interest of people to think of establishing the life beneath the sea water. Inspite of the challenges involved in constructing the building beneath the sea level, this creative concept has attracted the present generation. This made the builders to focus their business towards the construction of smart cities beneath the sea level. When the topic is about smart city, then sensors and internet of things(also called as infrastructure of the information society) will also add to the concept. This part of the chapter dicusses about the IoT related components involved in the construction of smart cities.

The Internet of Things(IoT) was invented in the year 1985 and internet of underwater things (IoUT) was first discussed in the year 2012. The IoUT is also called as "the network of smart interconnected underwater objects". The IoUT components used in the under water projects can be in various areas like

1. Autonomous surface vehicles (ASVs),
2. Underwater sensors
3. Buoys,
4. Autonomous underwater vehicles (AUVs)
5. Ships

The network architecture (Chien-Chi Kao et al.,2017) of Under Water Sensor Network is shown in the figure 1.

Figure 1. Under water sensor network architecture

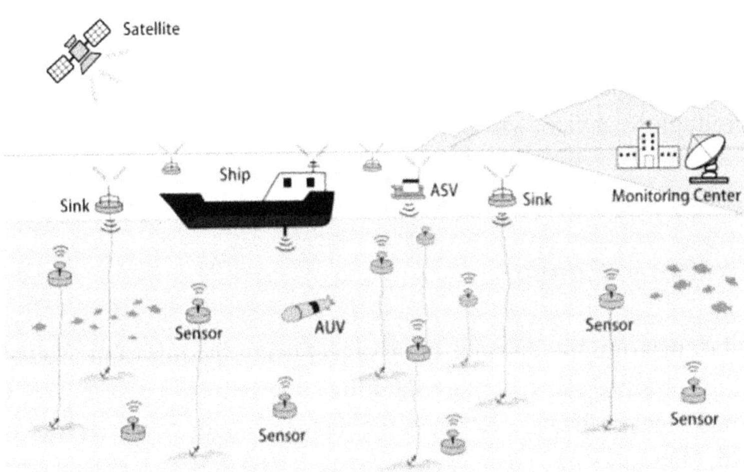

The UWSN architecture has many components and the main component is the sensor. The sensors are called as nodes with acoustic modems. They are distributed either in deep or in shallow water. These sensors are used to sense, forward and relay the information. The sensor are used to sense difffent ecological information like pressure, water quality, biological element,chemical element,temperature etc. The information from the sensors are forwarded to the components on the water surface called as sinks. The sinks can be ships,ASVs and buoys. These sinks are euipped with both acoustic and radio modems. The information received by the sinks via acoustic channels are tranfered to the remote monitoring center via radio channels. The monitroing center is at the seashore and this will monitor the water areas. The monitoring center will collect and analyze the information whereas AUVs collects and forwards the information.

UNDERWATER SENSORS

LM35 Sensor

The LM35 is a waterproof sensor to sense the temperature beneath the sea water. The voltage in this sensor will differ when there is a slight change in the temperature. It can sense temperatures ranging from -55 degree Celsius to +150 degree Celsius. There will be an increase of 10mV for every one degree Celsius change in temperature.

pH Sensor

The pH sensor is used to measure the quality of the water. This sensor is used to sense the acidity and basic nature of the water. Water with more hydrogen content is called as acidic in nature and water with more hydroxyl is alkaline in nature. The pH sensor can be used to measure the quality of the drinking water beneath the sea. This sensor will also be useful while recycling the sea water for domestic use.

ESP8266

This module is connected to the Arduino for transferring the data from one Wifi module to another Wifi module. This module is connected with the LM35 sensor or the pH sensor for collecting the data from the sensor in the digital format. The wifi networking part of the module will work only on the land. So when this microcontroller is used under the water the range of internet coverage will be limited. This module plays a significant role as different sensors can be interfaced with the module .

Waterproof UltraSonic Sensor

This device is used to measure the distance of an object using the ultrasonic sound. The transducer in the sensor is used for sending and receiving the ultrasonic pulses. These pulses will return back the information about the objects distance. Some of the reputed brands of waterproof ultrasonic sensor are xcluma, NeitKaarsh, Questquo and Voyager .

Coral Reef Monitoring Sensor

This sensor is used for sending reports of the status of coral reef beneath the sea water. The coral reef monitoring sensor identifies the locations of the corals and its patterns. The coral reef plays an important role in the ecosystem. 0.1% of the earth surface is occupied by the reef. The coral reef provides shelter for 25% of the marine species. Sea bird is a reputed manufacturer of coral reef monitoring sensors.

UnderWater Pressure Sensor

The pressure sensor is use for measuring the pressure beneath the sea water. This sensor produces a signal based on the pressure enforced by the liquid. Meacon automation, Shenzhen Xinjingcheng technologies, Xi'an Zhongming Electrical limited are some of the reputed manufactures of pressure sensors.

Underwater Depth Sensor

This sensor is used to find the depth under the sea level. The sensor is integrated with temperature and altitude measurement. The depth sensor reflects a infra red light and measures the depth based on the time it takes to reflect light.This sensor assures accuracy at any level of depth. Some of the reputed manufacturer of underwater depth sensor are Hunan Mac Sensor Ltd and Hefei WNK Smart Technology .

Under Water Proximity Sensor

A proximity sensor is used to sense the occurrence of neighboring objects without the physical contact. This sensor emits an electromagnetic radiation and observes the changes in the return signal. Terabee and dukane seacom are the leading manufactures of underwater proximity sensors.

Architecture Of Underwater Smart City

The system of interrelated underwater computing devices, processing the data from the underwater sesnor and transfering them over the network without the intervention of human is called Internet of Underwater Things(IoUT). The IoT devices situated in the Under Water smart cities broadcast the data to the cluster head. The IoUT device with high processing power and high memory is nominated as the cluster head. The Cluster head communicates and monitored by the base station. The IoUT devices are registered with the smart Gateway agent. The security issues (Goyal et al, 2020) are taken care by the Gateway agent. The Gateway agent coordinates with the cloud for processing and security.

FUTURE CONCEPT OF UNDERWATER CITY

Sunken Cities of Ancient History

The idea of underwater city is not a new idea. Underwater construction has been tried by our ancestors. The lion city in china and the Port Royal in Jamaica are best examples for ancient sunken cities. The sunken cities of the ancient history laid the path for the concept of underwater life.

Figure 2. Under water smart city- architecture

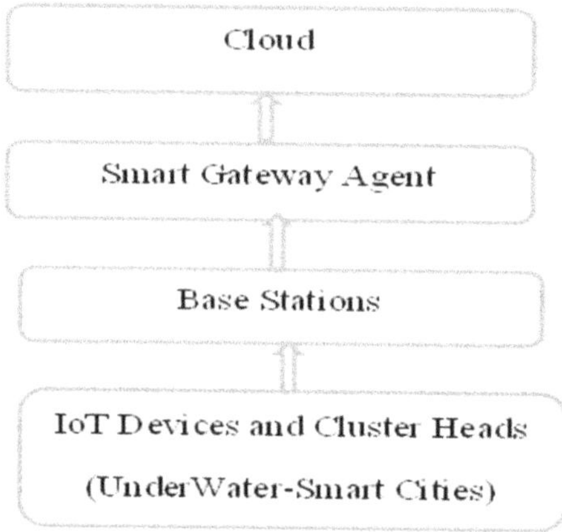

Port Royal, Jamaica

The Port named Royal in Jamaica was called as the "wickedest city in the west". It was also considered as the heaven for the pirates .The earthquake in 1962 devastated two third of the city. Today only a portion of the city is preserved and it is considered as archaeological treasure by the UNESCO. Some pieces from the city are available at the National Museum of Jamaica in Kingston.

Figure 3. Lion City, China (Lost Underwater Lion City: Rediscovery of China's 'Atlantis'. (2015, July 23). Retrieved from http://www.lovethesepics.com/2012/06/lost-underwater-lion-city-rediscovery-of-chinas-atlantis/)

Submerged in 1959 when the manmade reservoir Qiandao Lake was created to provide hydroelectric power for the nearby, fast-growing city of Hangzhou, China's Lion City (or Shi Cheng) was founded almost 1,400 years ago and not 'rediscovered' until 2001. The subaquatic city, around 85-131 feet below the surface, is home to some astoundingly well preserved examples of Ming and Qing Dynasty architecture including pagodas, temples and countless archways featuring intricately carved dragons, lions and phoenixes, while Qiandao Lake is home to another submerged city – He Cheng – thought to date back even further. Due to the depth, temperature and visibility of Lion City's location, only experienced divers are permitted to visit the ancient underwater city.

Dwarka, India

The ancient city, Dwarka, is considered to be built by Lord Krishna . This palace is made of silver, gold and precious stones. The palace is discovered by the national institute of Ocean Technology 131 feet under the gulf of khambhat.

Yonaguni Monument, Japan

The yonaguni monument, was discovered by the local diver on the coast of the yonaguni island which is the westernmost point of japan. It is a pyramid shaped structure 90 feet above the seabed. This structure was believed to be built some 3000 years ago.

Figure 4. Dwarka, Khambhat
(Image result for dwarka underwater city: Underwater city, Under water ruins, Sunken city. (n.d.). Retrieved from https:// in.pinterest.com/pin/ 4076462 03761029698 / ?lp=true)

Figure 5.Yonaguni,Japan. (Yonaguni Underwater Ruins: Okinawa Attractions: Travel Japan: JNTO. (n.d.). Retrieved from https://www.japan.travel/en/spot/564/)

EXISTING MASTERPIECE ARCHITECTURE CONSTRUCTED UNDERWATER

Ocean Spiral, a Deep Sea Future City Concept

Shimizu Corporation is a construction company in japan. It is the first construction company that came up with the innovative idea of deep sea future city. This project is considered as the first underwater city .According to the company's plan, the underwater city can accommodate five thousands permanent residents. The project is named as ocean spiral and the design involves a sphere of wide five hundred meter. This sphere is connected to a source center through a fifteen kilometer helix shaped channel. According to the company this project concept can be used for both residential and corporate zones. The scientists are exploring ways to use the energy resources available at the seabed. The professionals from shimizu believe that the technology needed to construct the planned structure and make life possible under the sea water might be available near future i.e. 2030.

Villa Epecuén, Argentina

Villa Epecuén was a popular Argentinian tourist village nestled on the edges of Laguna Epecuén, This village existed until the heavy rain during the mid-1980s.This rain caused the lake to overflow consuming the once thriving town. Technically no longer totally underwater, since 2009 the floodwaters have receded exposing a haunting landscape of half-submerged houses, trees and roads but it's one of the world's few underwater cities that visitors don't need scuba diving skills to explore. One former resident,

Figure 6. Ocean Spiral by Shimizu Corporation
(Deep Sea future city concept: Retrieved from https://www.dezeen.com/2017/11/06/video-ocean-spiral-shimizu-corporation)

85-year-old Pablo Novak, decided to return to Villa Epecuén when the waters receded: he is the subject of Australian filmmaker Matthew Salleh's 2013 documentary, *Pablo's Villa*.

Underwater Sculpture Park, Grenada

The artist Jason de Caires Taylor created the sea bed sculpture park in 2006. This park consists of 65 statues of humans sited permanently in the sea water of Grenada, covering the area of about 800 meters. This sculpture has become the base for natural ecosystem. This park also helped the visitors to divert them from the endangered island that is available nearby. This park is located two miles away from the St Goerges capital on the west of the island. The international directory of sculpture park have established similar types of underwater parks in various locations like Africa, Caribbean, Cuba, Jamaica, Europe, United states of America, Australia and new Zealand.

Heracleion, Egypt

Located off the coast of the Egyptian town of Abu-Qir just northwest of Alexandria, lies the sunken city of Heracleion. Discovered in 2000 thirty feet under the sea, excavations of the ancient site – formerly the main trading point of the Mediterranean during its heyday in the 6th century BC – have unearthed artifacts including giant statues, pottery, jewelry, wrecked ships and coins dating back to the Ptolemaic dynasty. Currently there are plans in the works to construct an underwater museum that would not only

Figure 7. Underwater sculpture park
(Phillip, R. (2017, August 21). WATCH: The Underwater Sculpture Park is an ethereal must-see. Retrieved from http://www.loopslu.com/content/watch-underwater-sculpture-park-ethereal-must-see-3)

allow visitors to see Heracleion's sunken treasures up close, but also protect the ancient city from pollution and poaching.

Kihavah Huravalhi Island, Maldives

The island situated in Maldives is surrounded by natural reef and blue lagoons. The restaurant in this island is the only restaurant in the world with underwater wine cellar. The bar is located around two meters below the sea. This restaurant is constructed with 360 degree view. It has catering facility for 18 persons at a time. People can have their breakfast and enjoy watching the underwater sea world. The restaurant was circular in shape and was encased with thick transparent glass through which the sea life can been seen. It has octagonal shaped ceiling which is also transparent to view the marine life above the head.

Atlantis The Palm, Dubai

The Dubai's is located in the Jumeirah Island and it is famous for its isolated sandy beach. The underwater aquarium, the aqua venture waterpark and the chambers aquarium are the highlights of 5 star atlantics. The free Wi-Fi facility is available for the entire property, including the underwater aquarium. The room decorated with Arabian styles and a balcony with beautiful view, are inspired by many people. Each room is facilitated with flat screen satellite TV, mini bar and tea/coffee maker.

Figure 8. The lost city of heracleion
(Thuras, D. (2016, July 21). The Lost City of Heracleion. Retrieved from https://www.atlasobscura.com/places/the-lost-city-of-heracleion-al-maadeyah-egypt)

The Muraka, Conrad Maldives

The Conrad Maldives Rangali Island proudly announced the opening ceremony of the worlds first undersea hotel residence. It is a two storey villa sixteen feet below the sea level. This villa is named as "Muraka" which means "Coral" in the Maldives language. This sunken hotel consists of gym, private bar, butler's quarters and infinity pool. The bathtub, shower and the bed have the beautiful view of the ocean. The top floor of the two storey villa rests above the water. This suite costs around fifty thousand dollar per night which includes private boat and a personal chef. The muraka was constructed bot environmentally conscious and innovative .The materials needed for the structure was built in Singapore and then shipped to the Conrad. The chief architect of this master piece consideres Muraka Hotel as the life time achievement.

Figure 9. Underwater wine cellar, Maldives
(Underwater Wedding Maldives: Underwater Restaurant Weddings. Retrieved from https://www.anantara.com/en/kihavah-/offers/ underwater-wine-cellar-and-restaurant-wedding)

Figure 10.The palm, Dubai
(About Atlantis, The Palm: Luxury Hotel & 5 Star Resort In Dubai. (n.d.). Retrieved from https://www.atlantis.com/dubai/atlantis-the-palm)

Niyama Private Islands, Maldives

Niyama Private Property is another masterpiece in Maldives island for underwater night club. A group of Sub six is treated as the world's first underwater venture with nightclub, a 24-hour spa, bathhouses with private pools, and a private seaside. Even though the light of day will afford you flawless panoramas of the Turkish blue waters, the actual scenery begins when the sunlight goes down. This property is situated 6 meters below the sea level. An separate seaplane is required to reach the place. The icicle-style ceiling, glowing neon lighting, the nightclub and fluorescent blues added feathers of the restaurant.

Radisson Blu in Berlin, Germany

It's a unique hotel in Germany and the world's largest cylindrical aquarium. This cylinder is twenty five meters high with 1500 fishes and almost 1,000,000 liters of salt water. Thought it is not underwater hotel, but they have given the impression of being underwater.

Resorts World Sentosa, Singapore

This resort is well known as ocean suite, situated in Singapore. This suite consist of two storey construction with the upper level above the sea level and the ground floor located below the sea level. Over 40000 marine lives are visible through the transparent glass. It is located 16 feet below the sea and the outside area is covered with acrylic.

Figure 11.Underwater Hotel, Germany
(World's first underwater hotel opens in the Maldives. (2019, June 21). Retrieved from https://archpaper.com/2018/11/maldives-underwater-hotel-conrad-maldives-rangali-island-muraka/#gallery-0-slide-4)

Figure 12.Ocean Suite, Singapore (zaran et al., 2017)

Underwater Theme Park, Behrain

The world largest underwater theme park with an area of about 100,000 square meters is established in behrain in 2017.This is project of behrain tourism in association with Supreme Council for Environment (SCE) and Exhibitions Authority (BTEA).

The eco-friendly park will be useful for the researchers with high source of data on aquatic ecologies, and will improve ecological awareness on the significance of conserving marine life. After the successful submersion of the project, specialized divers from registered government dive centers have inspected the project to certify the park's safety to both leisure divers and professional divers.

CONCLUSION

This chapter discuss about the construction of building under the sea level. Establishing the cities beneath the sea is an innovative idea and it is considered as the future concept to manage the environmental issues. This chapter has highlighted the challenges in underwater construction and discussed the IoT related components involved in the construction. This chapter gives the examples of the master piece architect that is already established under the sea water. The concept of theme park, amusement park, sculpture park under the sea level will pull the constructors and builders to move towards the construction of smart cities and migrating the people under the water in near future.

REFERENCES

Goyal, N., Dave, M., & Verma, A. K. (2020). SAPDA: Secure Authentication with Protected Data Aggregation Scheme for Improving QoS in Scalable and Survivable UWSNs. *Wireless Personal Communications*, *113*(1), 1–15. Advance online publication. doi:10.100711277-020-07175-8

Guleria, K., & Verma, A. K. (2018a, December). Ant Colony Optimization Techniques–An Enhancement to Energy Efficiency in WSNs. In *International conference on Computer Networks, Big data and IoT* (pp.670-677). Springer.

Kao, C. C., Lin, Y. S., Wu, G. D., & Huang, C. J. (2017, May). A study of applications, challenges, and channel models on the Internet of Underwater Things. In *2017 International Conference on Applied System Innovation (ICASI)* (pp. 1375-1378). IEEE. 10.1109/ICASI.2017.7988162

Kieth, K. M. (1977). Floating cities: A new challenge for transnational law. *Marine Policy, 1*, 190–204. Available on: https://www.sciencedirect.com/science/article/pii/0308597X77900264

Kim, H., Mokdad, L., & Ben-Othman, J. (2018). Designing uav surveillance frameworks for smart city and extensive ocean with differential perspectives. *IEEE Communications Magazine*, *56*(4), 98–104. doi:10.1109/MCOM.2018.1700444

Liou, E. C., Kao, C. C., Chang, C. H., Lin, Y. S., & Huang, C. J. (2018, April). Internet of Underwater Things: Challenges and routing protocols. In *2018 IEEE International Conference on Applied System Invention (ICASI)* (pp. 1171-1174). IEEE. 10.1109/ICASI.2018.8394494

Saira Banu Atham, Prabaharan, & Sangeetha. (2019). Method for Providing Efficient Real-Time Multimedia Communication using VoIP Over Communication Channels. *International Journal of Advanced Science and Technology*, 1832-1843.

Vince, G. (2013). *BBC - Future - The rising and sinking threats to our cities* Available on: http://www.bbc.com/future/story/20130613-the-rising-threat-to-ourcities. https://www.bbc.com/future/article/20171130-the-ambitious-plan-to-stop-the-ground-from-sinking

Compilation of References

Abrudan, T., Kypris, O., Trigoni, N., & Markham, A. (2016). Impact of rocks and minerals on underground magneto-inductive communication and localization. *IEEE Access: Practical Innovations, Open Solutions*, *4*, 3999–4010. doi:10.1109/ACCESS.2016.2597641

Ahmed, M., Salleh, M., & Channa, M. I. (2017). Routing protocols based on node mobility for Underwater Wireless Sensor Network (UWSN): A survey. *Journal of Network and Computer Applications*, *78*, 242–252. doi:10.1016/j.jnca.2016.10.022

Ahmed, M., Salleh, M., & Channa, M. I. (2018). Routing protocols based on protocol operations for underwater wireless sensor network: A survey. *Egyptian Informatics Journal*, *19*(1), 57–62. doi:10.1016/j.eij.2017.07.002

Akkaya, K., & Newell, A. (2009). Self-deployment of sensors for maximized coverage in underwater acoustic sensor networks. *Computer Communications*, *32*(7), 1233–1244. doi:10.1016/j.comcom.2009.04.002

Akram, V. K., & Dagdeviren, O. (2016). On k-connectivity problems in distributed systems. In N. Meghanathan (Ed.), *Advanced Methods for Complex Network Analysis* (pp. 30–57). IGI Global. doi:10.4018/978-1-4666-9964-9.ch002

Akyildiz, I. F., Pompili, D., & Melodia, T. (2004). Challenges for efficient communication in underwater acoustic sensor networks. *SIGBED Review*, *1*(2), 3–8. doi:10.1145/1121776.1121779

Akyildiz, I. F., Pompili, D., & Melodia, T. (2005). Underwater acoustic sensor networks: Research challenges. *Ad Hoc Networks*, *3*(3), 257–279. doi:10.1016/j.adhoc.2005.01.004

Akyildiz, I. F., Pompili, D., & Melodia, T. (2006). State-of-the-art in protocol research for underwater acoustic sensor networks. *Proceedings of the 1st ACM International Workshop on Underwater Networks*, 7–16. 10.1145/1161039.1161043

Akyildiz, I. F., Su, W., Sankarasubramaniam, Y., & Cayirci, E. (2002). subramaniamSankara, Y., and Cayirci, E. (2002). Wireless sensor networks: A survey. *Computer Networks*, *38*(4), 393–422. doi:10.1016/S1389-1286(01)00302-4

Akyildiz, I. F., & Vuran, M. C. (2010). Error Control. In *Wireless Sensor Networks*. John Wiley & Sons Ltd. doi:10.1002/9780470515181.ch6

Al Salti, F., Alzeidi, N., & Arafeh, B. R. (2017). EMGGR: An energy-efficient multipath grid-based geographic routing protocol for underwater wireless sensor networks. *Wireless Networks*, *23*(4), 1301–1314. doi:10.100711276-016-1224-0

Albaladejo, C., Sanchez, P., Iborra, A., Soto, F., Lopez, J. A., & Torres, R. (2010). Wireless sensor networks for oceanographic monitoring: A systematic review. *Sensors (Basel)*, *10*(7), 6948–6968. doi:10.3390100706948 PMID:22163583

Al-Dharrab, S., Uysal, M., & Duman, T. (2013). Cooperative underwater acoustic communications. *IEEE Communications Magazine*, *51*(7), 146–153. doi:10.1109/MCOM.2013.6553691

Alghamdi, R., Saeed, N., Dahrouj, H., Alouini, M. S., & Al-Naffouri, T. Y. (2019, September). Towards Ultra-Reliable Low-Latency Underwater Optical Wireless Communications. In *2019 IEEE 90th Vehicular Technology Conference (VTC2019-Fall)* (pp. 1-6). IEEE. 10.1109/VTCFall.2019.8891506

Alhumyani, H., Ammar, R., Albarakati, H., & Alharbi, A. (2016). Deployment strategies for underwater sensing and processing networks. *IEEE Symposium on Computers and Communication (ISCC)*, 358-363, 10.1109/ISCC.2016.7543766

Ali, M. T., Rahim, S. S., Jan, M. A., Ishtiaq, A., Ahmed, S., Ahmad, M., ... Khan, M. A. (2018). Dist-Coop: Distributed cooperative transmission in UWSNs using optimization congestion control and opportunistic routing. *Int. J. Adv. Comput. Sci. Appl*, *9*(6), 356–368. doi:10.14569/IJACSA.2018.090649

Ali, M., Khan, A., Mahmood, H., & Bhatti, N. (2019). Cooperative, reliable, and stability-aware routing for underwater wireless sensor networks. *International Journal of Distributed Sensor Networks*, *15*(6), 1550147719854249. doi:10.1177/1550147719854249

Alippi, C., Camplani, R., Galperti, C., & Roveri, M. (2011). A robust adaptive solar-powered WSN framework for aquatic environmental monitoring. *IEEE Sensors Journal*, *11*(1), 45–55. doi:10.1109/JSEN.2010.2051539

Ali, T., Jung, L. T., & Faye, I. (2014). Classification of routing algorithms in volatile environment of underwater wireless sensor networks. *International Journal of Communication Networks and Information Security*, *6*(2), 129.

Ali, T., Jung, L. T., & Faye, I. (2014). Diagonal and vertical routing protocol for underwater wireless sensor network. *Procedia: Social and Behavioral Sciences*, *129*, 372–379. doi:10.1016/j.sbspro.2014.03.690

Al-Karaki, J. N., & Kamal, A. E. (2004). Routing techniques in wireless sensor networks: A survey. *IEEE Wireless Communications*, *11*(6), 6–28. doi:10.1109/MWC.2004.1368893

Almalkawi, I. T., Guerrero Zapata, M., Al-Karaki, J. N., & Morillo-Pozo, J. (2010). Wireless multimedia sensor networks: Current trends and future directions. *Sensors (Basel)*, *10*(7), 6662–6717. doi:10.3390100706662 PMID:22163571

Alraje. (2013). *Localization Techniques in Wireless Sensor Networks*. Academic Press.

Alsalih, W., Hassanein, H., & Akl, S. (2008a). Placement of multiple mobile data collectors in underwater acoustic sensor networks. *Wireless Communications and Mobile Computing*, *8*(8), 1011–1022. doi:10.1002/wcm.656

Alsalih, W., Hassanein, H., & Akl, S. (2008b). Delay constrained placement of mobile data collectors in underwater acoustic sensor networks. *Proceedings of 33rd IEEE Conference on Local Computer Networks*, 91-97. 10.1109/LCN.2008.4664156

Alsheikh, M. A., Lin, S., Niyato, D., & Tan, H. (2014). Machine learning in wireless sensor networks: Algorithms, strategies, and applications. *IEEE Communications Surveys and Tutorials*, *16*(4), 1996–2018. doi:10.1109/COMST.2014.2320099

Alves, J., Potter, J., Zappa, G., Guerrini, P., & Been, R. (2012, October). A testbed for collaborative development of underwater communications and networking. In *MILCOM 2012-2012 IEEE Military Communications Conference* (pp. 1–8). IEEE. doi:10.1109/MILCOM.2012.6415691

Aman, W., Rahman, M. M. U., & Qadir, J. (2018). *Secure Routing in OFDM based Multi-Hop Underwater Acoustic Sensor Networks*. arXiv; arXiv:1807.01556

Arachchilage, N. A. G. (2015). *User-Centred Security Education: A Game Design to Thwart Phishing Attacks*. arXiv preprint arXiv:1511.03459

Arora, A., Gupta, S., & Kumar, A. (2018). Survey of Machine Learning Methods, What's More, Their Sensor and IoT Applications for Future World. *AJCS Journal, 21*(2).

Ateniese, G., Capossele, A., Gjanci, P., Petrioli, C., & Spaccini, D. (2015). Security framework for underwater acoustic sensor networks. *Proceedings of the MTS/IEEE OCEANS*, 1–9.

Avizienis, L., Laprie, J.-C., Randell, B., & Landwehr, C. (2004). Basic concepts and taxonomy of dependable and secure computing. *IEEE Transactions on Dependable and Secure Computing*, *1*(1), 11–33. doi:10.1109/TDSC.2004.2

Awan, J. H., Memon, S., Pathan, S. M., Usman, M., Khan, R. A., Abbasi, S., ... Hussain, Z. (2017). A user friendly security framework for the protection of confidential information. *Int. J. Comput. Sci. Netw. Secur*, *17*(04), 215–223.

Awan, K. M., Shah, P. A., Iqbal, K., Gillani, S., Ahmad, W., & Nam, Y. (2019). Underwater wireless sensor networks: A review of recent issues and challenges. *Wireless Communications and Mobile Computing*, *2019*, 2019. doi:10.1155/2019/6470359

Ayaz, M., & Abdullah, A. (2009). Underwater wireless sensor networks: Routing issues and future challenges. *Proceedings of 7th International Conference on Advances in Mobile Computing and Multimedia*, 370–375. 10.1145/1821748.1821819

Ayaz, M., Abdullah, A., & Faye, I. (2010, November). Hop-by-hop reliable data deliveries for underwater wireless sensor networks. In *2010 International Conference on Broadband, Wireless Computing, Communication and Applications* (pp. 363-368). IEEE. 10.1109/BWCCA.2010.97

Ayaz, M., Jung, L. T., Abdullah, A., & Ahmad, I. (2012). Reliable data deliveries using packet optimization in multi-hop underwater sensor networks. *Journal of King Saud University-Computer and Information Sciences*, *24*(1), 41–48. doi:10.1016/j.jksuci.2011.09.001

Ayodele, T.O. (2010). *Introduction to Machine Learning*. Academic Press.

Azar, A. T., & Vaidyanathan, S. (Eds.). (2015). *Computational intelligence applications in modeling and control*. Springer International Publishing. doi:10.1007/978-3-319-11017-2

Babcock, C. (2015). *Preparing for the Cyber Battleground of the Future*. 50th Space Communications Squadron Schriever AFB United States.

Bainbridge, S., Eggeling, D., & Page, G. (2016). Lessons from the field—Two years of deploying operational wireless sensor networks on the great barrier reef. *Sensors (Basel)*, 6842–6855. PMID:22163988

Bakr, M. A., & Lee, S. (2017). Distributed multisensor data fusion under unknown correlation and data inconsistency. *Sensors (Basel)*, *17*(11), 2472. doi:10.339017112472 PMID:29077035

Baladrón, C., Aguiar, J. M., Calavia, L., Carro, B., Sánchez-Esguevillas, A., & Hernández, L. (2012). Performance study of the application of artificial neural networks to the completion and prediction of data retrieved by underwater sensors. *Sensors (Basel)*, *12*(2), 1468–1481. doi:10.3390120201468 PMID:22438720

Balakrishnan, G., Yang, M., Jiang, Y., & Kim, Y. (2007). Performance analysis of error control codes for wireless sensor networks. *Int. Conf. Information Technology*, 876-879. 10.1109/ITNG.2007.149

Barbeau, M., Blouin, S., Cervera, G., Garcia-Alfaro, J., & Kranakis, E. (2014). *Appendix to simulation of mobile underwater communications*. Academic Press.

Barborak, D., Dahbura, A., & Malek, M. (1993). The consensus problem in fault-tolerant computing. *ACM Computing Surveys*, *25*(2), 171–220. doi:10.1145/152610.152612

Basagni, S., Petrioli, C., Petroccia, R., & Stojanovic, M. (2010). *Optimizing network performance through packet fragmentation in multi-hop underwater communications. In IEEE Sydney, OCEANSSYD 2010*. IEEE.

Basagni, S., Petrioli, C., Petroccia, R., & Stojanovic, M. (2012). Optimized packet size selection in underwater wireless sensor network communications. *IEEE Journal of Oceanic Engineering*, *37*(3), 321–337. doi:10.1109/JOE.2012.2197271

Basagni, S., Valerio, D., Gjanci, P., & Petrioli, C. (2017). Finding MARLIN: Exploiting multi-modal communications for reliable and low-latency underwater networking. In *IEEE Conference on Computer Communications (INFOCOM)* (pp. 1701-1709). Atlanta, GA: IEEE. 10.1109/INFOCOM.2017.8057132

Bengio, Y. (2009). Learning deep architectures for AI. *Foundations and Trends in Machine Learning*, *2*(1), 1–127. doi:10.1561/2200000006

Benson, B. (2010). *Design of a Low-cost Underwater Acoustic Modem for Short-Range SensorNetworking Applications* (Ph.D. Thesis). University of California, San Diego, CA, USA.

Berger, C. R., Zhou, S., Willett, P., & Liu, L. (2008). Stratification effect compensation for improved underwater acoustic ranging. *IEEE Transactions on Signal Processing*, *56*(8), 3779–3783. doi:10.1109/TSP.2008.924801

Berlian, M. H., & Sahputra, T. E. R. (2016). Implementation of smart environment monitoring and analytics in real-time system framework based on the internet of underwater things and big data. *Proceedings of the IEEE International Electronics Symposium (IES)*. 10.1109/ELECSYM.2016.7861040

Bevilacqua, M., Tsourdos, A., Starr, A., & Durazo-Cardenas, I. (2015, February). Data fusion strategy for precise vehicle location for intelligent self-aware maintenance systems. In *2015 6th International Conference on Intelligent Systems, Modelling and Simulation* (pp. 76-81). IEEE. 10.1109/ISMS.2015.37

Bharamagoudra, M. R., & Manvi, S. K. S. (2016). Deployment scheme for enhancing coverage and connectivity in underwater acoustic sensor networks. *Wireless Personal Communications*, *89*(4), 1265–1293. Advance online publication. doi:10.100711277-016-3315-8

Bhardwaj, M. (2017). Faulty Link Detection in Cluster based Energy Efficient Wireless Sensor Networks. *International Journal of Scientific Research in Network Security and Communication*, *5*(3), 1–8.

Bhatti, G. (2018). Machine learning based localization in large-scale wireless sensor networks. *Sensors (Basel)*, *18*(12), 1–20. doi:10.339018124179 PMID:30487457

Birnbaum, M. (2017). Russian submarines are prowling around vital undersea cables. It's making NATO nervous. *The Washington Post*, 22.

Björnemo, E. (2009). Error correcting codes in static channels. Energy Constrained Wireless Sensor Networks.

Björnemo, E., Ahlén, A., & Gidlund, M. (2010). WiComPI: Wireless control for the process industri. *VINNOVA Workshop on Smarter, Faster, and Converging Solutions*.

Blackman, S. S., & Broida, T. J. (1990). Multiple sensor data association and fusion in aerospace applications. *Journal of Robotic Systems*, *7*(3), 445–485. doi:10.1002/rob.4620070309

Blough, B., & Brown, H. W. (1999). The broadcast comparison model for on-line fault diagnosis in multicomputer systems: Theory and implementation. *IEEE Transactions on Computers*, *48*(5), 470–493. doi:10.1109/12.769431

Bondavalli, C., Chiaradonna, S., Di Giandomenico, F., & Grandoni, F. (2000, March). A., Di Giandomenico, S., Grandoni, F., & Fabrizio. (IEEE). Threshold-based mechanisms to discriminate transient from intermittent faults. *IEEE Transactions on Computers*, *49*(3), 230–245. doi:10.1109/12.841127

Branch, J. W., Szymanski, B., Giannella, C., Wolff, R., & Kargupta, H. (2009). In-network outlier detection in wireless sensor networks. *Knowledge and Information Systems*, *34*(1), 1–28.

Burnett, D. R. (2018). OSPAR and Coastal State Encroachment on High Seas Submarine Cable Freedoms. *Sustainable Ocean Resource Governance: Deep Sea Mining, Marine Energy and Submarine Cables*, 234.

Byun, A. R., Son, K., Cho, H. S., & Jang, Y. S. (2013, July). Channel reserved MAC protocol for underwater sensor networks. In *2013 Fifth International Conference on Ubiquitous and Future Networks (ICUFN)* (pp. 110-111). IEEE.

Cai, S., Gao, Z., Yang, D., & Yao, N. (2013). A network coding based protocol for reliable data transfer in underwater acoustic sensor. *Ad Hoc Networks, 11*(5), 1603–1609. doi:10.1016/j.adhoc.2013.02.001

Caiti, A., Calabro, V., Dini, G., Lo Duca, A., & Munafo, A. (2012). Secure cooperation of autonomous mobile sensors using an underwater acoustic network. *Sensors (Basel), 12*(2), 1967–1989. doi:10.3390120201967 PMID:22438748

Capossele, A., Petrioli, C., Saturni, G., Spaccini, D., & Venturi, D. (2017). Securing Underwater Communications: Key Agreement based on Fully Hashed MQV. *Proceedings of the ACM International Conference on Underwater Networks & Systems*, 12. 10.1145/3148675.3152760

Cardarelli, E., Sabattini, L., Secchi, C., & Fantuzzi, C. (2014, September). Multisensor data fusion for obstacle detection in automated factory logistics. In *2014 IEEE 10th International Conference on intelligent computer communication and processing (ICCP)* (pp. 221-226). IEEE. 10.1109/ICCP.2014.6937000

Casey, K., Lim, A., & Dozier, G. (2008). A sensor network architecture for tsunami detection and response. *International Journal of Distributed Sensor Networks, 4*(1), 27–42. doi:10.1080/15501320701774675

Catipovic, J. (1991). Underwater acoustic local area network for ROV and instrument communications. *Proc. AUVS'91*, 446-460.

Cato, D. (2008). Ocean ambient noise: Its measurement and its significance to marine animals. *Proceedings of the Institute of Acoustics. Institute of Acoustics (Great Britain)*, 30.

Cayirci, E., Tezcan, H., Dogan, Y., & Coskun, V. (2006). Wireless sensor networks for underwater surveillance systems. *Ad Hoc Networks, 4*(4), 431–446. doi:10.1016/j.adhoc.2004.10.008

Celik, N., & Saeed, T. Y. (2018). Modeling and performance analysis of multihop underwater optical wireless sensor networks. *IEEE Wireless Communications and Networking Conference (WCNC)*, 1-6. 10.1109/WCNC.2018.8377388

Chaitanya, D. E., Sridevi, V., & Rao, G. S. B. (2011). Path loss analysis of underwater communications. *Proceeding of the 2011 IEEE Students' Technology Symposium*, 65-70. 10.1109/TECHSYM.2011.5783803

Chakraborty, U., Tewary, T., & Chatterjee, R. P. (2009). Exploiting the Loss-Frequency Relationship using RF Communication in Underwater Communication Networks. *2009 International Conference on Computers and Devices for Communication*, 1-4.

Chan, A. B., Liang, Z. S. J., & Vasconcelos, N. (2008, June). Privacy preserving crowd monitoring: Counting people without people models or tracking. In *2008 IEEE Conference on Computer Vision and Pattern Recognition* (pp. 1-7). IEEE. 10.1109/CVPR.2008.4587569

Chander, B., Kumar, P., & Kumaravelan. (2018). A analysis of machine learning in wireless sensor network. *International Journal of Engineering & Technology, 7*(4), 185-192.

Chandrasekhar, V., & Winston, K. G. (2006). Localization in underwater sensor networks-survey and challenges. *Proceedings of the ACM International Conference on Mobile Computing and Networking*, 33–40.

Channels, M. O. F., & Abdelzaher, T. (2012). *Mining Diverse Opinions Abo, Gavin High-Efficiency Ferrite Meander Antenna (HEMA) for LTE Applications*. Academic Press.

Chen, J., Wu, X., & Chen, G. (2008, October). REBAR: a reliable and energy balanced routing algorithm for UWSNs. In *2008 Seventh international conference on grid and cooperative computing* (pp. 349-355). IEEE. 10.1109/GCC.2008.12

Chen, K. J., Somani, S., & Arun. (2006). Distributed fault detection of wireless sensor networks. In *Proceedings of the 2006 workshop on Dependability issues in wireless ad hoc networks and sensor networks* (pp. 65-72). ACM. 10.1145/1160972.1160985

Cheng, E., Wu. L., Yuan, F., Gao, C., & Yi, A.D. (2019). *Node selection algorithm for underwater acoustic sensor network based on particle swarn optimization*. Academic Press.

Chen, H., & Lou, W. (2015). on protecting end-to-end location privacy against local eavesdropper in wireless sensor networks. *Pervasive and Mobile Computing, 16*, 36–50. doi:10.1016/j.pmcj.2014.01.006

Chessa, S. S., & Paolo. (2001). Comparison-based system-level fault diagnosis in ad hoc networks. In *Proceedings 20th IEEE Symposium on Reliable Distributed Systems* (pp. 257-266). IEEE. 10.1109/RELDIS.2001.970776

Che, X., Wells, I., Dickers, G., Kear, P., & Gong, X. (2010). Re-Evaluation of RF electromagnetic communication in underwater sensor networks. *IEEE Communications Magazine, 48*(12), 143–151. doi:10.1109/MCOM.2010.5673085

Chiang, M. (2005). Balancing transport and physical layers in wireless multihop networks: Jointly optimal congestion control and power control. *IEEE Journal on Selected Areas in Communications, 23*(1), 104–116. doi:10.1109/JSAC.2004.837347

Choi, Y., J.-Y., Huh, S.-J., Choi, Y., & Yoon-Hwa. (2009). A distributed adaptive scheme for detecting faults in wireless sensor networks. *WSEAS Transactions on Communications, 8*(2), 269-278.

Cho, S. (2000). Adaptive error control scheme for multimedia applications in integrated terrestrialsatellite wireless networks. *Wireless Communications and Networking Conference*. 10.1109/WCNC.2000.903926

Chouiten, M., Domingues, C., Didier, J. Y., Otmane, S., & Mallem, M. (2012, March). Distributed mixed reality for remote underwater telerobotics exploration. In *Proceedings of the 2012 Virtual Reality International Conference* (p. 1). ACM. 10.1145/2331714.2331716

Climent, S., Capella, J. V., Meratnia, N., & Serrano, J. J. (2012). Underwater sensor networks: A new energy efficient and robust architecture. *Sensors (Basel), 12*(1), 704–731. doi:10.3390120100704 PMID:22368492

Conde, M. E., Cruz, S., Munoz, D. M., Llanos, C. H., & Fortaleza, E. L. (2013, February). An efficient data fusion architecture for infrared and ultrasonic sensors, using FPGA. In *2013 IEEE 4th Latin American Symposium on Circuits and Systems (LASCAS)* (pp. 1-4). IEEE. 10.1109/LASCAS.2013.6519059

Cong, Y., Yang, G., Wei, Z., & Zhou, W. (2010). Security in underwater sensor network. *Communications and Mobile Computing, International Conference on, 1*, 162–168.

Connell, I. T. W., Connell, T. W., & Amand, J. P. S. (2011). *U.S. Patent No. 8,063,737*. Washington, DC: U.S. Patent and Trademark Office.

Coutinho, R. W., Boukerche, A., Vieira, L. F., & Loureiro, A. A. (2018). Underwater wireless sensor networks: A new challenge for topology control–based systems. *ACM Computing Surveys, 51*(1), 19. doi:10.1145/3154834

Coutinho, R., Boukerche, A., Vieira, L., & Loureiro, A. (2017). EnRO: Energy balancing routing protocol for underwater sensor networks. In *IEEE International Conference on Communications (ICC)*, (pp. 1-6). Paris: IEEE.

Cui hong, J., Kong, J., Gerla, M., & Zhou, S. (2005). *Challenges building scalable and distributed underwater UWSN*. Academic Press.

Cui hong, J., Kong, J., Gerla, M., & Zhou, S. (2006). Challenges: Building scalable mobile underwater wireless sensor networks for aquatic applications. *IEEE Network, 20*.

Cui, J.H., Kong, J., Gerla, M. Zhou, S. (2005) *Challenges: Building scalable and distributed Underwater Wireless Sensor Networks (UWSNs) for aquatic applications*. Academic Press.

Dalberg, E., Lauberts, A., Lennartsson, R. K., Levonen, M. J., & Persson, L. (2006, July). Underwater target tracking by means of acoustic and electromagnetic data fusion. In *2006 9th International Conference on Information Fusion* (pp. 1-7). IEEE 10.1109/ICIF.2006.301613

Das, A. P., & Thampi, S. M. (2017). Fault-resilient localization for underwater sensor networks. *Ad Hoc Networks, 2017*(55), 132–142. doi:10.1016/j.adhoc.2016.09.003

Dasarathy, B. V. (1997). Sensor fusion potential exploitation-innovative architectures and illustrative applications. *Proceedings of the IEEE, 85*(1), 24–38. doi:10.1109/5.554206

Dasarathy, S., & Merli, M. (2016). Sarcopenia from mechanism to diagnosis and treatment in liver disease. *Journal of Hepatology, 65*(6), 1232–1244. doi:10.1016/j.jhep.2016.07.040 PMID:27515775

David. (2011). *Energy consumption analysis of underwater acoustic sensor networks*. Academic Press.

Davidovic, S., Loukoianova, M. E., Sullivan, C., & Tourpe, H. (2019). *Strategy for Fintech Applications in the Pacific Island Countries*. International Monetary Fund. doi:10.5089/9781498326735.087

Davis, A., & Chang, H. (2012, October). *Underwater wireless sensor networks. In 2012 Oceans*. IEEE.

De Biasi, M., Snickars, C., Landernäs, K., & Isaksson, A. J. (2008). Simulation of Process Control with WirelessHART Networks Subject to Packet Losses. In *4th IEEE Conference on Automation Science and Engineering*, (pp. 548-553). Washington, DC: IEEE. 10.1109/COASE.2008.4626509

Dee, J. S. (1995). *U.S. Patent No. 5,408,740*. Washington, DC: U.S. Patent and Trademark Office.

Dey, A. (2016). Machine learning algorithms: A review. *International Journal of Computer Science and Information Technologies, 7*(3), 1174–1179.

Ding, M., Chen, D., Xing, K., & Cheng, X. (2005). Localized fault-tolerant event boundary detection in sensor networks. In *IEEE 24th Annual Joint Conference of the IEEE Computer and Communications Societies.* (pp. 902-913). IEEE.

Ding, Y., Li, N., Song, B., & Yang, Y. (2017). The mobile node deployment algorithm for underwater wireless sensor networks. *Chinese Automation Congress (CAC)*, 456-460. doi: 10.1109/CAC.2017.8242811

Dini, G., & Lo Duca, A. (2012). *A secure communication suite for underwater acoustic sensor networks*. Academic Press.

Dini, G., & Duca, A. L. (2011). A secure network discovery protocol for Underwater Acoustic Networks. *Proceedings of the 2011 IEEE Symposium on Computers and Communications (ISCC)*, 636–638. 10.1109/ISCC.2011.5983910

Domingo, M. C. (2009). Optimal placement of wireless nodes in underwater wireless sensor networks with shadow zones. Proceedings of IEEE 2nd IFIP Wireless Days (WD). doi:10.1109/WD.2009.5449707

Domingo, M. C. (2008). Overview of channel models for underwater wireless communication networks. *Elsevier Physical Communications Journal, 1*(3), 163–182. doi:10.1016/j.phycom.2008.09.001

Domingo, M. C. (2013). Marine communities based congestion control in underwater wireless sensor networks. *Information Sciences, 228*, 203–221. doi:10.1016/j.ins.2012.11.011

Domingo, M. C., & Prior, R. (2007). Design and analysis of a gps-free routing protocol for underwater wireless sensor networks in deep water. *International Conference on Sensor Technologies and Applications, SENSORCOMM '07*, 215–220. 10.1109/SENSORCOMM.2007.4394923

Domingo, M. C., & Prior, R. (2008). Energy analysis of routing protocols for underwater wireless sensor networks. *Computer Communications*, *31*(6), 1227–1238. doi:10.1016/j.comcom.2007.11.005

Dong, T., Hu, W., & Liao, X. (2016). Dynamics of the congestion control model in underwater wireless sensor networks with time delay. *Chaos, Solitons, and Fractals*, *92*, 130–136. doi:10.1016/j.chaos.2016.09.019

Dong, Y., Dong, H., & Zhang, G. (2014). Study on denial of service against underwater acoustic networks. *Journal of Communication*, *9*(2), 135–143. doi:10.12720/jcm.9.2.135-143

Dong, Y., Li, Z., Wang, R., & Zhang, K. (2017). Range-based localization in underwater wireless sensor networks using deep neural network: poster abstract. *Proceedings of the 16th ACM/IEEE International Conference on Information Processing in Sensor Networks*, 321-322. 10.1145/3055031.3055069

Dosaranian-Moghadam, M., & Amo-Rahimi, Z. (2018). Energy efficiency and reliability in underwater wireless sensor networks using Cuckoo Optimizer Algorithm. *AUT Journal of Electrical Engineering*, *50*(1), 93–100. doi:10.22060/eej.2018.13347.5154

Durrant-Whyte, H. F. (1990). Sensor models and multisensor integration. In *Autonomous robot vehicles* (pp. 73–89). Springer. doi:10.1007/978-1-4613-8997-2_7

Du, X. J., Huang, K. J., Ian, S. L., Feng, Z. X., & Fan, L. I. U. (2014). LB-AGR: Level-based adaptive geo-routing for underwater sensor network. *Journal of China Universities of Posts and Telecommunications*, *21*(1), 54–59. doi:10.1016/S1005-8885(14)60268-5

Elghazel, W., Medjaher, K., Zerhouni, N., Bahi, J., Farhat, A., Guyeux, C., & Hakem, M. (2015). Random forests for industrial device functioning diagnostics using wireless sensor networks. *IEEE Aerospace Conference*, 1–9. 10.1109/AERO.2015.7119275

Elhadef, B. M., Elkadiki, A., & Hisham. (2006). Diagnosing mobile ad-hoc networks: two distributed comparison-based self-diagnosis protocols. In Diagnosing mobile ad-hoc networks: two distributed comparison-based self-diagnosis protocols (pp. 18-27). ACM.

Elhadef, Boukerche, M., & Elkadi, A. (2008). A distributed fault identification protocol for wireless and mobile ad hoc networks. *Journal of Parallel and Distributed Computing,* (3), 321-335.

Elhadef, B. M., & Azzedine. (2007). A failure detection service for large-scale dependable wireless ad-hoc and sensor networks. In *The Second International Conference on Availability, Reliability and Security (ARES'07)* (pp. 182-189). IEEE. 10.1109/ARES.2007.3

Elhoseny, M., Tharwat, A., Farouk, A., & Hassanien, A. E. (2017). K-Coverage model based on genetic algorithm to extend WSN lifetime. *IEEE Sensors Letters*, *1*(4), 1–4. doi:10.1109/LSENS.2017.2724846

Ellis, R., & Mohan, V. (Eds.). (2019). *Rewired: Cybersecurity Governance*. John Wiley & Sons.

Elmenreich, W. (2002). *An introduction to sensor fusion*. Vienna University of Technology.

Erol, M., Vieira, L. F., & Gerla, M. (2007, September). Localization with Dive'N'Rise (DNR) beacons for underwater acoustic sensor networks. In *Proceedings of the second workshop on Underwater networks* (pp. 97-100). ACM. 10.1145/1287812.1287833

Fan, X., Zhang, F. B., Zhang, Y. Q., & Wang, G. (2011). Information Fusion Technology and its Application to Multi-sensor Integrated Navigation System for Autonomous Underwater Vehicle. *Huoli yu Zhihui Kongzhi, 36*(3), 78-81

Farhat, A., Guyeux, C., Makhoul, A., Jaber, A., Tawil, R., & Hijazi, A. (2019). Impacts of wireless sensor networks strategies and topologies on prognostics and health management. *Journal of Intelligent Manufacturing*, *30*(5), 2129–2155. doi:10.100710845-017-1377-4

Felamban, M., Shihada, B., & Jamshaid, K. (2013). Optimal Node Placement in Underwater Wireless Sensor Networks. *IEEE 27th International Conference on Advanced Information Networking and Applications (AINA),* 492-499. doi: 10.1109/AINA.2013.40

Feldstein, M. (2008). How to Help People Whose Home Values Are Underwater—The Economic Spiral Will Get Worse Unless We Do Something about Negative Equity. *Wall Street Journal*, 18.

Felemban, E. (2013). Advanced border intrusion detection and surveillance using wireless sensor network technology, *International Journal of Communications. Network and System Sciences*, *6*(5), 251–259. doi:10.4236/ijcns.2013.65028

Felemban, E., Shaikh, F. K., Qureshi, U. M., Sheikh, A. A., & Qaisar, S. B. (2015). Underwater sensor network applications: A comprehensive survey. *International Journal of Distributed Sensor Networks*, *11*(11), 896832. doi:10.1155/2015/896832

Foote, K. E., & Huebner, D. J. (1995). *The Geographer's Craft Project*. Department of Geography, University of Texas.

Forrest, A. L., Laval, B. E., Lim, D. S. S., Williams, D. R., Trembanis, A. C., Marinova, M. M., Shepard, R., Brady, A. L., Slater, G. F., Gernhardt, M. L., & McKay, C. P. (2010). Performance evaluation of underwater platforms in the context of space exploration. *Planetary and Space Science*, *58*(4), 706–716. doi:10.1016/j.pss.2009.08.007

Freitag, L., Stojanovic, M., Grund, M., & Singh, S. (2002). Acoustic communications for regional undersea observatories. Proceedings of Oceanology Inti, 1-10.

Gaba, A., & Goyal, N. (2013). Review over Diverse Location Aided Routing. *Global Journal for Current Engineering Research*, *2*(2), 141–144.

Galan, B., & Grossmann, I. E. (1998). Optimal design of distributed wastewater treatment networks. *Industrial & Engineering Chemistry Research*, *37*(10), 4036–4048. doi:10.1021/ie980133h

Gao, M., Foh, C. H., & Cai, J. (2012). On the selection of transmission range in underwater acoustic sensor networks. *Sensors (Basel)*, *12*(4), 4715–4729. doi:10.3390120404715 PMID:22666054

Garcia, M., Sendra, S., Atenas, M., Lloret, J. (2012). Underwater Wireless Ad-hoc Networks: A Survey. *Mobile Ad Hoc Networks: Current Status and Future Trends Survey*. doi:10.1201/b11447

Garon, J. M. (2018). Cyber-World War III. *Journal of Law & Cyber Warfare*, *7*(1), 1–60.

Ge, W. (2016). Numerical simulation on the deploying and recovering process for a node of underwater wireless sensor network. *OCEANS 2016*, 1-4. doi: . doi:10.1109/OCEANSAP.2016.7485602

Georgy, J., Noureldin, A., & Mellema, G. R. (2011). Clustered mixture particle filter for underwater multitarget tracking in multistatic active sonobuoy systems. *IEEE Transactions on Systems, Man and Cybernetics. Part C, Applications and Reviews*, *42*(4), 547–560. doi:10.1109/TSMCC.2011.2154329

Ghafoor, H., & Noh, A. Y. (2019). An overview of next generation underwater target detection and tracking: An integrated underwater architecture. *IEEE Access: Practical Innovations, Open Solutions*, *7*, 98841–98853. doi:10.1109/ACCESS.2019.2929932

Ghoreyshi, S. M., Shahrabi, A., & Boutaleb, T. (2017). Void-handling techniques for routing protocols in underwater sensor networks: Survey and challenges. *IEEE Communications Surveys and Tutorials*, *19*(2), 800–827. doi:10.1109/COMST.2017.2657881

Gitzendanner, R., Puglia, F., Martin, C., Carmen, D., Jones, E., & Eaves, S. (2004). High power and high energy lithium-ion batteries for under-water applications. *Journal of Power Sources*, *136*(2), 416–418. doi:10.1016/j.jpowsour.2004.03.032

Gjanci, P., Petrioli, C., Basagni, S., Phillips, C., Boloni, L., & Turgut, D. (2018). Path finding for maximum value of information in multi-modal underwater wireless sensor networks. *IEEE Transactions on Mobile Computing*, *17*(2), 404–418. doi:10.1109/TMC.2017.2706689

Gkikopouli, A., Nikolakopoulos, G., & Manesis, S. (2012). A survey on underwater wireless sensor networks and applications. *Proceedings of 20th Mediterranean Conference on Control & Automation (MED)*, 1147-1154. 10.1109/MED.2012.6265793

Goldsmith, A. (2005). *Wireless Communications*. Cambridge University Press. doi:10.1017/CBO9780511841224

Goyal, N., Dave, M. & Verma, A. K. (2017a). Improved data aggregation for cluster based underwater wireless sensor networks. *Proceedings of the National Academy of Sciences, India Section A: Physical Sciences, 87*(2), 235-245.

Goyal, N., Dave, M. & Verma, A. K. (2020). SAPDA: Secure Authentication with Protected Data Aggregation Scheme for Improving QoS in Scalable and Survivable UWSNs. *Wireless Personal Communication*, Vol 113, Issue 1, pp 1-15. Doi:10.100711277-020-07175-8

Goyal, N., Dave, M., & Verma, A. K. (2014). Fuzzy based clustering and aggregation technique for under water wireless sensor networks. In *2014 International Conference on Electronics and Communication Systems (ICECS)* (pp. 1-5). IEEE. 10.1109/ECS.2014.6892804

Goyal, N., Dave, M., & Verma, A. K. (2017). Data aggregation in underwater wireless sensor network: Recent approaches and issues. *Journal of King Saud University-Computer and Information Sciences.*

Goyal, N., Dave, M., & Verma, A. K. (2017). Improved data aggregation for cluster based underwater wireless sensor networks. *Proceedings of the National Academy of Sciences, India Section A: Physical Sciences, 87*(2), 235-245. 10.100740010-017-0344-y

Goyal, N., Dave, M., & Verma, A. K. (2017b). Data aggregation in an underwater wireless sensor network: Recent approaches and issues. *Journal of King Saud University-Computer and Information Sciences.*

Goyal, N., Dave, M., & Verma, A. K. (2017b). Data aggregation in underwater wireless sensor network: Recent approaches and issues. *Journal of King Saud University-Computer and Information Sciences.*

Goyal, N., Sandhu, J. K., & Verma, L. (2019). Machine Learning based Agglomeration in Underwater Wireless Sensor Network. Vol. 9, Issue 6, pp 240-245 DOI:16.10089.IJMTE.2019.V9I6.19.28495

Goyal, N., Sandhu, J. K., & Verma, L. (2019). *Machine Learning based Data Agglomeration in Underwater Wireless Sensor Networks*. Vol. 9, Issue 6, pp. 240-245, IJMTE. Academic Press.

Goyal, N., Sandhu, J. K., & Verma, L. (2019b). *Machine Learning based Data Agglomeration in Underwater Wireless Sensor Networks*. Academic Press.

Goyal, N., Sandhu, J. K., &Verma, L. (2019). *Machine Learning based Data Agglomeration in Underwater Wireless Sensor Networks*. Vol. 9, Issue 6, pp. 240-245. Academic Press.

Goyal, K. N. (2016). An Optimal Scheme for Minimizing Energy Consumption in WSN. *Global Research and Development Journal for Engineering*, *1*(8), 1–7.

Goyal, N., Dave, M.,, & Verma, A. K. (2018). Adaptive Error Control Technique for Cluster-Based Underwater Wireless Sensor Networks. In *International Conference on Wireless Intelligent and Distributed Environment for Communication* (pp. 269-280). Springer. 10.1007/978-3-319-75626-4_20

Goyal, N., Dave, M., & Verma, A. K. (2016). Congestion control and load balancing for cluster based underwater wireless sensor networks. In *2016 Fourth International Conference on Parallel, Distributed and Grid Computing (PDGC)* (pp. 462-467). IEEE. 10.1109/PDGC.2016.7913240

Goyal, N., Dave, M., & Verma, A. K. (2016). Energy efficient architecture for intra and inter cluster communication for underwater wireless sensor networks. *Wireless Personal Communications*, *89*(2), 687–707. doi:10.100711277-016-3302-0

Goyal, N., Dave, M., & Verma, A. K. (2017). Trust model for cluster head validation in underwater wireless sensor networks. *Underwater Technology*, *34*(3), 106–113. doi:10.3723/ut.34.107

Goyal, N., Dave, M., & Verma, A. K. (2018). A novel fault detection and recovery technique for cluster-based underwater wireless sensor networks. *International Journal of Communication Systems*, *31*(4), e3485. doi:10.1002/dac.3485

Goyal, N., Dave, M., & Verma, A. K. (2019a). Data aggregation in underwater wireless sensor network: Recent approaches and issues. *Journal of King Saud University-Computer and Information Sciences.*, *31*(3), 275–286. doi:10.1016/j.jksuci.2017.04.007

Goyal, N., Dave, M., & Verma, A. K. (2019a). Protocol stack of underwater wireless sensor network: Classical approaches and new trends. *Wireless Personal Communications*, *104*(3), 995–1022. doi:10.100711277-018-6064-z

Goyal, N., & Gaba, A. (2013). A new approach of location aided routing protocol using minimum bandwidth in mobile ad-hoc network. *International Journal of Computer Technology and Applications*, *4*(4), 653.

Goyal, N., Sandhu, J. K., & Verma, L. (2019b). Machine learning based data agglomeration in underwater wireless sensor networks. International Journal of Management. *Technology And Engineering*, *9*(6), 240–245.

Green, D. (2007). *Underwater Acoustic Communication and Modem-Based Navigation Aids*. Springer. doi:10.1007/978-3-540-77090-9_43

Guleria, K., & Verma, A. K. (2018b, December). Unequal Clustering-An Enhancement to Network Longevity in WSNs. In *2018 4th International Conference on Computing Communication and Automation (ICCCA)* (pp. 1-4). IEEE. 10.1109/CCAA.2018.8777605

Guleria, K., Kumar, S., & Verma, A. K. (2019 c). Energy Aware Location Based Routing Protocols in Wireless Sensor Networks. *World Scientific News*, *124*(2), 326–333.

Guleria, K., & Verma, A. K. (2018). An energy efficient load balanced cluster-based routing using ant colony optimization for WSN. *International Journal of Pervasive Computing and Communications*, *14*(3/4), 233–246. doi:10.1108/IJPCC-D-18-00013

Guleria, K., & Verma, A. K. (2018a, December). Ant Colony Optimization Techniques–An Enhancement to Energy Efficiency in WSNs. In *International conference on Computer Networks, Big data and IoT* (pp. 670-677). Springer.

Guleria, K., & Verma, A. K. (2019a). Comprehensive review for energy efficient hierarchical routing protocols on wireless sensor networks. *Wireless Networks*, *25*(3), 1159–1183. doi:10.100711276-018-1696-1

Guleria, K., & Verma, A. K. (2019b). Meta-heuristic Ant Colony Optimization Based Unequal Clustering for Wireless Sensor Network. *Wireless Personal Communications*, *105*(3), 891–911. doi:10.100711277-019-06127-1

Gulli, A., & Pal, S. (2017). *Deep learning with Keras*. Packt Publishing Limited.

Gungor, V. C., & Hancke, G. P. (2009). Industrial Wireless Sensor Networks: Challenges, Design Principles, and Technical Approaches. *IEEE Transactions on Industrial Electronics*, *56*(10), 4258–4265. doi:10.1109/TIE.2009.2015754

Guo, S., Zhong, Z., & He, T. (2009). FIND: faulty node detection for wireless sensor networks. In *Proceedings of the 7th ACM conference on embedded networked sensor systems* (pp. 253-266). ACM. 10.1145/1644038.1644064

Guo, Y., & Liu, Y. (2013). Localization for anchor-free underwater sensor networks. *Computers & Electrical Engineering*, *39*(6), 1812–1821. doi:10.1016/j.compeleceng.2013.02.001

Guo, Z., Colombi, G., Wang, B., Cui, J. H., Maggiorini, D., & Rossi, G. P. (2008, January). Adaptive routing in underwater delay/disruption tolerant sensor networks. In *2008 Fifth Annual Conference on Wireless on Demand Network Systems and Services* (pp. 31-39). IEEE. 10.1109/WONS.2008.4459352

Haeberlen, Kouznetsov, A., Druschel, P., & Peter. (2006). The Case for Byzantine Fault Detection. *HotDep*.

Hall, D. L., & Llinas, J. (1997). An introduction to multisensor data fusion. *Proceedings of the IEEE*, *85*(1), 6–23. doi:10.1109/5.554205

Hall, D. L., McMullen, S. A., & Hall, C. M. (2015, September). New perspectives on level-5 information fusion: The impact of advances in information technology and user behavior. In *2015 IEEE International Conference on Multisensor Fusion and Integration for Intelligent Systems (MFI)* (pp. 214-219). IEEE. 10.1109/MFI.2015.7295811

Hamid, M. A., Abdullah-Al-Wadud, M., Hassan, M. M., Almogren, A., Alamri, A., Kamal, A. R. M., & Mamun-Or-Rashid, M. (2018). A key distribution scheme for secure communication in acoustic sensor networks. *Future Generation Computer Systems*, *86*, 1209–1217. doi:10.1016/j.future.2017.07.025

Han, Jiang, Shu, Xu, & Wang. (2012). Localization Algorithms of Underwater Wireless Sensor Networks: a survey. *MDPI*.

Han, G., Jiang, J., Shu, L., Xu, Y., & Wang, F. (2012). Localization algorithms of underwater wireless sensor networks: A survey. *Sensors (Basel)*, *12*(2), 2026–2061. doi:10.3390120202026 PMID:22438752

Han, G., Jiang, J., Sun, N., & Shu, L. (2015). Secure communication for underwater acoustic sensor networks. *IEEE Communications Magazine*, *53*(8), 54–60. doi:10.1109/MCOM.2015.7180508

Han, G., Liu, L., Jiang, J., Shu, L., & Rodrigues, J. J. (2016). A collaborative secure localization algorithm based on trust model in underwater wireless sensor networks. *Sensors (Basel)*, *16*(2), 229. doi:10.339016020229 PMID:26891300

Han, G., Zhang, C., Shu, L., & Rodrigues, J. J. P. C. (2015). Impacts of deployment strategies on localization performance in underwater acoustic sensor networks. *IEEE Transactions on Industrial Electronics*, *62*(3), 1725–1733. doi:10.1109/TIE.2014.2362731

Han, J., Al-Bzoor, M., Zuba, M., Cui, J. H., Ammar, R. A., & Rajasekaran, S. (2012). PADP: Prediction assisted dynamic surface gateway placement for mobile underwater networks. *IEEE Symposium on Computers and Communications (ISCC)*, 139-144.

Han, S. Y., Chen, Y. H., & Tang, G. Y. (2017). Fault diagnosis and fault-tolerant tracking control for discrete-time systems with faults and delays in actuator and measurement. *Journal of the Franklin Institute*, *354*(12), 4719–4738. doi:10.1016/j.jfranklin.2017.05.027

Harinath, K., Vishalakshi, R., Kumar, R. R., & Shanmukhi, M. (2019). Exploring machine learning techniques for localization in wireless sensor networks. *Journal of Information and Computational Science*, *9*(11), 568–575.

Harland, E., & Richards, S. (2006). *SEA 7 Technical report: Underwater ambient noise (Non-technical summary).* Report by QINETIQ.

Harris, C. J., Bailey, A., & Dodd, T. J. (1998). Multi-sensor data fusion in defense and aerospace. *Aeronautical Journal, 102*(1015), 229–244.

HART Communication Foundation. (2007). *2.4GHz DSSS O-QPSK Physical Layer Specification*. HART Communication Foundation.

Headrick, R., & Freitag, L. (2009). Growth of underwater communication technology in the US Navy. *IEEE Communications Magazine, 1*(47), 80–82. doi:10.1109/MCOM.2009.4752681

Heidemann, J., Li, Y., Syed, A., Wills, J., & Ye, W. (2005). Underwater sensor networking: Research challenges and potential applications. *Proceedings of the Technical Report ISI-TR-2005-603, USC/Information Sciences Institute.*

Heidemann, J., Stojanovic, M., & Zorzi, M. (1958). Stojanovic, M., & Zo, M. (2012). Underwater sensor networks: Applications, advances and challenges. *Philosophical Transactions of the Royal Society of London. Series A, Mathematical and Physical Sciences, 370*(1958), 158–175. doi:10.1098/rsta.2011.0214

Heidemann, J., Ye, W., Wills, J., Syed, A., & Li, Y. (2006, April). Research challenges and applications for underwater sensor networking. In *IEEE Wireless Communications and Networking Conference.WCNC 2006* (Vol. 1, pp. 228-235). IEEE. 10.1109/WCNC.2006.1683469

He, J., Chen, J., Cheng, P., & Cao, X. (2014). Secure time synchronization in wireless sensor networks: A maximum consensus-based approach. *IEEE Transactions on Parallel and Distributed Systems, 25*(4), 1055–1065. doi:10.1109/TPDS.2013.150

Hollinger, G. A., Choudhary, S., Qarabaqi, P., Murphy, C., Mitra, U., Sukhatme, G. S., Stojanovic, M., Singh, H., & Hover, F. (2012). Underwater data collection using robotic sensor networks. *IEEE Journal on Selected Areas in Communications, 30*(5), 899–911. doi:10.1109/JSAC.2012.120606

Howard, S. L., Schlegel, C., Iniewski, K., & Iniewski, K. (2006). Error Control Coding in Low-Power Wireless Sensor Networks: When is ECC Energy-Efficient. *EURASIP Journal on Wireless Communications and Networking, 2006*(1), 1–14. doi:10.1155/WCN/2006/74812

Huang, C. J., Wang, Y. W., Liao, H. H., Lin, C. F., Hu, K. W., & Chang, T. Y. (2011). A power-efficient routing protocol for underwater wireless sensor networks. *Applied Soft Computing, 11*(2), 2348–2355. doi:10.1016/j.asoc.2010.08.014

Hui, W., Liu, M., & Lin, Z.S. (2015). An Efficient Depth adjustment Deployment Scheme for Underwater Wireless Sensor Network. *Proceedings of the 34th Chinese Control Conference*, 7771-7776.

Hummelholm, A. (2019, July). Undersea Optical Cable Network and Cyber Threats. In *European Conference on Cyber Warfare and Security*. Academic Conferences International Limited.

Hu, Q., & Hancke, G. P. (2017, March). A session hijacking attack on physical layer key generation agreement. In *2017 IEEE International Conference on Industrial Technology (ICIT)* (pp. 1418-1423). IEEE. 10.1109/ICIT.2017.7915573

Hu, T., & Fei, Y. (2010). QELAR: A machine-learning-based adaptive routing protocol for energy-efficient and lifetime-extended underwater sensor networks. *IEEE Transactions on Mobile Computing, 9*(6), 796–809. doi:10.1109/TMC.2010.28

Hwang, D., & Kim, D. (2008, September). DFR: Directional flooding-based routing protocol for underwater sensor networks. *Oceans, 2008*, 1–7.

Ibragimov, M., Lee, J. H., Kalyani, M., Namgung, J. I., Park, S. H., Yi, O., & Lim, Y. K. (2016). CCM-UW security modes for low-band underwater acoustic sensor networks. *Wireless Personal Communications, 2016*(89), 479–499. doi:10.100711277-016-3283-z

Ibrahim, S., Ammar, R. & Cui, J-H. (2010). Surface gateway placement strategy for maximizing underwater sensor network lifetime. *The IEEE symposium on computers and communications*. DOI: . doi:10.1109/ISCC.2010.5546562

Ibrahim, S., Jun-Hong, C., & Reda, A. (2007). Surface-Level Gateway Deployment for Underwater Sensor Networks. *Proceedings of IEEE Military Communications Conference*, 1 - 7. 10.1109/MILCOM.2007.4454996

IEEE. (1964). Table of generators for Bose-Chaudhuri codes. *IEEE Transactions on Information Theory*, 390.

IEEE. (2006). *IEEE Std 802.15.4*. Institute of Electrical and Electronics Engineers, Inc.

Inam, J. A., Ahmed, S., Ullah, N., Raza, A., Rehman, A. U., Khattak, A., & hahid Rahim, S. S. (2018). Congestion Control Reliability and Adaptive Cooperation for Efficient Underwater Sensor Networks. *EAI Endorsed Transactions on Scalable Information Systems, 5*(16).

Ismail, N.-S., & Hussein, L. (2010). Analyzing the performance of acoustic channel in underwater wireless sensor network (UWSN). In *2010 Fourth Asia International Conference on Mathematical/Analytical Modelling and Computer Simulation* (pp. 550-555). IEEE. 10.1109/AMS.2010.111

Iyer, S., & Rao, D. V. (2015, Feb.). Genetic algorithm based optimization technique for underwater sensor network positioning and deployment. In *Proceedings of the IEEE Underwater Technology (UT '15)*. Chennai, India: IEEE. 10.1109/UT.2015.7108229

Jacka, L., & Booth, K. (2012, December). What about the firewall? Creating virtual worlds in a public primary school using Sim-on-a-Stick. *Australian Computers in Education Conference*.

Jaikaeo, C., Srisathapornphat, C., & Shen, C. (2001, June). Diagnosis of sensor networks. In *IEEE International Conference on Communications. Conference Record (Cat. No. 01CH37240)* (pp. 1627-1632). IEEE.

Jain, S., Pilli, E. S., Govil, M. C., & Rao, D. V. (2015, February). Performance evaluation of congestion-aware routing protocols for underwater sensor networks with multimedia data. In 2015 IEEE Underwater Technology (UT) (pp. 1-6). IEEE. doi:10.1109/UT.2015.7108271

James, E. P. (2017). *Russia a 'risk' to undersea cables, defense chief warns, BBC*. Retrieved from https://www.bbc.com/news/uk-42362500

James, E. P. (2018). *Australia to cement internet deal with Solomon as leaders meets in Canberra, SBS News*. Retrieved from https://www.sbs.com.au/news/australia-to-cement-internet-deal-with-solomons-as-leaders-meet-in-canberra

Javaid, N., Ejaz, M., Abdul, W., Alamri, A., Almogren, A., Niaz, I., & Guizani, N. (2017). Cooperative position aware mobility pattern of AUVs for avoiding void zones in underwater WSNs. *Sensors (Basel), 17*(3), 580. doi:10.339017030580 PMID:28335377

Jayaraman, P. P., Zaslavsky, A., & Delsing, J. (2010). Intelligent processing of k-nearest neighbors queries using mobile data collectors in a location aware 3D wireless sensor network. *Proceeding of International conference on Industrial, Engineering and other Applications of Applied Intelligent Systems, Trends in Applied Intelligent Systems*, 6098, 260–270. 10.1007/978-3-642-13033-5_27

Jiang & Peng. (2009). A new method for node fault detection in wireless sensor networks. *Sensors, 9*(2), 1282-1294.

Jiang, S.M. (2018). On securing underwater acoustic networks: A survey. *IEEE Commun. Surv. Tutor.*

Jiang, J., Han, G., Zhu, C., Chan, S., & Rodrigues, J. J. (2017). A trust cloud model for underwater wireless sensor networks. *IEEE Communications Magazine, 55*(3), 110–116. doi:10.1109/MCOM.2017.1600502CM

Jiang, S., & Georgakopoulos, S. (2011). Electromagnetic Wave Propagation into Fresh Water. *Journal of Electromagnetic Analysis and Applications*, *3*(07), 261–266. doi:10.4236/jemaa.2011.37042

Jin, Z., Ma, Y., Su, Y., Li, S., & Fu, X. (2017). A Q-learning-based delay-aware routing algorithm to extend the lifetime of underwater sensor networks. *Sensors (Basel)*, *17*(7), 1660. doi:10.339017071660 PMID:28753951

Jin, Z., Zhao, Q., & Su, Y. (2019). RCAR: A reinforcement-learning-based routing protocol for congestion-avoided underwater acoustic sensor networks. *IEEE Sensors Journal*, *19*(22), 10881–10891. doi:10.1109/JSEN.2019.2932126

Johnson, E. N., DeBitetto, P. A., Trott, C. A., & Bosse, M. C. (1996). *The 1996 Draper Laboratory small autonomous aerial vehicle*. Project Documentation.

Jonathan, H. (2015). *Are Russian Submarines Attacking the Internet? Telegeography*. Retrieved from https://blog.telegeography.com/can-russian-submarines-attack-submarine-cables

Jones, M. (2015). *The effects of conformity and training in a phishing context: Conforming to the school of phish*. The University of Alabama in Huntsville.

Jouhari, M., & Ibrahimi, K., andBenattou, M. (2015, October). Topology control through depth adjustment and transmission power control for UWSN routing protocols. In *2015 International Conference on Wireless Networks and Mobile Communications (WINCOM)* (pp. 1-5). IEEE. 10.1109/WINCOM.2015.7381310

Jr, Duarte, Ziwich, E., Albini, R., & Albini. (2011). A survey of comparison-based system-level diagnosis. *ACM Computing Surveys*, *43*(3), 22.

Jung, L. T., & Abdullah, A. B. (2011, June). Underwater wireless network energy efficiency and optimal data packet size. In *International Conference on Electrical, Control and Computer Engineering 2011 (InECCE)* (pp. 178-182). IEEE. 10.1109/INECCE.2011.5953871

Jun-Hong, C., Jiejun, K., Gerla, M., & Shengli, Z. (2006). The challenges of building mobile underwater wireless networks for aquatic applications. *IEEE Network*, *20*(3), 12–18. doi:10.1109/MNET.2006.1637927

Kamoun-Abid, F., Meddeb-Makhlouf, A., Zarai, F., & Guizani, M. (2019). DVF-fog: Distributed virtual firewall in fog computing based on risk analysis. *International Journal of Sensor Networks*, *30*(4), 242–253. doi:10.1504/IJSNET.2019.101242

Kang, J., Zhang, Y., & Nath, B. (2007). TARA: Topology-aware resource adaptation to alleviate congestion in sensor networks. *IEEE Transactions on Parallel and Distributed Systems*, *18*(7), 919–931. doi:10.1109/TPDS.2007.1030

Kao, C. C., Lin, Y. S., Wu, G. D., & Huang, C. J. (2017, May). A study of applications, challenges, and channel models on the Internet of Underwater Things. In *2017 International Conference on Applied System Innovation (ICASI)* (pp. 1375-1378). IEEE. 10.1109/ICASI.2017.7988162

Khaledi, S., Mann, H., Perkovich, J., & Zayed, S. (2013). Design of an underwater mine detection system. *Proceedings of the IEEE Systems and Information Engineering Design Symposium (SIEDS)*.

Khaleghi, B., Khamis, A., Karray, F. O., & Razavi, S. N. (2013). Multisensor data fusion: A review of the state-of-the-art. *Information Fusion*, *14*(1), 28–44. doi:10.1016/j.inffus.2011.08.001

Khan, A., & Jenkins, L. (2008). Undersea wireless sensor network for ocean pollution prevention. *Proceedings of the IEEE International Conference on Communication Systems Software and Middleware (COMSWARE)*. 10.1109/COMSWA.2008.4554369

Khan, A., Khan, M., Ahmed, S., Mohd, A. A. R., & Khan, M. (2019). Energy harvesting based routing protocol for underwater sensor networks. *PLoS One*, *14*(7). doi:10.1371/journal.pone.0219459 PMID:31314772

Kheirabadi, M. T., & Mohamad, M. M. (2013). Greedy routing in underwater acoustic sensor networks: A survey. *International Journal of Distributed Sensor Networks*, *9*(7), 701834. doi:10.1155/2013/701834

Khurana, K., & Goyal, N. (2016). A Survey on Deployment Strategies and Energy Efficiency of Wireless Sensor Networks. *International Journal of Electrical Electronics & Computer Science Engineering*, *3*(3), 16–20.

Khurana, T., Singh, S., & Goyal, N. (2012). An Evaluation of Ad-hoc Routing Protocols for Wireless Sensor Networks. *International Journal of Advanced Research in Computer Science and Electronics Engineering*, *1*(5).

Kieth, K. M. (1977). Floating cities: A new challenge for transnational law. *Marine Policy*, *1*, 190–204. Available on: https://www.sciencedirect.com/science/article/pii/0308597X77900264

Kim, H., Mokdad, L., & Ben-Othman, J. (2018). Designing uav surveillance frameworks for smart city and extensive ocean with differential perspectives. *IEEE Communications Magazine*, *56*(4), 98–104. doi:10.1109/MCOM.2018.1700444

Kleinschmidt, J. H., Borelli, W. C., & Pellenz, M. E. (2007). An analytical model for energy efficiency of error control schemes in sensor networks. In *International Conference on Communications 2007*, (pp. 3895-3900). 10.1109/ICC.2007.642

Krishnamachari, B., Estrin, D., & Wicker, S. B. (2002). The impact of data aggregation in wireless sensor networks. *Proceeding of 22nd International Conference on Distributed Computing Systems Workshops*, 575– 578. 10.1109/ICDCSW.2002.1030829

Kumar, A., & Sharma, S. (2019). Enhanced Energy-efficient heterogeneous Routing Protocols in WSNs for IoT Application. *IJEAT*, *9*(1).

Kumar, K., & Kumar, A. (2018). Wireless Sensor Networks: A review on "Challenges and Opportunities for the Future world-LTE". *AJCS*, *1*(2).

Kumar, D. P., Amgoth, T., & Annavarapu, C. S. R. (2019). Machine learning algorithms for wireless sensor networks: A survey. *INFORM FUSION*, *49*, 1–25. doi:10.1016/j.inffus.2018.09.013

Kumari, V. S., & Dondeti, V. (2019). Survey: Wireless sensor networks in node localization by techniques of machine learning. *International Journal of Research and Analytical Reviews*, *6*(1), 728–737.

Kumar, M., & Goyal, N. (2014). Reviewing underwater acoustic wireless sensing networks. *International Journal of Computer Science and Technology*, *5*(2), 95–98.

Kumar, P., & Kumar, P., & Priyadarshini, S. P. (2012). Underwater acoustic sensor network for early warning generation. *Proceedings of the IEEE OCEANS*. 10.1109/OCEANS.2012.6405009

Kumar, R., & Singh, N. (2014). A survey on data aggregation and clustering schemes in underwater sensor networks. *International Journal of Grid and Distributed Computing*, *7*(6), 29–52. doi:10.14257/ijgdc.2014.7.6.04

Lal, C., Petroccia, R., Conti, M., & Alves, J. (2016, August). Secure underwater acoustic networks: Current and future research directions. In *2016 IEEE third underwater communications and networking conference (UComms)* (pp. 1-5). IEEE.

Lanbo, L., Shengli, Z., & Jun-Hong, C. (2008). Prospects and problems of wireless communication for underwater sensor networks. *Wireless Communications and Mobile Computing*, *8*(8), 977–994. doi:10.1002/wcm.654

Langley, P., & Simon, H. A. (1995). Applications of machine learning and rule induction. *Communications of the ACM*, *38*(11), 54–64. doi:10.1145/219717.219768

Law, Y. W., Doumen, J., & Hartel, P. (2006). Survey and benchmark of block ciphers for wireless sensor networks. *ACM Transactions on Sensor Networks*, *2*(1), 65–93. doi:10.1145/1138127.1138130

Le, T. T., & Kim, D. S. (2013, April). Distributed cooperative transmission for underwater acoustic sensor networks. In 2013 IEEE wireless communications and networking conference workshops (WCNCW) (pp. 205-210). IEEE.

Lee, W. L., Datta, A., & Cardell-Oliver, R. (2006). *Winms: Wireless sensor network-management system, an adaptive policy-based management for wireless sensor networks*. Academic Press.

Lee, D., & Chung, K. (2010). Adaptive duty-cycle based congestion control for home automation networks. *IEEE Transactions on Consumer Electronics*, *56*(1), 42–47. doi:10.1109/TCE.2010.5439124

Lefort, R., Real, G., & Dremeau, A. (2017). Direct regressions for underwater acoustic source localization in fluctuating oceans. *Applied Acoustics*, *116*, 303–310. doi:10.1016/j.apacoust.2016.10.005

Lehto, M., Hummelholm, A., Iida, K., Jakstas, T., Kari, M. J., Minami, H., & Saunavaara, J. (2019). Arctic Connect Project and Cyber Security Control, ARCY. *Informaatioteknologiantiedekunnanjulkaisuja*, (78/2019).

Li, C. S., & Hu, Y. (2012). Study on State-based Remote Detection Technology for SQL Injection Vulnerability. *Information Security and Communications Privacy, 5*.

Li, X., Fang, S., & Zhang, Y. (2007). The study on clustering algorithm of the underwater acoustic sensor networks. *Proceedings of the 14th IEEE International Conference on Mechatronics and Machine Vision in Practice (M2VIP 2007)*, 78–81.

Li, X., Nayak, A., Simplot-Ryl, D., & Stojmenovic, I. (2009). *Sensor placement in sensor and actuator networks in wireless and actuator network: algorithm and protocol*. Academic Press.

Li, X., Song, Y., Guo, J., Feng, C., Li, G., Yan, T., & He, B. (2017). Sensor fault diagnosis of autonomous underwater vehicle based on extreme learning machine. *IEEE International Symposium on Underwater Technology*, 1-5.

Liam, F. (2018). *Australia, Solomon Islands, PNG sign undersea cable deal amid criticism from China, ABC*. Retrieved from https://www.abc.net.au/news/2018-07-12/australia-solomon-islands-png-sign-undersea-cable-deal/9983102

Li, B., Zhou, S., Stojanovic, M., Freitag, L., & Willett, P. (2008). Multicarrier communication over underwater acoustic channels with nonuniform Doppler shifts. *IEEE Journal of Oceanic Engineering*, *33*(2), 198–209. doi:10.1109/JOE.2008.920471

Liebe, H. J., Hufford, G. A., & Manabe, T. (1991). A model for the complex permittivity of water at frequencies below 1THz. *International Journal of Infrared and Millimeter Waves*, *12*(7), 659–675. doi:10.1007/BF01008897

Liggins, M. II, Hall, D., & Llinas, J. (Eds.). (2017). *Handbook of multisensor data fusion: theory and practice*. CRC press. doi:10.1201/9781420053098

Lin, H., Wei, W., Zhao, P., Ma, X., Zhang, R., Liu, W., Deng, T., & Peng, K. (2015). Energy-efficient compressed data aggregation in underwater acoustic sensor networks. *Wireless Networks*, *22*(06), 1–13.

Liou, E. C., Kao, C. C., Chang, C. H., Lin, Y. S., & Huang, C. J. (2018, April). Internet of Underwater Things: Challenges and routing protocols. In *2018 IEEE International Conference on Applied System Invention (ICASI)* (pp. 1171-1174). IEEE. 10.1109/ICASI.2018.8394494

Lisa, G. (2013). The great underwater firewall. *The World of Chinese*, *3*(6), 71–74.

Liu, B., Garcin, F., Ren, F., & Lin, C. (2008, June). A study of forward error correction schemes for reliable transport in underwater sensor networks. In *2008 5th Annual IEEE Communications Society Conference on Sensor, Mesh and Ad Hoc Communications and Networks* (pp. 197-205). IEEE. 10.1109/SAHCN.2008.33

Liu, J., & Wang, J. (2016, January). A MACA-based collision avoidance MAC protocol for underwater acoustic sensor networks. In *2016 IEEE/OES China Ocean Acoustics (COA)* (pp. 1-4). IEEE.

Liu, L. (2010). A qos-based topology control algorithm for underwater wireless sensor networks. *International Journal of Distributed Sensor Networks.*

Liu, W., Weaver, J., Weaver, L., Whelan, T., Bagrodia, R., Forero, P. A., & Chavez, J. (2018, May). APOLL: Adaptive Polling for Reconfigurable Underwater Data Collection Systems. In *2018 OCEANS-MTS/IEEE Kobe Techno-Oceans (OTO)* (pp. 1-9). IEEE.

Liu, J., Wang, Z., Zuba, M., Peng, Z., Cui, J. H., & Zhou, S. (2014). DA-Sync: A Doppler-assisted time-synchronization scheme for mobile underwater sensor networks. *IEEE Transactions on Mobile Computing*, *13*(3), 582–595. doi:10.1109/TMC.2013.13

Liu, J., Zhou, Z., Peng, Z., Cui, J. H., Zuba, M., & Fiondella, L. (2013). Efficient time synchronization for mobile underwater sensor networks. *IEEE Transactions on Parallel and Distributed Systems*, *24*(2), 406–416. doi:10.1109/TPDS.2012.71

Liu, L. (2011). A deployment algorithm for underwater sensor networks in ocean environment. *Journal of Circuits, Systems, and Computers*, *20*(6), 1051–1066. doi:10.1142/S021812661100775X

Liu, Y. C., Liu, S. Y., & Wang, N. (2016). Fully-tuned fuzzy neural network based robust adaptive tracking control of unmanned underwater vehicle with thruster dynamics. *Neurocomputing*, *196*, 1–13. doi:10.1016/j.neucom.2016.02.042

Li, Y. B., Li, L. L., Xu, B. B., Wu, W., Wu, R. Y., Wan, X., Cheng, Q., & Cui, T. J. (2016). Transmission-type 2-bit programmable metasurface for single-sensor and single-frequency microwave imaging. *Scientific Reports*, *6*(1), 23731. doi:10.1038rep23731 PMID:27025907

Li, Y. F., & Zhou, Z. H. (2014). Towards making unlabeled data never hurt. *IEEE Transactions on Pattern Analysis and Machine Intelligence*, *37*(1), 175–188. PMID:26353217

Li, Y., Dai, W., Ming, Z., & Qiu, M. (2015). Privacy protection for preventing data over-collection in smart city. *IEEE Transactions on Computers*, *65*(5), 1339–1350. doi:10.1109/TC.2015.2470247

Li, Y., & Wang, Y. (2018). Synergistic strategy for the geographical traceability of wild Boletus tomentipes by means of data fusion analysis. *Microchemical Journal*, *140*, 38–46. doi:10.1016/j.microc.2018.04.001

Llinas, J., Bowman, C., Rogova, G., Steinberg, A., Waltz, E., & White, F. (2004). *Revisiting the JDL data fusion model II*. Space and Naval Warfare Systems Command.

Lloret, J. (2013). *Underwater sensor nodes and networks*. Academic Press.

Lloret, J. (2013). Underwater sensor nodes and networks. *Sensors (Basel)*, *13*(9), 11782–11796. doi:10.3390130911782 PMID:24013489

Lloret, J., Sendra, S., Ardid, M., & Rodrigues, J. J. (2012). Underwater wireless sensor communications in the 2.4 GHz ISM frequency band. *Sensors (Basel)*, *12*(4), 4237–4264. doi:10.3390120404237 PMID:22666029

Llor, J., & Malumbres, M. P. (2012). Underwater wireless sensor networks: How do acoustic propagation models impact the performance of higher-level protocols? *Sensors (Basel)*, *12*(2), 1312–1335. doi:10.3390120201312 PMID:22438712

Lu, F., Mirza, D., & Schurgers, C. (2010, September). D-sync: Doppler-based time synchronization for mobile underwater sensor networks. In *Proceedings of the Fifth ACM International Workshop on UnderWater Networks* (p. 3). ACM. 10.1145/1868812.1868815

Lu, J. (2012). *Reinforcement learning based energy efficient routing protocols for underwater acoustic wireless sensor networks*. University of Western Australia.

Luo, J., Fan, L., Wu, S., & Yan, X. (2018). Research on localization algorithms based on acoustic communication for underwater sensor networks. *Sensors (Basel)*, *18*(2), 67. doi:10.339018010067 PMID:29301369

Luo, R. C., & Kay, M. G. (1988, August). Multisensor integration and fusion: issues and approaches. In *Sensor Fusion* (Vol. 931, pp. 42–49). International Society for Optics and Photonics. doi:10.1117/12.946646

Luo, R. C., Yih, C. C., & Su, K. L. (2002). Multisensor fusion and integration: Approaches, applications, and future research directions. *IEEE Sensors Journal*, *2*(2), 107–119. doi:10.1109/JSEN.2002.1000251

Luo, Y., Pu, L., Peng, Z., & Shi, Z. (2016). RSS-based secret key generation in underwater acoustic networks: Advantages, challenges, and performance improvements. *IEEE Communications Magazine*, *54*(2), 32–38. doi:10.1109/MCOM.2016.7402258

Lu, Y., He, R., Chen, X., Lin, B., & Yu, C. (2020). Energy-efficient depth-based opportunistic routing with q-learning for underwater wireless sensor networks. *Sensors (Basel)*, *20*(4), 1–25. doi:10.339020041025 PMID:32074989

Lv, C., Wang, S., Tan, M., & Chen, L. (2014). UA-MAC: An underwater acoustic channel access method for dense mobile underwater sensor networks. *International Journal of Distributed Sensor Networks*, *10*(2), 374028. doi:10.1155/2014/374028

Magalhaes, F. A., Vannozzi, G., Gatta, G., & Fantozzi, S. (2015). Wearable inertial sensors in swimming motion analysis: A systematic review. *Journal of Sports Sciences*, *33*(7), 732–745. doi:10.1080/02640414.2014.962574 PMID:25356682

Majumder, S., Scheding, S., & Durrant-Whyte, H. F. (2001). Multisensor data fusion for underwater navigation. *Robotics and Autonomous Systems*, *35*(2), 97–108. doi:10.1016/S0921-8890(00)00126-3

Malek & Miroslaw. (1980). A comparison connection assignment for diagnosis of multiprocessor systems. In *Proceedings of the 7th annual symposium on Computer Architecture* (pp. 31-36). ACM.

Manjula, R. B., & Manvi, S. S. (2013). Coverage optimization based sensor deployment by using PSO for anti-submarine detection in UWASNs. *Proceedings of the International Symposium on Ocean Electronics*. 10.1109/SYMPOL.2013.6701906

Maqsood, H., Javaid, N., Zain-ul-Abidin, M., Ejaz, M., & Shakeel, U. (2015, November). Energy balanced interference aware energy efficient depth base routing protocol for UWSNs. In *2015 10th international conference on broadband and wireless computing, communication and applications (BWCCA)* (pp. 13-19). IEEE. 10.1109/BWCCA.2015.85

Matsakis, L. (2018). *What would really happen if Russia Attacked Undersea Internet Cables, Wired*. Retrieved from https://www.wired.com/story/russia-undersea-internet-cables/

Mauldin, A. (2017). *Cable Breakage: When and How Cables Go Down, Telegeography*. Retrieved from https://blog.telegeography.com/what-happens-when-submarine-cables-break

McGillivary, P. (2018). Why Maritime Cybersecurity Is an Ocean Policy Priority and How It Can Be Addressed. *Marine Technology Society Journal*, *52*(5), 44–57. doi:10.4031/MTSJ.52.5.11

McGlynn, M. J., & Borbash, S. A. (2001). Birthday protocols for low energy deployment and flexible neighbor discovery in ad hoc wireless networks. *Proceedings of the 2nd ACM International Symposium on Mobile Ad Hoc Networking & Computing*, 137–145. 10.1145/501416.501435

Menon, K. A. U., Divya, P., & Ramesh, M. V. (2012). Wireless sensor network for river water quality monitoring in India. *Proceedings of the IEEE International Conference on Computing, Communication and Networking Technologies (ICCCNT)*. 10.1109/ICCCNT.2012.6512437

Min, H., Cho, Y., & Heo, J. (2012). Enhancing the reliability of head nodes in underwater sensor networks. *Sensors (Basel)*, *12*(2), 1194–1210. doi:10.3390120201194 PMID:22438707

Min, R., Bhardwaj, M., Cho, S.-H., Ickes, N., Shih, E., Sinha, A., & .... (2002). Energy-Centric Enabling Technologies for Wireless Sensor Networks. *IEEE Wireless Communications*, *9*(4), 28–39. doi:10.1109/MWC.2002.1028875

Mitchell, H. B. (2007). *Multi-sensor data fusion: an introduction*. Springer Science & Business Media.

Mitchell, T. M. (1997). *Machine Learning* (1st ed.). McGraw-Hill, Inc.

Mohamed, N. J., Jawhar, I., Al-Jaroodi, J., & Zhang, L. (2011). Sensor network architectures for monitoring underwater pipelines. *Sensors (Basel)*, *11*(11), 10738–10764. doi:10.3390111110738 PMID:22346669

Moradi, M., Rezazadeh, J., & Ismail, A. S. (2012). A reverse localization scheme for underwater acoustic sensor networks. *Sensors (Basel)*, *12*(4), 4352–4380. doi:10.3390120404352 PMID:22666034

Mortazavi, E., Javidan, R., Dehghani, M. J., & Kavoosi, V. (2017). A robust method for underwater wireless sensor joint localization and synchronization. *Ocean Engineering*, *137*, 276–286. doi:10.1016/j.oceaneng.2017.04.006

Murphy, J. B., Gutierrez-Alonso, G., Nance, R. D., Fernandez-Suarez, J., Keppie, J. D., Quesada, C., & Dostal, J. (2006). Origin of the Rheic Ocean: Rifting along a Neoproterozoic suture. *Geology*, *34*(5), 325–328. doi:10.1130/G22068.1

Nadeem, J., Naveed, I., Ashfaq, A., Nabil, A., & Umar, Q. (2015). An Efficient Data-Gathering Routing Protocol for Underwater Wireless Sensor Networks. Sensors. *Sensors (Basel)*, *15*(11), 29149–29181. doi:10.3390151129149 PMID:26593924

Nasir, H., Javaid, N., Ashraf, H., Manzoor, S., Khan, Z. A., Qasim, U., & Sher, M. (2014, November). CoDBR: cooperative depth based routing for underwater wireless sensor networks. In *2014 Ninth International Conference on Broadband and Wireless Computing, Communication and Applications* (pp. 52-57). IEEE. 10.1109/BWCCA.2014.45

Nasridinov, A., Ihm, S.-Y., & Park, Y.-H. (2013). Skyline-based aggregator node selection in wireless sensor networks. *International Journal of Distributed Sensor Networks*, *9*(9), 1–7. doi:10.1155/2013/356194

Nasri, N., Mnasri, S., & Val, T. (2019). 3D node deployment strategies prediction in wireless sensors network. *International Journal of Electronics*, 1–19.

Nguyen, V. D., Luong, N. S., & Patzold, M. (2017). A method to estimate the path gains and propagation delays of underwater acoustic channels using the arrival phase information of the multipath components. *International Journal of Electronics and Communications*, *73*, 129–138. doi:10.1016/j.aeue.2017.01.003

Nicosevici, T., Garcia, R., Carreras, M., & Villanueva, M. (2004, November). A review of sensor fusion techniques for underwater vehicle navigation. In Oceans' 04 MTS/IEEE Techno-Ocean'04 (IEEE Cat. No. 04CH37600) (Vol. 3, pp. 1600-1605). IEEE. doi:10.1109/OCEANS.2004.1406361

Nightingale, D., Troxel, S., & Libby, B. (2002). Use of the internet in maintaining and operating the Intermediate Scale Measurement System (ISMS) underwater test facility. *The Journal of the Acoustical Society of America*, *112*(5), 2328–2328. doi:10.1121/1.4779397

Otnes, R., Asterjadhi, A., Casari, P., Goetz, M., Husøy, T., Nissen, I., … Zorzi, P. (2012). Underwater Acoustic Networking Techniques. Springer.

Ovaliadis, K., Savage, N., & Kanakaris, V. (2010). Energy efficiency in underwater sensor networks: A research review. *Journal of Engineering Science and Technology Review*, *3*(1), 151–156. doi:10.25103/jestr.031.25

Oxenham, M. G., Kewley, D. J., & Nelson, M. J. (1996, November). Performance assessment of data fusion systems. In *Proceeding of 1st Australian Data Fusion Symposium* (pp. 36-41). IEEE. 10.1109/ADFS.1996.581078

Paltrinieri, N., Landucci, G., & Salvo Rossi, P. (2017, June). Real-Time Data for Risk Assessment in the Offshore Oil and Gas Industry. In *ASME 2017 36th International Conference on Ocean, Offshore and Arctic Engineering*. American Society of Mechanical Engineers Digital Collection. 10.1115/OMAE2017-61486

Pandarinath, P. (2011) Secure localization with defense against selective forwarding attacks in wireless sensor networks. *Proceedings of the 2011 IEEE 3rd International Conference on Electronics Computer Technology (ICECT)*, 5, 112–117. 10.1109/ICECTECH.2011.5941968

Pant, N. (2017). A comparative study of congestion control in wireless sensor networks using efficient resource management. In *Proc. the 2nd Int. Conf. Advancement in Engineering, Applied Science and Management (ICAEASM)* (pp. 320-325). Academic Press.

Paradis, S., Roy, J., & Treurniet, W. (1998). Integration of all data fusion levels using a blackboard architecture. *Proceedings of the 1998 International Conference on Data Fusion-EuroFusion'98, October 1998, Great Malvern, UK,* 195-202.

Paradis, L., & Han, Q. (2007). A survey of fault management in wireless sensor networks. *Journal of Network and Systems Management*, *15*(2), 171–190. doi:10.100710922-007-9062-0

Park, M. K., & Rodoplu, V. (2007). UWAN-MAC: An energy-efficient MAC protocol for underwater acoustic wireless sensor networks. *IEEE Journal of Oceanic Engineering*, *32*(3), 710–720. doi:10.1109/JOE.2007.899277

Pau, L. F. (1988). Sensor data fusion. *Journal of Intelligent & Robotic Systems*, *1*(2), 103–116. doi:10.1007/BF00348718

Peach, C., & Yarali, A. (2013). An overview of underwater sensor networks. *Proceedings of the routing techniques regarding network layer*, 31–36.

Pelizzari, U., & Tovaglieri, S. (2004). *Manual of Freediving: Underwater on a single breath*. Idelson-Gnochi Limited Publishers.

Peng, Z., Le, S., Zuba, M., Mo, H., Zhu, Y., Pu, L., . . . Cui, J. H. (2011, June). Aqua-TUNE: A testbed for underwater networks. In OCEANS 2011 IEEE-Spain (pp. 1-9). IEEE. doi:10.1109/Oceans-Spain.2011.6003661

Peng, C., & Du, X. (2017). SDBR: A Secure Depth-Based Anonymous Routing Protocol in Underwater Acoustic Networks. *International Journal of Performability Engineering*, *13*, 731–741. doi:10.23940/ijpe.17.05.p16.731741

Peng, C., Du, X., Li, K., & Li, M. (2016). An ultra-lightweight encryption scheme in underwater acoustic networks. *Journal of Sensors*, *2016*, 8763528. doi:10.1155/2016/8763528

Perez, R. M., Pintado, J. G., & Gómez, A. S. (2012). A real-time measurement system for long-life flood monitoring and warning applications. *Sensors (Basel)*, *12*(4), 4213–4236. doi:10.3390120404213 PMID:22666028

Perrine, K., Nieman, K., Henderson, T., Lent, K., & Evans, B. (2010). Doppler estimation and correction for shallow underwater acoustic communications. In *2010 Conference Record of the Forty Fourth Asilomar Conference on Signals, Systems and Computers* (pp. 746-750). IEEE. 10.1109/ACSSC.2010.5757663

Plate, R., & Wakayama, C. (2015). Utilizing kinematics and selective sweeping in reinforcement learning-based routing algorithms for underwater networks. *Ad Hoc Networks*, *34*, 105–120. doi:10.1016/j.adhoc.2014.09.012

Pompili, D., Melodia, T., & Akyildiz, I. F. (2009). Three-dimensional and two-dimensional deployment analysis for underwater acoustic sensor networks. *Ad Hoc Networks*, *7*(4), 778–790. doi:10.1016/j.adhoc.2008.07.010

Porkodi, K.P., & Zubair Rahman, A.M.J.M. (2018). *Enhanced Underwater Wireless Sensor Networks Security with Tic-Tac-Toe Ai-Minimax Algorithm in Game Theory*. Academic Press.

Postolache, O., Pereira, J. D., & Girao, P. S. (2014). Wireless sensor network based solution for environmental monitoring: Water quality assessment case study,'' IET Sci. *Measurement Techniques*, *8*(6), 610–616.

Prasan, U. D., & Murugappan, D. S. (2012). S.A. Combinatorial Multi-Objective Trust Model for Efficient and Secured Routing in UWSN. *Int. J. Eng. Trends Technol.*, *22*(6), 259–264. doi:10.14445/22315381/IJETT-V22P255

Prasan, U. D., & Murugappan, D. S. (2012). Underwater Sensor Networks: Architecture, Research Challenges and Potential Applications. *International Journal of Engineering Research and Applications*, *2*(2), 251–256.

Pratap, N., & Arya, S. (2015). A Review of Devnagari Character Recognition from Past to Future. *International Journal of Computer Science and Telecommunications*, *3*(6), 77–82.

Preparata, M., Metze, G., & Chien, R. T. (1967). On the connection assignment problem of diagnosable systems. *IEEE Transactions on Electronic Computers*, *6*(6), 848–854. doi:10.1109/PGEC.1967.264748

Proakis, J. G. (2001). Block and Convolutional Channel Codes. In *Digital Communications*. McGraw-Hill.

Qiu, W. (2011). *Submarine cables cut after Taiwan earthquake in Dec 2006*. Submarine Cable Networks Website.

Qiu, J., Xing, Z., Zhu, C., Lu, K., He, J., Sun, Y., & Yin, L. (2019). Centralized fusion based on interacting multiple model and adaptive Kalman filter for target tracking in underwater acoustic sensor networks. *IEEE Access: Practical Innovations, Open Solutions*, *7*, 25948–25958. doi:10.1109/ACCESS.2019.2899012

Radha, S., Josemin, G., & Bala, J. (2017). *Flooding in static and free deployment for underwater wireless sensor network. International Conference on Signal Processing and Communication (ICSPC'17)*, 28-32.

Rahman, Z. U. (2016). On Utilizing Static Courier Nodes to Achieve Energy Efficiency with Depth Based Routing for Underwater Wireless Sensor Networks. *IEEE 30th International Conference on Advanced Information Networking and Applications (AINA)*, 1184-1191. doi: 10.1109/AINA.2016.170

Rana, A., & Salau, A. (2018). Recent Trends in IoT, its Requisition with IoT Built Engineering: A Review. Springer.

Rao, M., & Kamila, N. K. (2018). Bayesian network based energy efficient ship motion monitoring. *Karbala International Journal of Modern Science*, *4*(1), 69–85. doi:10.1016/j.kijoms.2017.11.001

Rauchenstein, L. T., Vishnu, A., Li, X., & Deng, Z. D. (2018). Improving underwater localization accuracy with machine learning. *The Review of Scientific Instruments*, *89*(7), 1–13. doi:10.1063/1.5012687 PMID:30068124

Reiber, J. (2018). The Fastest Way Across the Seas: Cyberspace Operations and Cybersecurity in the Indo-Pacific. In Eurasia's Maritime Rise and Global Security (pp. 83-94). Palgrave Macmillan.

Roach, J. (2018a). *Under the sea, Microsoft tests a datacenter that's quick to deploy, could provide internet connectivity for years, Microsoft*. Retrieved from https://news.microsoft.com/features/under-the-sea-microsoft-tests-a-datacenter-thats-quick-to-deploy-could-provide-internet-connectivity-for-years/

Roach, J. (2018b). *Project Natick, Microsoft*. Retrieved from https://natick.research.microsoft.com/

Romer, K., & Mattern, F. (2004). The design space of wireless sensor networks. *IEEE Wireless Communications*, *11*(6), 54–61. doi:10.1109/MWC.2004.1368897

Ruiz, L. B. S. I., Wong, H. C., Nogueira, J. M., & Loureiro, A. A. (2004). Fault management in event-driven wireless sensor networks. In ACM international symposium on Modeling, analysis and simulation of wireless and mobile systems (pp. 149-1567). ACM.

Saeed, N., & Alouini, M. (2019). On achievable accuracy of localization in magnetic induction-based internet of underground things for oil and gas reservoirs. CoRR, vol. abs/1803.02442

Saeed, N., & Alouini, M. (2019). Towards the internet of underground things: A systematic survey. CoRR, vol. abs/1902.03844

Saeed, N., & Celik, A. (2018). Underwater optical wireless communications, networking, and localization: A survey. CoRR, vol. abs/1803.02442

Saeed, H., Ali, S., Rashid, S., Qaisar, S., & Felemban, E. (2014). Reliable monitoring of oil and gas pipelines using wireless sensor network (WSN): REMONG. *Proceedings of the IEEE International Conference on System of Systems Engineering (SOSE)*. 10.1109/SYSOSE.2014.6892493

Saeed, N., & Alouini, M. (2019). Outlier detection and optimal anchor placement for 3-D underwater optical wireless sensor network localization. *IEEE Transactions on Communications*, *67*(1), 611–622. doi:10.1109/TCOMM.2018.2875083

Saeed, N., & Alouini, M. (2019). Performance analysis of connectivity and localization in multi-Hop underwater optical wireless sensor networks. *IEEE Transactions on Mobile Computing*, *18*(11), 2604–2615. doi:10.1109/TMC.2018.2878672

Saeed, N., & Alouini, M. (2019). *Towards the Internet of X-things: New Possibilities for Underwater*. Underground, and Outer Space Exploration.

Saeed, N., Celik, A., Al-Naffouri, T. Y., & Alouini, M. S. (2019). Underwater optical wireless communications, networking, and localization: A survey. *Ad Hoc Networks*, *94*, 101935. doi:10.1016/j.adhoc.2019.101935

Sahoo, M. N., & Khilar, P. M. (2014). Diagnosis of wireless sensor networks in presence of permanent and intermittent faults. *Wireless Personal Communications*, *78*(2), 1571–1591. doi:10.100711277-014-1836-6

Saifullah, A., Xu, Y., Lu, C., & Chen, Y. (2010). Real-Time Scheduling for WirelessHART Networks. In *2010 31st IEEE Real-Time Systems Symposium*, (pp. 150-159). San Diego, CA: IEEE.

Saira Banu Atham, Prabaharan, & Sangeetha. (2019). Method for Providing Efficient Real-Time Multimedia Communication using VoIP Over Communication Channels. *International Journal of Advanced Science and Technology*, 1832-1843.

Sajid, M., Wahid, A., Pervaiz, K., Khizar, M., Khan, Z. A., Qasim, U., & Javaid, N. (2016, July). SMIC: Sink mobility with incremental cooperative routing protocol for underwater wireless sensor networks. In *2016 10th International Conference on Complex, Intelligent, and Software Intensive Systems (CISIS)* (pp. 256-263). IEEE.

Salama, E. M., & Abdelrahman, A. (2019). *Security of Critical Cyber-Physical Systems: Fundamentals and Optimization* (Doctoral dissertation). Virginia Tech.

Salva-Garau, F., & Stojanovic, M. (2003). Multi-cluster protocol for adhoc mobile underwater acoustic networks. In *OCEANS 2003* (Vol. 1, pp. 91–98). Academic Press.

Sandeep, D. N., & Kumar, V. (2017). Review on clustering, coverage and connectivity in underwater wireless sensor networks: A communication techniques perspective. *IEEE Access: Practical Innovations, Open Solutions*, *5*, 11176–11199. doi:10.1109/ACCESS.2017.2713640

Sangwan, A., & Singh, R. P. (2015). Survey on coverage problems in wireless sensor networks. *Wireless Personal Communications*, *80*(4), 1475–1500. doi:10.100711277-014-2094-3

Sankarasubramaniam, Y., Akan, Ö. B., & Akyildiz, I. F. (2003, June). ESRT: event-to-sink reliable transport in wireless sensor networks. In *Proceedings of the 4th ACM international symposium on Mobile ad hoc networking & computing* (pp. 177-188). ACM. 10.1145/778415.778437

Sankarasubramaniam, Y., Akyildiz, I. F., & McLaughlin, S. W. (2003). Energy efficiency based packet size optimization in wireless sensor networks. *IEEE Int. Workshop on Sensor Network Protocols and Applications*, 1-8. 10.1109/SNPA.2003.1203351

Sarao, P., Chattu, K., & Swapna, C. (2018). *Routing issues and challenges in Underwater Wireless Sensor Networks.* Academic Press.

Seah, W. K. G., & Tan, H.-X. (2007). Multipath virtual sink architecture for underwater sensor networks. *Proceedings of the OCEANS—Asia Pacific*, 1–6.

Seders, L. A., Shea, C. A., Lemmon, M. D., Maurice, P. A., & Talley, J. W. (2007). LakeNet: An integrated sensor network for environmental sensing in lakes. *Environmental Engineering Science*, *24*(2), 183–191. doi:10.1089/ees.2006.0044

Sendra, S., Lamparero, J. V., Lloret, J., & Ardid, M. (2011). Underwater Communications in wireless sensor networks using WLAN at 2.4GHz. *2011 Eighth IEEE International Conference on Mobile Ad-Hoc and Sensor Systems*, 892-897. 10.1109/MASS.2011.106

Sendra, S., Lloret, J., Garcia, M., & Toledo, J. F. (2011). Power saving and energy optimization techniques for Wireless Sensor Networks. *Journal of Communication*, *6*(6), 439–459.

Senel, F. (2016). Coverage-aware connectivity-constrained unattended sensor deployment in underwater acoustic sensor networks. *Wireless Communications and Mobile Computing*, *16*(14), 2052–2064. doi:10.1002/wcm.2667

Senel, F., Melike, K., Kantarci, E., & Yilmaz, T. (2014). Self-deployment of mobile underwater acoustic sensor networks for maximized coverage and guaranteed connectivity. *Ad Hoc Networks*, 1–25.

Shah, K., & Kumar, M. (2007). Distributed independent reinforcement learning (DIRL) approach to resource management in wireless sensor networks. *Proceeding of International Conference on Mobile Adhoc and Sensor Systems*, 1-9. 10.1109/MOBHOC.2007.4428658

Shams Shamsabad Farahani, S. (2018). Congestion Control Approaches Applied to Wireless Sensor Networks: A Survey. *Journal of Electrical and Computer Engineering Innovations*, *6*(2), 125–144.

Shareef, A., Zhu, Y., & Musavi, M. (2008). Localization using neural networks in wireless sensor networks. *Proceedings of the 1st International Conference on Mobile Wireless Middleware, Operating Systems, and Applications, 4*, 1-7.

Shchurov, V. A. (1991). Coherent and diffusive fields of underwater acoustic ambient noise. *The Journal of the Acoustical Society of America*, *90*(2), 991–1001. doi:10.1121/1.401913

Shen, J., Tan, H. W., Wang, J., Wang, J. W., & Lee, S. Y. (2015). A novel routing protocol providing good transmission reliability in underwater sensor networks. *J. Internet Technol.*, *16*, 171–178.

Shih, E., Cho, S., Lee, F. S., Calhoun, B. H., & Chandrakasan, A. (2004). Design Considerations for Energy-Efficient Radios in Wireless Microsensor Networks. *The Journal of VLSI Signal Processing*, *37*(1), 77–94. doi:10.1023/B:VLSI.0000017004.57230.91

Shin, D., & Kim, D. (2009, December). FRT: Fast and reliable transport protocol for underwater wireless sensor networks. In *2009 IEEE Asia-Pacific Services Computing Conference (APSCC)* (pp. 402-407). IEEE. 10.1109/APSCC.2009.5394096

Shin, S. Y., & Park, S. H. (2011). A Cost Effective Block Framing Scheme for Underwater Communication. *Sensors (Basel), 11*(12), 11717–11735. doi:10.3390111211717 PMID:22247689

Shiozaki, A., Okuno, K., Suzuki, K., & Segawa, T. (1991). A Hybrid ARQ Scheme with Adaptive Forward Error Correction for Satellite Communications. *IEEE Transactions on Communications, 39*(4), 482–484. doi:10.1109/26.81734

Shu, J., Liu, S., Liu, L., Zhan, L., & Hu, G. (2017). Research on link quality estimation mechanism for wireless sensor networks based on support vector machine. *Chinese Journal of Electronics, 26*(2), 377–384. doi:10.1049/cje.2017.01.013

Silverajan, B., Ocak, M., & Nagel, B. (2018, July). Cybersecurity Attacks and Defences for Unmanned Smart Ships. In *2018 IEEE International Conference on Internet of Things (iThings) and IEEE Green Computing and Communications (GreenCom) and IEEE Cyber, Physical and Social Computing (CPSCom) and IEEE Smart Data (SmartData)* (pp. 15-20). IEEE. 10.1109/Cybermatics_2018.2018.00037

Simeone, O. (2018). A very brief introduction to machine learning with applications to communication systems. *IEEE Transactions on Cognitive Communications and Networking, 4*(4), 648–664. doi:10.1109/TCCN.2018.2881442

Simola, J., & Rajamäki, J. (2018). *Improving Cyber Situational Awareness in Maritime Surveillance*. Academic Conferences and Publishing International Limited.

Simon, A., Deo, M., Selvam, V., & Babu, R. (2016). An overview of machine learning and its applications. *International Journal of Electrical Sciences & Engineering, 1*(1), 22–24.

Simplicio, M. A. Jr, Silva, M. V., Alves, R. C. A., & Shibata, T. K. C. (2017). Lightweight and escrow-less authenticated key agreement for the internet of things. *Computer Communications, 98*, 43–51. doi:10.1016/j.comcom.2016.05.002

Singer, A., Yang, S., & Oelze, M. (2017). Acoustic communications: Through soils, sands, water, and tissue. *The Journal of the Acoustical Society of America, 141*(5), 3986–3987. doi:10.1121/1.4989117

Singh, D. A. A. G., & Leavline, E. J. (2013). Eercm: energy efficient and reliable communication model for achieving qos in underwater sensor networks. *Communications, 4*(5).

Singh, T., Bedi, R., & Marwaha, M. (2013). Need and role of underwater acoustic sensor networks. *International Journal of Electronics Communication and Computer Engineering, 4*(4), 1155–1163.

Sklar, B. (2003). Modulation and Coding Trade-Offs. In *Digital Communications*. Prentice Hall PTR.

Song, J., Han, S., & Mok, A. K. (2008). WirelessHART: Applying Wireless Technology in Real-Time Industrial Process Control. *IEEE Real-Time and Embedded Technology and Applications Symposium*, (pp. 377-386). 10.1109/RTAS.2008.15

Song, Y., & Deng, Y. (2019). A new method to measure the divergence in evidential sensor data fusion. *International Journal of Distributed Sensor Networks, 15*(4), 1550147719841295. doi:10.1177/1550147719841295

Soreide, N. N., Woody, C. E., & Holt, S. M. (2004). Overview of ocean based buoys and drifters: Present applications and future needs, *Proceedings of 16th International Conference on Interactive Information and Processing Systems (IIPS) for Meteorology, Oceanography and Hydrology.*

Sovacool, B. K., & Bulan, L. C. (2011). Behind an ambitious megaproject in Asia: The history and implications of the Bakun hydroelectric dam in Borneo. *Energy Policy, 39*(9), 4842–4859. doi:10.1016/j.enpol.2011.06.035

Sozer, E. M., Stojanovic, M., & Proakis, J. G. (2000). Underwater acoustic networks. *IEEE Journal of Oceanic Engineering, 25*(1), 72–83. doi:10.1109/48.820738

Spiegel, M. R. (1992). *Theory and Problems of Probability and Statistics*. McGraw-Hill.

Srinivas, S., Ranjitha, P., Ramya, R., & Narendra, K. G. (2012). Investigation of oceanic environment using large-scale UWSN and UANETs. *Proceedings of the IEEE International Conference onWireless Communications, Networking and Mobile Computing (WiCOM)*. 10.1109/WiCOM.2012.6478552

Srinivas, K., & Dua, M. (2019, August). Fog Computing and Deep CNN Based Efficient Approach to Early Forest Fire Detection with Unmanned Aerial Vehicles. In *International Conference on Inventive Computation Technologies* (pp. 646-652). Springer.

Srinivas, K., & Dua, M. (2019, June). Early Discovery of Disaster Events from Sensor Data Using Fog Computing. In *International Conference on Intelligent Computing, Information and Control Systems* (pp. 114-121). Springer.

Stojanovic, M. (1996). Recent advances in high-speed underwater acoustic communications. *Oceanic Engineering IEEE Journal of*, *21*(2), 125–136. doi:10.1109/48.486787

Stojanovic, M., & Preisig, J. (2009). Underwater acoustic communication channels: Propagation models and statistical characterization. *IEEE Communications Magazine*, *47*(1), 84–89. doi:10.1109/MCOM.2009.4752682

Sun, Y., Wang, Z., & China, P. R. (2018). Fault diagnosis method of autonomous underwater vehicle based on deep learning. *Proceeding of IOP Conf. Series: Materials Science and Engineering*, 1-6.

Syed, A. A., Ye, W., & Heidemann, J. (2008, April). T-Lohi: A new class of MAC protocols for underwater acoustic sensor networks. In *IEEE INFOCOM 2008-The 27th Conference on Computer Communications* (pp. 231-235). IEEE. 10.1109/INFOCOM.2008.55

Syed, A. B., Andrabi, & Manoj, K. (2017). A comparative study of routing protocols in underwater wireless sensor networks. *International Journal of Computer Science and Communication*, *8*(1), 27–29.

Tan, X., Sun, Z., & Akyildiz, I. F. (2015). Wireless underground sensor networks: MI-based communication systems for underground applications. *IEEE Antennas & Propagation Magazine*, *57*(4), 74–87. doi:10.1109/MAP.2015.2453917

Tayal, S., & Nandi, A. (2017). Effect of FIBL in-conjunction with channel parameters on analog and RF FOM of FinFET. *Superlattices and Microstructures*, *105*, 152–162. doi:10.1016/j.spmi.2017.03.018

Teguig, E. H., Touati, Y., & Ali-Cherif, A. (2017). ECC Based-Approach for Keys Authentication and Security in WSN. *Proceedings of the 2017 9th IEEE-GCC Conference and Exhibition (GCCCE)*, 1–4. 10.1109/IEEEGCC.2017.8447901

Telegeography. (2018). *Submarine Cable Map*. Retrieved from https://www.submarinecablemap.com/#/

Thayer, C. A. (2012, February). The rise of China and maritime security in southeast Asia. In *Presentation to 11th IDE Forum*, (p. 11). Academic Press.

The Economist. (2008). *Of cables and conspiracies*. Retrieved from https://www.economist.com/international/2008/02/07/of-cables-and-conspiracies

The Royal Society. (2017). *Machine learning: the power and promise of computers that learn by example*. Author.

Thomopoulos, S. C. A. (1991). Theories in distributed decision fusion. *IFAC Proceedings Volumes, 24*(5), 195-200.

Tilak, S. N., Abu-Ghazaleh, B., & Heinzelman, W. (2002). Infrastructure tradeoffs for sensor networks. In *Proceedings of the 1st ACM international workshop on Wireless sensor networks and applications, WSNA '02*, (pp. 49–58). ACM. 10.1145/570738.570746

Tsai. (2017). *Hybrid localization approach for underwater sensor network*. Academic Press.

Tuna, G., & Gungor, V. C. (2017). A survey on deployment techniques, localization algorithms, and research challenges for underwater acoustic sensor networks. *International Journal of Communication Systems*, *30*(17), 1–21. doi:10.1002/dac.3350

U.S. (2018). *Treasury Sanctions Russian Federal Security Service Enablers, U.S. Department of the Treasure*. Retrieved from https://home.treasury.gov/news/press-releases/sm0410

Ubaldi, B. (2013). *Open government data*. Academic Press.

Umrao. G. (2020). *Undersea Cables and the Challenges of Protecting Seabed Lines of Communication.* Center for International Maritime Security. Retrieved 9 May 2020, from http://cimsec.org/undersea-cables-challenges-protecting-seabed-lines-communication/35889

Vancouver, C. A. S. I. S. (2018). Russian Deep-Sea Operations and Canadian Cybersecurity Issues. *The Journal of Intelligence, Conflict, and Warfare*, *1*(2), 6–6. doi:10.21810/jicw.v1i2.641

Varadharajan, K. (2018). Secure localization using coordinated gradient descent technique for underwater wireless sensor networks. *ICTACT J. Commun. Technol.*, *9*(1), 1716–1720. doi:10.21917/ijct.2018.0252

Vaseghi, B., Pourmina, M.A., & Mobayen, S. (2017). Finite-time chaos synchronization and its application in wireless sensor networks. *Trans. Inst. Meas. Control.*

Vaseghi, B., Pourmina, M. A., & Mobayen, S. (2017). Secure communication in wireless sensor networks based on chaos synchronization using adaptive sliding mode control. *Nonlinear Dynamics*, *89*(3), 1689–1704. doi:10.100711071-017-3543-9

Vieira, L. F. M., Almiron, M. G., & Loureiro, A. A. F. (2011). 3D MANETs: Link Probability, Node Degree, Network Coverage and applications. *IEEE Wireless Communications and Networking Conference*, 2042-2047, 10.1109/WCNC.2011.5779443

Vikas Juneja & Gupta. (2018). *Security against Vampire Attack in ADHOC Wireless Sensor Network: Detection and Prevention Techniques*. Academic Press.

Vince, G. (2013). *BBC - Future - The rising and sinking threats to our cities* Available on: http://www.bbc.com/future/story/20130613-the-rising-threat-to-ourcities. https://www.bbc.com/future/article/20171130-the-ambitious-plan-to-stop-the-ground-from-sinking

Vuran, M. C., & Akyildiz, I. F. (2009). Error Control in Wireless Sensor Networks: A Cross Layer Analysis. *IEEE/ACM Transactions on Networking*, *17*(4), 1186–1199. doi:10.1109/TNET.2008.2009971

Vuran, M., Salam, A., Wong, R., & Irmak, S. (2018). Internet of underground things in precision agriculture: Architecture and technology aspects. *Ad Hoc Networks*, *81*, 160–173. doi:10.1016/j.adhoc.2018.07.017

Wahid, A., & Dongkyun, K. (2010). Analyzing routing protocols for underwater wireless sensor networks. *International Journal of Communication Networks and Information Security*, *2*(3), 253.

Wahid, A., & Kim, D. (2012). An energy efficient localization-free routing protocol for underwater wireless sensor networks. *International Journal of Distributed Sensor Networks*, *8*(4), 307246. doi:10.1155/2012/307246

Waldmeyer, M., Tan, H. P., & Seah, W. K. G. (2011). Multi-stage AUV-aided localization for underwater wireless sensornetworks. *Proceedings of the IEEE Workshops of Advanced Information Networking and Applications (WAINA)*.

Walikar, G. A. (2018, February). Reliable and stable route construction in mobile ad hoc networks using R- learning techniques. *International Journal of Engineering Science Invention*, *2*(2), 58–63.

Waltz, E. L., & Buede, D. M. (1986). Data fusion and decision support for command and control. *IEEE Transactions on Systems, Man, and Cybernetics*, *16*(6), 865–879. doi:10.1109/TSMC.1986.4309005

Wan, C. Y., Eisenman, S. B., & Campbell, A. T. (2003, November). CODA: congestion detection and avoidance in sensor networks. In *Proceedings of the 1st international conference on Embedded networked sensor systems* (pp. 266-279). ACM. Retrieved from https://www.defencetalk.com/ntus-sense-ational-invention-helps-underwater-vessels-navigate-with-ease-45958/

Wanglai, G. (2018). BRICS cybersecurity cooperation: Achievements and deepening paths. *China Int'l Stud.*, *68*, 124.

Wang, P., Li, C., & Zheng, J. (2007). Distributed minimum-cost clustering protocol for underwater sensor networks (UWSNs). *Proceedings of the IEEE International Conference on Communications (ICC'07)*, 3510–3515. 10.1109/ICC.2007.580

Wang, Q., Dai, H. N., Li, X., Wang, H., & Xiao, H. (2016). On modeling eavesdropping attacks in underwater acoustic sensor networks. *Sensors (Basel)*, *16*(5), 721. doi:10.339016050721 PMID:27213379

Wang, S., & Shin, Y. (2019). 3D-Deployment of Magnetic Induction Relays in Underwater Sensor Networks. *International Conference on Information Networking (ICOIN)*, 222-226. 10.1109/ICOIN.2019.8718105

Wan, J., Chen, M., Xia, F., Di, L., & Zhou, K. (2013). From machine-to-machine communications towards cyber-physical systems. *Computer Science and Information Systems*, *10*(3), 1105–1128. doi:10.2298/CSIS120326018W

Winston, K. G. S., Xian, T. H., Zheng, L., & Marcelo, H. (2005). Multiple-UUV approach for enhancing connectivity in underwater ad-hoc sensor networks. *Proceedings of MTS/IEEE on Oceans*. DOI: 10.1109/OCEANS.2005.1640102

Winter, J., Xu, Y., & Lee, W.-C. (2005). Energy efficient processing of k nearest neighbor queries in location-aware sensor networks. *Proceeding of 2nd Annual International Conference on Mobile and Ubiquitous Systems: Networking and Services*, 281-292. 10.1109/MOBIQUITOUS.2005.28

Wong, Y. F., Ngoh, L. H., Wong, W. C., & Winston, K. G. (2006). Wakeup Scheme for Ocean Monitoring Underwater Sensor Networks (UWSN). *OCEANS 2006 - Asia Pacific:ScholarBank@NUS Repository*. doi:10.1109/OCEANSAP.2006.4393874

Wu, H., Chen, M., & Guan, X. (2012). A network coding based routing protocol for underwater sensor networks. *Sensors (Basel)*, *12*(4), 4559–4577. doi:10.3390120404559 PMID:22666045

Xiao, S., Gong, W., & Towsley, D. (2014). Communication Security and Key Safety. In *Dynamic Secrets in Communication Security* (pp. 5–11). Springer. doi:10.1007/978-1-4614-7831-7_2

Xie, P., & Cui, J. H. (2007, August). An FEC-based reliable data transport protocol for underwater sensor networks. In *2007 16th International Conference on Computer Communications and Networks* (pp. 747-753). IEEE.

Xie, P., Cui hong, J., & Lao, L. (2005). Vbf: Vector-based forwarding protocol for underwater sensor networks. *Proceeding of IFIP Networking*, 1216–1221.

Xie, P., Cui, J. H., & Lao, L. (2006, May). VBF: vector-based forwarding protocol for underwater sensor networks. In *International conference on research in networking* (pp. 1216-1221). Springer. 10.1007/11753810_111

Xie, P., Zhou, Z., Peng, Z., Cui, J. H., & Shi, Z. (2010). SDRT: A reliable data transport protocol for underwater sensor networks. *Ad Hoc Networks*, *8*(7), 708–722. doi:10.1016/j.adhoc.2010.02.003

Xu. M., Liu, L. (2016). Sender-receiver role-based energy-aware scheduling for Internet of Underwater Things. *IEEE Trans*.

Xu, M., Liu, G., & Guan, J. (2015). Towards a secure medium access control protocol for cluster-based underwater wireless sensor networks. *International Journal of Distributed Sensor Networks*, *11*(5), 325474. doi:10.1155/2015/325474

Yadav, S., & Kumar, V. (2019). Hybrid Compressive Sensing Enabled Energy Efficient Transmission of Multi-Hop Clustered UWSNs. *International Journal of Electronics and Communications*. doi:10.1016/j.aeue.2019.152836

Yang, B., Lei, Y., & Yan, B. (2016). Distributed multi-human location algorithm using Naive Bayes classifier for a binary pyroelectric infrared sensor tracking system. *IEEE Sensors Journal*, *16*(1), 216–223. doi:10.1109/JSEN.2015.2477540

Yang, G., Dai, L., & Wei, Z. (2019). Challenges, Threats. *Security Issues and New Trends of Underwater Wireless Sensor Networks*, *147*, 210–216.

Yang, Y., Gao, Z., Zhou, H., & Qiu, X. (2014). An Uncertainty-Based Distributed Fault Detection Mechanism for Wireless Sensor Networks. *Sensors (Basel)*, *14*(5), 7655–7683. doi:10.3390140507655 PMID:24776937

Yan, H., Shi, Z. J., & Cui, J. H. (2008, May). DBR: depth-based routing for underwater sensor networks. In *International conference on research in networking* (pp. 72-86). Springer. 10.1007/978-3-540-79549-0_7

Yick, J., Mukherjee, B., & Ghosal, D. (2008). Wireless sensor network survey. *Computer Networks*, *52*(12), 2292–2330. doi:10.1016/j.comnet.2008.04.002

Yingying, Z., Xia, L., & Shiliang, F. (2011). Deployment analysis in two-dimensional Underwater Acoustic Wireless Sensor Networks. *IEEE International Conference on Signal Processing, Communications and Computing (ICSPCC)*, 1-5. 10.1109/ICSPCC.2011.6061819

Yoon, S., Azad, A. K., Oh, H., & Kim, S. (2012). AURP: An AUV-aided underwater routing protocol for underwater acoustic sensor networks. *Sensors (Basel)*, *12*(2), 1827–1845. doi:10.3390120201827 PMID:22438740

Young, M. (2018). *How an Australian spy stopped China from growing internet influence South Pacific, News Limited.* Retrieved from https://www.news.com.au/technology/online/security/how-an-australian-spy-stopped-china-from-growing-internet-influence-south-pacific/news-story/4eb83151f54c66a50917a95096015666

Yuan, C., Chen, W., & Li, D. (2017). A Hierarchical Identity-Based Signcryption Scheme in Underwater Wireless Sensor Network. *Proceedings of the China Conference on Wireless Sensor Networks*, 44–54.

Yuan, Y., Liang, C., Kaneko, M., Chen, X., & Hogrefe, D. (2019). Topology Control for Energy-Efficient Localization in Mobile Underwater Sensor Networks Using Stackelberg Game. *IEEE Transactions on Vehicular Technology, 68*(2), 1487-1500. doi:10.1109/TVT.2018.2886904

Yu, L., Wang, N., & Meng, X. (2005). Real-time forest fire detection with wireless sensor networks. *Proceeding of International Conference on Wireless Communications, Networking and Mobile Computing*, 2, 1214–1217.

Zawodniok, M., & Jagannathan, S. (2007). Predictive congestion control protocol for wireless sensor networks. *IEEE Transactions on Wireless Communications*, *6*(11), 3955–3963. doi:10.1109/TWC.2007.051035

Zeng, C., King, D. J., Richardson, M., & Shan, B. (2017). Fusion of multispectral imagery and spectrometer data in UAV remote sensing. *Remote Sensing*, *9*(7), 696. doi:10.3390/rs9070696

Zhang, Z., & Blanco, L. H. (2014, May). A self adaptive MAC layer protocol for delay-tolerant underwater wireless sensor networks. In *2014 23rd Wireless and Optical Communication Conference (WOCC)* (pp. 1-8). IEEE. 10.1109/WOCC.2014.6839944

Zhang, J., Qiao, G., & Zhou, F. (2014, April). A relative clock based and energy sense MAC protocol for underwater acoustic communication networks. In *OCEANS 2014-TAIPEI* (pp. 1–7). IEEE. doi:10.1109/OCEANS-TAIPEI.2014.6964341

Zhang, Q., Yang, L. T., Chen, Z., & Li, P. (2018). A survey on deep learning for big data. *Information Fusion*, *42*, 146–157. doi:10.1016/j.inffus.2017.10.006

Zhang, S., Chen, K., Fan, Z., Cheng, E., & Su, A. W. (2019). The localization algorithm based on symmetry correction for underwater acoustic networks. *IEEE Access: Practical Innovations, Open Solutions*, *7*, 121127–121135. doi:10.1109/ACCESS.2019.2937106

Zhang, X., Han, X., Yin, J., & Sheng, X.-l. (2013). Study on Doppler effects estimate in underwater acoustic communication. In *Proceedings of Meetings on Acoustics ICA2013* (p. 070062). ASA. 10.1121/1.4799512

Zhang, Y., Jin, Z. G., Luo, Y. M., & Du, X. (2013). Node secure localization algorithm in underwater sensor network based on trust mechanism. *Jisuanji Yingyong*, *33*(5), 1208–1211. doi:10.3724/SP.J.1087.2013.01208

Zhang, Y., Sun, H., & Yu, J. (2015). Clustered routing protocol based on improved K-means algorithm for underwater wireless sensor networks. *Proceeding of IEEE International Conference on CYBER Technology in Automation, Control, and Intelligent Systems (CYBER)*, 1304-1309. 10.1109/CYBER.2015.7288132

Zhao, Q., Lambert, A., & Benson, C. R. (2014, November). FS-MACA: a future scheduling mac protocol in underwater acoustic networks. In *Proceedings of the International Conference on Underwater Networks & Systems* (pp. 1-5). 10.1145/2671490.2674470

Zhong, S., Zhong, H., Huang, X., Yang, P., Shi, J., Xie, L., & Wang, K. (2019). Networking Cyber-Physical Systems: System Fundamentals of Security and Privacy for Next-Generation Wireless Networks. In *Security and Privacy for Next-Generation Wireless Networks* (pp. 1–32). Springer. doi:10.1007/978-3-030-01150-5_1

Zhou, B., Li, S., Wang, W., Wang, J., Cheng, Y., & Wu, J. (2019). An Efficient Authentication Scheme Based on Deployment Knowledge Against Mobile Sink Replication Attack in UWSNs. *IEEE Internet of Things Journal, 6*(6), 9738-9747. doi:10.1109/JIOT.2019.2930823

Zhou, P., Jiang, S., Irissappane, A., Zhang, J., Zhou, J., & Teo, J. C. M. (2015). Toward energy-efficient trust system through watchdog optimization for WSNs. *IEEE Transactions on Information Forensics and Security*, *10*(3), 613–625. doi:10.1109/TIFS.2015.2389145

Zhou, Z., Yao, B., Xing, R., Shu, L., & Bu, S. (2016). E-CARP: An energy efficient routing protocol for UWSNs in the internet of underwater things. *IEEE Sensors Journal*, *16*(11), 4072–4082. doi:10.1109/JSEN.2015.2437904

Zhu, F., & Wei, J. (2018, December 16). An energy efficient routing protocol based on layers and unequal clusters in underwater wireless sensor networks. *Journal of Sensors*, *2018*, 1–10. Advance online publication. doi:10.1155/2018/5835730

Zhu, J., Du, X., Han, D., Wang, L., & Li, M. (2019). LEER: layer-based and energy-efficient routing protocol for underwater sensor networks. *Proceeding of International Conference on Ad Hoc Networks*, 306, 26-39. 10.1007/978-3-030-37262-0_3

About the Contributors

Nitin Goyal is working as an Associate Professor in the department of Chitkara University Research and Innovation Network (CURIN), Chitkara University, Punjab. He is PhD in Computer Engineering from NIT Kurukshetra, in underwater wireless sensor networks (UWSN). He obtained his B.Tech in 2007 and M.Tech in 2009 from Computer Science and Engineering under Kurukshetra University. He worked on underwater wireless sensor network (UWSN), wireless sensor network (WSN), mobile ad-hoc network (MANET) etc. He is having around 11 years of teaching experience along with Academic work. He has published approximately 40 research papers in various International/National Journals, Books and Conferences out of which 6 are SCI, 4 are SCOPUS, 5 is SCOPUS indexed book chapter. Also he has attended 13 workshops and 3 FDP. He has guided 7 M.Tech candidates and 2 PhD candidates are in process. He has also delivered expert lecture on NS2, national level FDP on "Research and Innovation". He is GATE and UGC-NET qualifier too. He is currently working in the area of machine learning in communication networks to solve some real world problems. Also he has 5 patents published and 2 project in DST. His research interests include MANET, FANET, UWSN and WSN. He is an active member of IEEE, CSI. He is the reviewer, TPC member, advisory board member, associate editor of many Conferences and Journals of repute.

* * *

Laith Alzubaidi is currently a Ph.D. student at Queensland University of Technology / Faculty of Science and Engineering, Brisbane, Australia. He received his Master's degree in computer science from the University of Missouri/USA in 2016. In his Master's thesis, he worked on the detection and classification of breast cancer with deep learning. Laith's broad research area falls in artificial intelligence and the Internet of Things (IoT). In particular, his research interests include Deep Learning for medical applications and IoT.

Saira Atham has received her B.E., degree in Computer Science and Engineering from Madras University, Chennai. She received her post-graduation degree in Computer Science and Engineering from Anna University, Chennai. She is currently working as an Associate Professor in Chitkara University Institute of Engineering and Technology, Chitkara University, Punjab, India. She has got 15 years of teaching experience. Her area of Research is VoIP. She has contributed many papers for enhancing the performance of Voice using the VoIP. She has published more than twenty papers in the journals and in the conference related to her research.

Manisha Bharti has completed her B.Tech in Electronics & Communication Engineering from Guru Nanak Dev University, Amritsar in 2002, M.Tech in Electronics & Communication Engineering from Punjab Technical University, Jalandhar in 2007. She received her Doctorate in 2017 from IKG Punjab Technical University, Jalandhar in the field of Optical Communication and Networks. Her areas of research include Optical communication and networks, Analog, digital and broadband communication circuits, optical signal processing and VLSI. Dr. Bharti has over 16 years of experience in Academics. She worked as Assistant Professor for 7+ years at Ambedkar Institute of Advanced Communication Technologies & Research, Delhi, under Govt. of NCT of Delhi and is currently working as Assistant Professor at the National Institute of Technology (NIT), Delhi. At present, she is guiding 3 research scholars for their Ph.D. degree. She has also guided many M.Tech and B.Tech students for their projects in different areas. She has published more than 15 research papers in reputed International/National Journals & more than 35 publications in conferences proceedings (International and National).

Monika Chaudhary is B.Tech, M.Tech from Kurukshetra University. She is pursuing PhD from Chitkara University, Rajpura, Punjab. She is having 7 years of teaching experience. Her research interest include MANET, VANET, WSN, UWSN, etc. She has published around 4 papers in various conferences and journals.

Kanwal Garg presently working as Assistant Professor in Department Of Computer Science And Application, Kurukshetra University Kurukshetra. Apart from district topper in senior secondary examination and Kurukshetra University topper in under graduation examination, he had completed his post graduation & doctorate from GJU S&T, Hisar under the faculty of Engineering and Technology. Owe the credit of more than 45 research papers published in international & national journals, conference & seminar. He attended 12 workshops/faculty Development Programme/winter/summer school and 01 Orientation Programme to enhance his curriculum. His area of expertise is Data Bases, Data Mining, & warehousing. Approximately 11 year of experience in teaching industry and administration. During this tenure he is actively involved in organizing co-curricular & social activities.

Tanvika Garg is an M.Tech student at National Institute of Technology, Delhi. B.Tech from Dhirubhai Ambani Institute of Information and Communication Technology.

Jitender Grover is pursuing PhD from École de technologie supérieure (ÉTS), Montreal, Canada. He has completed his MS by Research from IIIT, Hyderabad, India. He is working in wireless networks research area for 7 years and has published many research papers in this field.

Kalpna Guleria is an Associate Professor in Chitkara University Institute of Engineering and Technology, Chitkara University, Punjab, India. She is PhD in Computer Science Engineering from Thapar University, Patiala, India. She has received her M. Tech and B. Tech. with honors, majoring in Computer Science Engineering. She has research publications in SCI - Indexed International journals of high repute. Her research interests include Wireless Sensor Networks, Ad hoc Networks, Swarm Optimization techniques and Machine Learning. She is also an active reviewer of various International journals of high repute.

Osho Gupta is working as Assistant Professor in Chitkara University, India. He is having 1 year of teaching and academic experience. He obtained B.Tech in CSE and pursuing M.Tech from Chitkara University, India. He has published approximately 2 research papers.

Shilpi Harnal is working as an Assistant Professor in the CSE Dept. of JMIT Radaur, Yamunanagar. She has published more than 11 papers in various conferences/journals. She is B.Tech, M.Tech and PhD from Kurukshetra University.

Vikas Juneja is an Assistant Professor, of Department of Information Technology in JMIT, Radaur (Distt. Yamunanagar) Haryana. He completed his M.E. in CSE at NITTTR, Chandigarh. His area of interest includes in Cryptography and Network Security.

Isha Kansal is working as an Assistant Professor in Department of Computer Science and Engineering, Chitkara University, Punjab. She has done her Ph.D in Image and video Processing. Her main interests include image and video processing, machine learning.

Savita Khurana is working as an Assistant Professor in the IT Dept. of JMIT Radaur, Yamunanagar. She has published more than 14 papers in various conferences/journals. She is B.Tech, M.Tech from Kurukshetra University and pursuing PhD from IKGPTU, Jalandhar.

Ashok Kumar is currently an Assistant Professor in Chitkara University Research and Innovation Network (CURIN) Department, Punjab, India. He is PhD in Computer Science and Engineering from Thapar University, Punjab, India. He has 15+ years of teaching and research experience. He has filed 3 patents and published a number of publications in International Journals and Conferences of repute. His current areas of research interest include Cloud Computing, Internet of Things, and Mist Computing. Further, he is supervising 4 PhD scholars and 2 ME students. His teaching interest includes Python, Haskell, Java, C/C++, Advanced Data structures and Data mining.

Manish Kumar is working as Assistant Professor in University Institute of Engineering and Technology, Kurukshetra University. He has received the Bachelor of Engineering and technology degree in 2004 from Kurukshetra University, Master of Engineering and Technology in 2011 from Kurukshetra University and doing Ph.D. from Department of Electronic Science Department, Kurukshetra University. His research interest includes on AI, Machine Learning and Deep Learning, embedded systems design, hardware/software co-design.

Manni Kumar is working as Assistant Professor in Chitkara University, India. He is having 1 year of teaching and academic experience. He obtained B.Tech in CSE and pursuing M.Tech from Chitkara University, India. He has published 1 research paper till now. His research interest include WSN, MANET, VANET, UWSN, etc.

Neha Miglani is working as an Assistant Professor in the CSE Dept. of NIT Kurukshetra. She has published more than 11 papers in various conferences/journals. She is B.Tech, M.Tech from Kurukshetra University.

Krishna Pandey is presently working as Assistant Professor in the Department of Electronics and Communication, University Institute of Engineering and Technology, Kurukshetra University, Kurukshetra. He has nearly 8 years of teaching experience. He has been involved in the research related to the Wireless communication and Networks, Antenna design, and Optical Networks. He received his Master's degree from Gautam Buddha University Greater Noida, India. His current research area mainly includes Wireless sensor network, Underwater wireless sensor network, cloud computing, IoT, D2D communication, 5G Networks, Optimization Techniques in 5G Communication. He has papers in the International Journals and conferences.

Renu Popli is an Assistant Professor, of Department of Computer Science & Engineering in Chitkara University Institute of Engineering and Technoogy, Chitkara University, Punjab. She completed her P.hD from Kurukshetra University, Kurukshetra. Her area of interest includes in WSN, UWSN, and MANETs.

Arun Kumar Rana has completed his B.tech degree from Kurukshetra University and M.Tech. Degree From MMEC MULLANE (HR). Mr. Rana is currently pursuing his Ph.D. from MMEC MULLANE University HARYANA. His area of interest includes Image Processing, WSN, IoT and Embedded systems. Mr. Rana is currently working as Assistant Professor in PIET College Samalkha with more than 12 years of experience. He has published around 40 papers in National and International Journals and also in conferences. He has also published 4 books and many time member of Sci Scopus indexed international conference/symposium.

Deepti Rani is currently working as Assistant Professor in Chandigarh University, Mohali (Punjab). She has total 10 years of teaching experience from different reputed organizations. She is pursuing Ph.D. from M. D. University, Rohtak (Haryana). She has published more than 15 research papers in reputed National and International Conference Proceedings and Journals. Her main research area is Digital Forensics in Internet of Things.

Esha Rani is an Assistant Professor, of Department of Computer Science in IIHS, KUK (Distt. Kurukshetra) Haryana. She completed her M.Phil in Computer Science from CDLU, Sirsa.Her Area of interest includes Computer Network and Security.

Rajneesh Rani has received the B.Tech and M.Tech degrees,both in Computer Science and Engineering, from Punjab Technical University, Jalandhar, India in 2001 and Punjabi University Patiala, India in 2003 respectively. She has done her Ph.D in computer Science and Engineering from NIT Jalandhar in 2015.From 2003 to 2005, she was a lecturer in Guru Nanak Dev Engineering College, Ludhiana. Currently, she has been working as an assistant professor in NIT Jalandhar since 2007. Her teaching and research include areas like Image Processing, Pattern Recognition, Machine Learning, Computer Programming and Document Analysis and Recognition.

Shikha Rani is working as Assistant Professor in NIT, Kurukshetra, India. She is having 6 years of teaching and academic experience. She obtained B.Tech and M.Tech in ECE from Kurukshetra University, Haryana, India. Her publications are approximately 6 in different journals, conferences, and seminars.

Ahmed Saihood is currently PhD student at Isfahan University. He obtained his undergraduate degree in Computer Engineering (2009) from Basra University, Iraq, and received his M.Tech degree in Information Technology (2013) from MM University, India. He worked on wireless sensor network (WSN), mobile ad-hoc network (MANET), evolutionary algorithms, etc. He is having around 5 years of teaching experience along with Academic work. He has published approximately 6 research papers in various International/National Journals. His research interests include MANET, UWSN, WSN, evolutionary algorithms, Deep learning, machine learning,IoT and medical imaging. He is the reviewer, TPC member, advisory board member of some Journals and conferences.

Anju Sangwan is currently working as an Assistant Professor in Department of Computer Science & Engineering, Guru Jambheshwar University of Science & Technology, Hisar, India. She has received the M.Sc. (Pure Mathematics) and M.Tech. (IT) degrees from the Banasthali University, Rajasthan, India in 2007 and 2009 respectively & Ph.D. from Guru Jambheshwar University of Science & Technology, Hisar, India. She has 6 paper in international journal and 2 papers in international conferences to her credit. Her area of interest is wireless communication and sensor networks.

Anupma Sangwan has done M.C.A. in 2008, M.Tech. in Computer Engineering in 2010 and Ph.D. in Computer Science & Engineering on MANETs in 2017. Presently, she is working as Assistant Professor in the Department of Computer Science & Engineering, Guru Jambheshwar University of Science & Technology, Hisar, Haryana since 2014. She has more than 9 years teaching experience. Her research interests are Networking, Data Science and Machine Learning.

Gaurav Sharma is working as an Associate Professor in the CSE Dept. of JMIT Radaur, Yamunanagar. He has published more than 40 papers in various conferences/journals at national and international level. He is M.Tech from GJUST University, Haryana and PhD from Punjabi University, Patiala.

Nonita Sharma is currently working as Assistant Professor in the Department of Computer Science Engineering at Dr. B.R. Ambedkar National Institute of Technology Jalandhar. Her research interests are Data Mining, Wireless Sensor Network, Blockchain Technology, Internet of Things (IoT). She has published more than 20 papers in peer reviewed journal and international conferences. She has published a book entitled Extreme Gradient Forecasting for Data Mining Applications. She has received Best Paper Award in the International Conference held at CDAC Mohali, India.

Sharad Sharma has a vast experience of more than 20 years of teaching and administrative work. He has received his B. Tech degree in Electronics Engineering from Nagpur University, Nagpur, India, M.Tech in Electronics and Communication Engineering from Thapar Institute of Engineering and Technology, Patiala, India and Ph.D. from National Institute of Technology, Kurukshetra, India. He has a teaching experience of more than 20 years. He has conducted many workshops on Soft Computing and its applications in engineering, Wireless Networks, Simulators etc. He has a keen interest in teaching and implementing the latest techniques related to wireless and mobile communications. He has opened up a student chapter of IEEE as Branch Counselor. His research interests are routing protocol design, performance evaluation and optimization for wireless mesh networks using nature inspired computing, Internet of Things and Space Communication etc. He is member of various professional bodies.

Rishi Pal Singh is a Professor in the Department of Computer Science & Engineering, Guru Jambheshwar University of Science & Technology, Hisar, India. He has done M.Tech. from BITS, Pilani, India and Ph.D. from Jawaharlal Nehru University, New Delhi, India. He has published more than 30 papers in international journals of repute and presented more than 15 papers in international conferences. His area of interest is adhoc network, sensor networks and wireless communication.

Sneha Singh has completed her B.E. in Electronics & Telecommunication Engineering from Savitribai Phule Pune University in 2018 and is currently pursuing her M.Tech in Electronics & Communication Engineering from National Institute of Technology Delhi, New Delhi. Her area of research includes speech processing, digital signal processing.

Sachin Umrao is currently working as post doctoral associate at Yale School of Medicine, Yale University, USA. Prior to this he was post doctoral fellow at Department of Electronics and Computer Engineering, Hanyang University, Seoul, South Korea. He received his Ph.D. degree in 2017 from Sungkyunkwan University, South Korea. His research interests are Artificial Intelligence, Application of AI in Health-care and wireless networks, Device-to-Device communication, 4G/5G wireless communications, C-V2X and network security.

Poonam Rani Verma is a Research Scholar, NIT, Delhi, pursuing M.Tech in electronics and communication, with interest area, signal processing.

Index

www.ingramcontent.com/pod-product-compliance
Lightning Source LLC
Chambersburg PA
CBHW081214180926
51613CB00003B/240
* 9 7 8 1 7 9 9 8 3 6 4 0 7 *